Applied Mathematical Sciences
Volume 106

Applied Mathematical Sciences

1. *John:* Partial Differential Equations, 4th ed.
2. *Sirovich:* Techniques of Asymptotic Analysis.
3. *Hale:* Theory of Functional Differential Equations, 2nd ed.
4. *Percus:* Combinatorial Methods.
5. *von Mises/Friedrichs:* Fluid Dynamics.
6. *Freiberger/Grenander:* A Short Course in Computational Probability and Statistics.
7. *Pipkin:* Lectures on Viscoelasticity Theory.
8. *Giacoglia:* Perturbation Methods in Non-linear Systems.
9. *Friedrichs:* Spectral Theory of Operators in Hilbert Space.
10. *Stroud:* Numerical Quadrature and Solution of Ordinary Differential Equations.
11. *Wolovich:* Linear Multivariable Systems.
12. *Berkovitz:* Optimal Control Theory.
13. *Bluman/Cole:* Similarity Methods for Differential Equations.
14. *Yoshizawa:* Stability Theory and the Existence of Periodic Solution and Almost Periodic Solutions.
15. *Braun:* Differential Equations and Their Applications, 3rd ed.
16. *Lefschetz:* Applications of Algebraic Topology.
17. *Collatz/Wetterling:* Optimization Problems.
18. *Grenander:* Pattern Synthesis: Lectures in Pattern Theory, Vol. I.
19. *Marsden/McCracken:* Hopf Bifurcation and Its Applications.
20. *Driver:* Ordinary and Delay Differential Equations.
21. *Courant/Friedrichs:* Supersonic Flow and Shock Waves.
22. *Rouche/Habets/Laloy:* Stability Theory by Liapunov's Direct Method.
23. *Lamperti:* Stochastic Processes: A Survey of the Mathematical Theory.
24. *Grenander:* Pattern Analysis: Lectures in Pattern Theory, Vol. II.
25. *Davies:* Integral Transforms and Their Applications, 2nd ed.
26. *Kushner/Clark:* Stochastic Approximation Methods for Constrained and Unconstrained Systems.
27. *de Boor:* A Practical Guide to Splines.
28. *Keilson:* Markov Chain Models—Rarity and Exponentiality.
29. *de Veubeke:* A Course in Elasticity.
30. *Shiatycki:* Geometric Quantization and Quantum Mechanics.
31. *Reid:* Sturmian Theory for Ordinary Differential Equations.
32. *Meis/Markowitz:* Numerical Solution of Partial Differential Equations.
33. *Grenander:* Regular Structures: Lectures in Pattern Theory, Vol. III.
34. *Kevorkian/Cole:* Perturbation Methods in Applied Mathematics.
35. *Carr:* Applications of Centre Manifold Theory.
36. *Bengtsson/Ghil/Källén:* Dynamic Meteorology: Data Assimilation Methods.
37. *Saperstone:* Semidynamical Systems in Infinite Dimensional Spaces.
38. *Lichtenberg/Lieberman:* Regular and Chaotic Dynamics, 2nd ed.
39. *Piccini/Stampacchia/Vidossich:* Ordinary Differential Equations in $\mathbf{R}^n$.
40. *Naylor/Sell:* Linear Operator Theory in Engineering and Science.
41. *Sparrow:* The Lorenz Equations: Bifurcations, Chaos, and Strange Attractors.
42. *Guckenheimer/Holmes:* Nonlinear Oscillations, Dynamical Systems and Bifurcations of Vector Fields.
43. *Ockendon/Taylor:* Inviscid Fluid Flows.
44. *Pazy:* Semigroups of Linear Operators and Applications to Partial Differential Equations.
45. *Glashoff/Gustafson:* Linear Operations and Approximation: An Introduction to the Theoretical Analysis and Numerical Treatment of Semi-Infinite Programs.
46. *Wilcox:* Scattering Theory for Diffraction Gratings.
47. *Hale et al:* An Introduction to Infinite Dimensional Dynamical Systems—Geometric Theory.
48. *Murray:* Asymptotic Analysis.
49. *Ladyzhenskaya:* The Boundary—Value Problems of Mathematical Physics.
50. *Wilcox:* Sound Propagation in Stratified Fluids.
51. *Golubitsky/Schaeffer:* Bifurcation and Groups in Bifurcation Theory, Vol. I.

(continued following index)

Carlo Cercignani Reinhard Illner
Mario Pulvirenti

The Mathematical Theory of Dilute Gases

With 34 Illustrations

Springer-Verlag
New York Berlin Heidelberg London Paris
Tokyo Hong Kong Barcelona Budapest

Carlo Cercignani
Dip. di Matematica
Politecnico di Milano
I-20133 Milano
Italy

Reinhard Illner
Dept. of Mathematics
University of Victoria
Victoria, B.C. V8W 3P4
Canada

Mario Pulvirenti
Dip. di Matematica
University of Roma
I-00185 Roma
Italy

Mathematics Subject Classification (1991): 76P05, 82C40, 82B40

Library of Congress Cataloging-in-Publication Data
Cercignani, Carlo
The mathematical theory of dilute gases/Carlo Cercignani,
Reinhard Illner, Mario Pulvirenti.
p. cm. – (Applied mathematical sciences)
Includes bibliographical references and index.
ISBN 0-387-94294-7. – ISBN 3-540-94294-7
1. Transport theory. 2. Gases, Kinetic theory of.
3. Mathematical physics. 4. Chemistry, Physical and theoretical.
I. Illner, Reinhard. II. Pulvirenti, Mario. III. Title.
IV. Series: Applied mathematical sciences (Springer-Verlag New York, Inc.)
QC175.2.C43 1994
530.4'3–dc20 94-10086

Printed on acid-free paper.

Production managed by Natalie Johnson; manufacturing supervised by Gail Simon.
Photocomposed using the authors' TeX files and Springer-Verlag macro svplain.sty.
Printed and bound by Edwards Brothers, Inc., Ann Arbor, MI.
Printed in the United States of America.

9 8 7 6 5 4 3 2 1

ISBN 0-387-94294-7 Springer-Verlag New York Berlin Heidelberg
ISBN 3-540-94294-7 Springer-Verlag Berlin Heidleberg New York

Contents

Introduction

The idea for this book was conceived by the authors some time in 1988, and a first outline of the manuscript was drawn up during a summer school on mathematical physics held in Ravello in September 1988, where all three of us were present as lecturers or organizers. The project was in some sense inherited from our friend Marvin Shinbrot, who had planned a book about recent progress for the Boltzmann equation, but, due to his untimely death in 1987, never got to do it.

When we drew up the first outline, we could not anticipate how long the actual writing would stretch out. Our ambitions were high: We wanted to cover the modern mathematical theory of the Boltzmann equation, with rigorous proofs, in a complete and readable volume. As the years progressed, we withdrew to some degree from this first ambition— there was just too much material, too scattered, sometimes incomplete, sometimes not rigorous enough. However, in the writing process itself, the need for the book became ever more apparent. The last twenty years have seen an amazing number of significant results in the field, many of them published in incomplete form, sometimes in obscure places, and sometimes without technical details. We made it our objective to collect these results, classify them, and present them as best we could.

The choice of topics remains, of course, subjective. There are some subjects hardly touched in this book: Little reference is made to discrete velocity models, a very lively branch of kinetic theory. We chose to ignore this topic in order to limit the size of the book. Also, we confine our attention mostly to hard sphere interactions (except for some approximations, where analytical reasons force us to change the collision kernel). The very active subject of numerical simulation of the Boltzmann equation is given only brief coverage in Chapter 10; no attempts are even made to discuss generalizations of the simulation procedures to physically relevant situations like gas mixtures, inner degrees of freedom, or chemical reactions. We refer the reader to the book by Bird and to the extensive literature on numerical experiments; the proceedings of the Biannual Symposia on Rarefied Gas Dynamics are a good source of information on such results. There are undoubtedly other related topics we had to ignore out of lack of expertise, time, or sheer ignorance.

The results with which we are concerned can be classified in essentially five categories: 1. foundations (derivation and validation of the Boltzmann equation from the laws of mechanics), 2. existence and uniqueness results, 3. qualitative behavior, 4. fluid dynamical limits, and 5. numerical simulation. Results on 2 and 3 follow usually (but not necessarily) hand in hand.

We begin, in Chapter 1, with a historical account of kinetic theory. The next two chapters contain the formal derivation of the Boltzmann equation from the BBGKY hierarchy and the main properties of the Boltzmann equation including results on invariants and the *H*-theorem. Most of this is well known and well documented in earlier books; we do, however, present some recent generalizations regarding the functional equations associated with the invariants. Chapter 4, the longest chapter in this book, is concerned with the rigorous derivation and validation of the Boltzmann equation from the BBGKY hierarchy. This was first done, in a famous paper, by Lanford in 1975. We go to some length here in order to fill in details that were left out in Lanford's original work and in later generalizations done by two of the present authors. In several appendixes, we explain why the validation fails for discrete velocity models (Uchiyama's counterexample), we give a rigorous derivation of the BBGKY hierarchy, and we address the pathologies of multiple collisions. A detailed discussion of the origin of irreversibility is also offered.

The next few chapters are all about existence theory. First, we repeat at the beginning of Chapter 5 the existence results that actually follow from the validation done in Chapter 4 (local existence and uniqueness, and global existence for a rare gas cloud in vacuum). The rest of Chapter 5 contains the formulation and proof of the general global existence (without uniqueness) theorem proved in 1988 by DiPerna and Lions. In Chapter 6, we present the existence and uniqueness theory for the spatially homogeneous Boltzmann equation, mostly relying on pioneering work by Carleman in the 1930s, Morgenstern in the 1950s, Povzner in the 1960s, and Arkeryd in 1972. Chapter 7 contains the lengthy and very technical proof of asymptotic stability of Maxwellian equilibria due to Ukai and Asano in 1976. Most of this chapter deals, by necessity, with properties of solutions of the linearized Boltzmann equation, as it is the decay in time of such solutions that implies the asymptotic stability of Maxwellian equilibria.

Chapter 8 contains a discussion of boundary conditions and is a preparation for Chapter 9, in which we present recent results on the initial–boundary value problem. All the chapters on existence and uniqueness theory also deliver results on qualitative properties of the solutions, such as the approach to equilibrium. Asymptotic convergence to a Maxwellian follows from the spectral properties of the linearized collision operator in Chapter 7, and from a careful analysis of the H-theorem in Chapters 5, 6, and 9.

In Chapter 10 we give an outline of the most widely used particle simulation techniques. We abstain from going into convergence proofs, even though the techniques discussed are now known to converge.

Chapter 11 contains a presentation of the few rigorous results on the fluid dynamical limit available. We explain how the compressible Euler and the incompressible Navier–Stokes equations arise in suitable limits from the Boltzmann equation, how the H-functional is related to the entropy concept for conservation laws, and we outline the proof of one of the rigorous results on the transition regime between the Boltzmann equation and the compressible Euler equations.

This book can only be a temporary reference point in a rapidly developing field such as kinetic theory, but we hope that it can at least serve this purpose. Our thanks must go to friends and colleagues for advice, valuable comments, and help. In particular, we appreciate encouragement and contributions by Leif Arkeryd, Raffaele Esposito, and Herbert Spohn. Rosalie Rutka and Georgina Smith typed part of the manuscript, and we appreciate their laying the seed for the final TEX file; Denton Hewgill and Maurizio Vianello helped whenever advice regarding TEX questions was needed.

Our wives Silvana, Leslie, and Silvia deserve a big acknowledgment for their patience while we spent endless hours over proofs and in front of word processors.

Milan, Victoria, and Rome, Spring 1994.

Numbering and References

In order to keep the notation simple, equations are numbered by section number in each chapter. For example, if you read in Section 2 of Chapter 3, the fourth equation in that section would be numbered (2.4), and would be referred to as (2.4) in the rest of Chapter 3, but as (3.2.4) in the rest of the book. Theorems, definitions, and lemmas, which are not as numerous as equations, are numbered by chapter and section, so, for example, Theorem 4.5.1 refers to the first theorem in the fifth section of the fourth chapter. Definitions, theorems, and lemmas are in this fashion numbered consecutively in each section of the book.

References are listed at the end of each chapter, but before any appendixes.

1
Historical Introduction

1.1 What is a Gas? From the Billiard Table to Boyle's Law

As early as 1738 Daniel Bernoulli advanced the idea that gases are formed of elastic molecules rushing hither and thither at large speeds, colliding and rebounding according to the laws of elementary mechanics. Of course, this was not a completely new idea, because several Greek philosophers asserted that the molecules of all bodies are in motion even when the body itself appears to be at rest. The new idea was that the mechanical effect of the impact of these moving molecules when they strike against a solid is what is commonly called the pressure of the gas. In fact if we were guided solely by the atomic hypothesis, we might suppose that the pressure would be produced by the repulsions of the molecules. Although Bernoulli's scheme was able to account for the elementary properties of gases (compressibility, tendency to expand, rise of temperature in a compression and fall in an expansion, trend toward uniformity), no definite opinion could be passed on it until it was investigated quantitatively. The actual development of the kinetic theory of gases was, accordingly, accomplished much later, in the nineteenth century.

Within the scope of this book, the molecules of a gas will be assumed to be perfectly elastic spheres that move according to the laws of classical mechanics. Thus, e. g., if no external forces, such as gravity, are assumed to act on the molecules, each of them will move in a straight line unless it happens to strike another sphere or a solid wall. Systems of this kind are usually called billiards, for obvious reasons.

Although the rules generating the dynamics of these systems are easy to prescribe, the phenomena associated with the dynamics are not so simple. They are actually rather difficult to understand, especially if one is interested in the asymptotic behavior of the system for long times (ergodic properties) or in the case when the number of spheres is very large (kinetic and hydrodynamical limits). Both aspects of the dynamics of hard spheres are relevant when dealing with a gas, but we shall now concentrate upon the problem of outlining the behavior of this system when the number of the particles is very large. This is because there are about $2.7 \cdot 10^{19}$ molecules in a cubic centimeter of a gas at atmospheric pressure and a temperature of 0°C.

Given the enormous number of particles to be considered, it would of course be a perfectly hopeless task to attempt to describe the state of the gas by specifying the so-called microscopic state, i. e. the position and velocity of every individual particle, and we must have recourse to statistics. This is possible because in practice all that our observation can detect are changes in the macroscopic state of the gas, described by quantities such as density, velocity, temperature, stresses, and heat flow, which are related to the suitable averages of quantities depending on the microscopic state. A simple example is provided by an elementary calculation of the pressure in a container at rest, which will be presently sketched.

Let P be a point of the wall of the vessel, assumed to be flat, and let us take the x-axis in the direction of the normal to the wall, pointing toward the wall. Then a molecule with mass m, hitting the wall with velocity ξ, having components ξ_1, ξ_2, ξ_3, $(\xi_1 > 0)$ will transfer a momentum $m\xi_1$ to the wall; and a molecule recoiling from the wall with velocity ξ, having components ξ_1, ξ_2, ξ_3 $(\xi_1 < 0)$ will transfer a momentum $m|\xi_1|$ to the wall. If one constructs a cylinder upon a piece of the wall of area ΔS and side $\xi \Delta t$ (Fig. 1), all molecules with velocity ξ in this cylinder will strike or have struck (according to the sign of ξ_1) that piece of wall in a time interval Δt. Since the volume of the cylinder is $|\xi_1|\Delta t \Delta S$, we conclude that if all the molecules had velocity ξ, the total amount of momentum transferred to the wall in time Δt would be $n(|\xi_1|\Delta t \Delta S)(m|\xi_1|) = nm\xi_1^2 \Delta S \Delta t$, where n denotes the number of molecules per unit volume. If the molecules have different velocities we must take an average over their distribution of velocities and obtain $nm\overline{\xi_1^2}\Delta S \Delta t$, where the superimposed bar denotes the average value of a quantity. The transfer of momentum equals the impulse exerted on the area ΔS in time Δt; hence a force per unit area $nm\overline{\xi_1^2}$ is exerted by the wall on the gas and hence by the gas on the wall. In order to proceed, at this point we need an assumption of symmetry: if the gas is in a statistical equilibrium in a container macroscopically at rest, all the velocity components have the same probability distribution and hence

$$\overline{\xi_1^2} = \overline{\xi_2^2} = \overline{\xi_3^2} \tag{1.1}$$

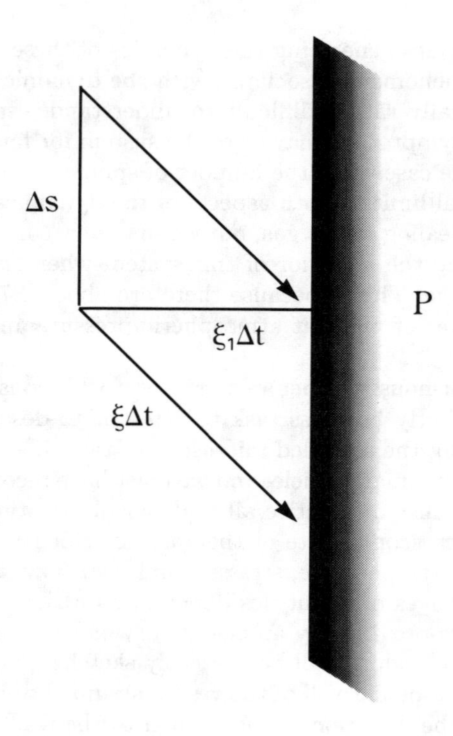

FIGURE 1.

Thus

$$3\overline{\xi_1^2} = \overline{\xi_1^2} + \overline{\xi_2^2} + \overline{\xi_3^2} = \overline{\xi_1^2 + \xi_2^2 + \xi_3^2} = \overline{\xi^2} \tag{1.2}$$

and the force per unit area, which is nothing other than the gas pressure p, will be given by:

$$p = \frac{1}{3} nm\overline{\xi^2}. \tag{1.3}$$

If V is the volume of the gas and N is the total number of molecules, it follows from this equation on multiplication by V, since $nV = N$, that

$$pV = \frac{1}{3} Nm\overline{\xi^2} = \frac{2}{3} Me \tag{1.4}$$

where e is the (kinetic) energy per unit mass and M is the total mass. Thus the product of pressure and volume depends only on the number of molecules and the average kinetic energy of a molecule. But according to an empirical law, the law of Boyle (1660) and Mariotte (1676), at constant temperature the product of the pressure and volume of a given amount of ideal gas is constant; we see that this law is reproduced if we make the reasonable assumption (which may be tantamount to a definition of temperature according to kinetic theory) that the average kinetic energy e only depends upon the (absolute) temperature T. In fact if we take into account the relation that combines the laws of Boyle-Mariotte and Gay-Lussac and Charles:

$$pV = MRT \tag{1.5}$$

where R is the gas constant, we conclude that

$$e = \frac{3}{2}RT. \tag{1.6}$$

We remark that R is constant for a given gas but is related to the molecular mass by $R = k/m$, where k is the Boltzmann constant ($k = 1.38 \cdot 10^{-23} J/^oK$); this follows, by considering a mixture of two different gases in the same vessel (Cercignani[8], 1988). We also remark that what has been said applies only to monatomic gases, which are well modeled by perfectly elastic spheres, and a gas sufficiently rarefied so that in a neighborhood of the wall we may neglect intermolecular collisions. A more careful analysis would show that the degree of rarefaction required for the argument to be valid is such that the product of the number density n by the cube of the molecular diameter σ must be negligible compared to unity.

At this point, however, a first question of principle must be considered. If we knew the exact position and velocity of every molecule of the gas at a certain time instant, the further evolution of the system would be completely determined, according to the laws of mechanics; even if we assume that at a certain moment the position and velocities of the molecules satisfy certain statistical laws, we are not entitled to expect that at any later time the state of the gas will conform to the same statistical assumptions, such as that embodied in Eq. (1.1), unless we prove that this is what mechanics predicts. In this case, it turns out that mechanics easily provides the required justification, but things are not so easy if we go further and ask how can we guarantee that the previous statistical assumption will be of practical importance, i.e., will actually be satisfied for a gas in equilibrium in a container. And questions become much more complicated if the gas is not in equilibrium, as is, e. g., the case for air around a flying vehicle.

Questions of this kind have been asked since the appearance of the kinetic theory of gases; today the matter is relatively well understood and a rigorous kinetic theory is emerging, as this book is trying to illustrate.

The importance of this matter stems from the need of a rigorous foundation of such a basic physical theory not only for its own sake, but also as a prototype of a mathematical construct central to the theory of nonequilibrium phenomena in large systems. Before describing the tools and results of rigorous kinetic theory, we shall first give a quick look at the history of the subject.

1.2 Brief History of Kinetic Theory

The first atomic theory is credited to Democritus of Abdera who lived in the fifth Century BC It was supported by other philosophers such as Leucippus (fifth Century BC) and through Epicurus (341–270 BC) it was transmitted to Romans. The most complete exposition of the view of the ancients is the famous poem of Lucretius (99–55 BC), *De Rerum Natura* ("On the Nature of the Things"). In medieval times some Arabian thinkers accepted the atomic theory, which was, however, fiercely fought by the scholastic theologians. During the Renaissance period, ideas related to atomism occur in the writings of Giordano Bruno (1548–1600), Galileo Galilei (1564–1642), and Francis Bacon (1561–1626). Later the French philosopher Petrus Gassendi (1592–1655) considered the idea of the atomic constitution of matter as a basic point of his philosophy. As mentioned in the previous section, it is only with Daniel Bernoulli (1700-1782) that this idea penetrates into the scientific domain, with an explanation of the gas pressure, and gives birth to the kinetic theory of gases. The same theory was afterward brought forward independently by George-Louis Lesage of Geneva (1724–1803), who devoted most of his work to the explanation of gravitation as due to the impact of atoms. Then John Herapath (1790–1869), in his *Mathematical Physics*, published in 1847, made a much more extensive application of the theory, and James Prescott Joule (1818–1889) estimated the average velocity of a molecule of hydrogen. A paper by K. Krönig (1822–1879) had the important role of drawing the attention of Rudolf Clausius (1822–1888) to the subject. With him, kinetic theory entered a mature stage, with the introduction of the concept of mean free path in 1858. In the same year, on the basis of this concept, James Clerk Maxwell (1831–1879) developed a preliminary theory of transport processes. In the same paper he gave a heuristic derivation of the velocity distribution function that bears his name. However, he almost immediately realized that the mean free-path method was inadequate as a foundation for kinetic theory, and in 1866, he developed a much more accurate method[21] based on the transfer equations, and he discovered the particularly simple properties of a model, according to which the molecules interact at distance with a force inversely proportional to the fifth power of the distance (now commonly called Maxwellian molecules). In the same paper he gave a better justification of his formula for the velocity distribution function for a gas in equilibrium.

With his transfer equations, Maxwell had come very close to an evolution equation for the distribution, but this step[3] must be credited to Ludwig Boltzmann (1844–1906). The equation under consideration is usually called the Boltzmann equation and sometimes the Maxwell-Boltzmann equation (to recognize the important role played by Maxwell in its discovery).

In the same paper, where he gave a heuristic derivation of his equation, Boltzmann deduced an important consequence from it, which later came to be known as the H-theorem. This theorem attempts to explain the irreversibility of natural processes in a gas, by showing how molecular collisions tend to increase entropy. The theory was attacked by several physicists and mathematicians in the 1890s, because it appeared to produce paradoxical results. However, within a few years of Boltzmann's suicide in 1906, the existence of atoms had been definitely established by experiments such as those on Brownian motion.

The paradoxes indicate, however, that some reinterpretation is necessary. Boltzmann himself had proposed that the H-theorem be interpreted statistically; later, Paulus Ehrenfest (1880–1933), together with his wife Tatiana, gave a brilliant analysis of the matter, which elucidated Boltzmann's ideas and made it highly plausible, at least from a heuristic standpoint. A rigorous analysis, however, was still to come.

In the meantime, the Boltzmann equation had become a practical tool for investigating the properties of dilute gases. In 1912 the great mathematician David Hilbert (1862–1943) indicated[16] how to obtain approximate solutions of the Boltzmann equation by a series expansion in a parameter, inversely proportional to the gas density. The paper is also reproduced as chapter XXII of his treatise entitled *Grundzüge einer allgemeinen Theorie der linearen Integralgleichungen.* The reasons for this are clearly stated in the preface of the book ("Neu hinzugefügt habe ich zum Schluss ein Kapitel über kinetische Gastheorie. [...] erblicke ich in der Gastheorie die glänzendste Anwendung der die Auflösung der Integralgleichungen betreffenden Theoreme").

In about the same year (1916–1917) Sidney Chapman[9] (1888–1970) and David Enskog[11] (1884–1947) independently obtained approximate solutions of the Boltzmann equation, valid for a sufficiently dense gas. The results were identical as far as practical applications were concerned, but the methods differed widely in spirit and detail. Enskog presented a systematic technique generalizing Hilbert's idea, while Chapman simply extended a method previously indicated by Maxwell to obtain transport coefficients. Enskog's method was adopted by S. Chapman and T. G. Cowling in their book *The Mathematical Theory of Non-uniform Gases* and thus became to be known as the Chapman-Enskog method.

Then for many years no essential progress in solving the equation came. Rather the ideas of kinetic theory found their way in other fields, such as radiative transfer, the theory of ionized gases, and, subsequently, in the theory of neutron transport. Almost unnoticed, however, the rigorous theory of the Boltzmann equation had started in 1933 with a paper[5] by Tage Gillis Torsten Carleman (1892–1949), who proved a theorem of global existence and uniqueness for a gas of hard spheres in the so-called space homogeneous case. The theorem was proved under the restrictive assumption that the initial data depend upon the molecular velocity only through its magnitude. This restriction is removed in a posthumous book by the same author[4].

In 1949 Harold Grad (1923–1986) wrote a paper[15], which became widely known because it contained a systematic method of solving the Boltzmann equation by expanding the solution into a series of orthogonal polynomials. In the same paper, however, Grad made a more basic contribution to the theory of the Boltzmann equation. In fact, he formulated a conjecture on the validity of the Boltzmann equation. In his words: "From the preceding discussion it is possible to see along what lines a rigorous derivation of the Boltzmann equation should proceed. First, from equilibrium considerations we must let the number density of molecules, N, increase without bound. At the same time we would like the macroscopic properties of the gas to be unchanged. To do this we allow m to approach zero in such a way that $mN = \rho$ is fixed. The Boltzmann equation for elastic spheres, (2.37) has a factor σ^2/m in the collision term. If σ is made to approach zero at such a rate that σ^2/m is fixed, then the Boltzmann equation remains unaltered. [...] In the limiting process described here, it seems likely that solutions of Liouville's equation attain many of the significant properties of the Boltzmann equation."

In the 1950s there were some significant results concerning the Boltzmann equation. A few exact solutions were obtained by C. Truesdell[25] in the U.S.A. and by V. S. Galkin[12,13] in the Soviet Union, while the existence theory was extended by D. Morgenstern[22], who proved a global existence theorem for a gas of Maxwellian molecules in the space homogeneous case. His work was extended by L. Arkeryd[1,2] in 1972.

In the 1960s, under the impact of the problems related to space research, the main interest was in the direction of finding approximate solutions of the Boltzmann equation and developing mathematical results for the perturbation of equilibrium[6,8]. Important methods developed by H. Grad[14] were brought to completion much later by S. Ukai, Y. Shizuta, and K. Asano[23,24,26].

The problem of proving the validity of the Boltzmann equation was still completely open. In 1972, C. Cercignani[7] proved that taking the limit indicated by Grad in the passage quoted above (now called the Boltzmann-Grad limit) produced, from a formal point of view, a perfectly consistent theory, i. e. the so-called Boltzmann hierarchy. This result clearly indicated that the difficulties of the rigorous derivation of the Boltzmann equation were not of a formal nature but were at least of the same order of difficulty as those of proving theorems of existence and uniqueness in the space inhomogeneous case. Subsequently, O. Lanford proved[20] that the formal derivation becomes rigorous if one limits oneself to a sufficiently short time interval. The problem of a rigorous, globally valid justification of the Boltzmann equation is still open, except for the case of an expanding rare cloud of gas in a vacuum, for which the difficulties were overcome by R. Illner and M. Pulvirenti[17,18], after that Illner and Shinbrot had provided the corresponding existence and uniqueness theorem for the Boltzmann equation[19].

Recently, R. Di Perna and P. L. Lions[10] have proved a global existence theorem for quite general data, but several important problems, such as proving that energy is conserved or controlling the growth of density are still open.

The last part of this historical sketch has come so close to current research that it would be inappropriate to continue it here. The rigorous theory developed so far and the open problems will be described in the next chapters.

References

1. L. Arkeryd, "On the Boltzmann equation. Part I: Existence," *Arch. Rat. Mech. Anal.* **45**, 1–16 (1972).
2. L. Arkeryd, "On the Boltzmann equation. Part II: The full initial value problem," *Arch. Rat. Mech. Anal.* **45**, 17–34 (1972).
3. L. Boltzmann, "Weitere Studien über das Wärmegleichgewicht unter Gasmolekülen," *Sitzungsberichte Akad. Wiss.*, Vienna, part II, **66**, 275–370 (1872).
4. T. Carleman, *Problèmes Mathématiques dans la Théorie Cinétique des Gaz*, Almqvist & Wiksell, Uppsala (1957).
5. T. Carleman, "Sur la théorie de l'équation intégro-differentielle de Boltzmann," *Acta Mathematica* **60**, 91–146 (1933).
6. C. Cercignani, *Mathematical Methods in Kinetic Theory*, Plenum Press, New York (1969), revised edition (1990).
7. C. Cercignani, "On the Boltzmann equation for rigid spheres," *Transp. Theory Stat. Phys.*, 211–225 (1972).
8. C. Cercignani, *The Boltzmann equation and its applications*, Springer, New York (1988).
9. S. Chapman, "The kinetic theory of simple and composite gases: Viscosity, thermal conduction and diffusion," *Proceedings of the Royal Society* (London) **A93**, 1–20 (1916/17).
10. R. Di Perna and P. L. Lions, "On the Cauchy problem for Boltzmann equations," *Ann. of Math.* **130**, 321–366 (1989).
11. D. Enskog, *Kinetische Theorie der Vorgänge in mässig verdünnten Gasen, I.Allgemeiner Teil*, Almqvist & Wiksell, Uppsala (1917).
12. V. S. Galkin, "Ob odnom resheniĭ kineticheskogo uravneniya," *Prikladnaya Matematika i Mekhanika* **20**, 445–446 (1956) (in Russian).
13. V. S. Galkin, "On a class of solutions of Grad's moment equations," PMM **22**, 532–536 (1958).
14. H. Grad, "Asymptotic equivalence of the Navier–Stokes and non-linear Boltzmann equation," *Proceedings of the American Mathematical Society Symposia on Applied Mathematics* **17**, 154–183 (1965).
15. H. Grad, "On the kinetic theory of rarified gases," *Comm. Pure Appl. Math.* **2**, 331–407 (1949).
16. D. Hilbert, "Begründung der kinetischen Gastheorie," *Mathematische Annalen 72*, 562–577 (1916/17).
17. R. Illner and M. Pulvirenti, "Global validity of the Boltzmann equation for a two-dimensional rare gas in a vacuum," *Commun. Math. Phys.* **105**, 189–203 (1986) .

18. R. Illner and M. Pulvirenti, "Global validity of the Boltzmann equation for two- and three-dimensional rare gas in vacuum: Erratum and improved result," *Comm. Math. Phys.* **121**, 143–146 (1989).
19. R. Illner and M. Shinbrot, "The Boltzmann equation: Global existence for a rare gas in an infinite vacuum," *Comm. Math. Phys.*, **95**, 217–226 (1984).
20. O. Lanford III, "The evolution of large classical systems," in *Dynamical systems, theory and applications*, J. Moser, ed., **LNP 35**, 1–111, Springer, Berlin (1975).
21. J. C. Maxwell, "On the dynamical theory of gases," *Philosophical Transactions of the Royal Society of London* **157**, 49–88 (1867).
22. D. Morgenstern, "General existence and uniqueness proof for spatially homogeneous solutions of the Maxwell-Boltzmann equation in the case of Maxwellian molecules," *Proceedings of the National Academy of Sciences* (U.S.A.) **40**, 719–721 (1954).
23. T. Nishida and K. Imai, "Global solutions to the initial value problem for the nonlinear Boltzmann equation," *Publications of the Research Institute for Mathematical Sciences, Kyoto University* **12**, 229–239 (1977).
24. Y. Shizuta and K. Asano, "Global solutions of the Boltzmann equation in a bounded convex domain," *Proceedings of the Japan Academy* **53**, 3–5 (1974).
25. C. Truesdell, "On the pressures and the flux of energy in a gas according to Maxwell's kinetic theory, II," *Jour. Rat. Mech. Anal.* **5**, 55–128 (1956).
26. S. Ukai, "On the existence of global solutions of mixed problem for non-linear Boltzmann equation," *Proceedings of the Japan Academy* **50**, 179–184 (1974).

2
Informal Derivation of the Boltzmann Equation

2.1 The Phase Space and the Liouville Equation

As indicated in the previous chapter, we shall investigate the hard sphere model of a gas. The reason for choosing such a simple model is based on the expectation that in the asymptotic regimes (hydrodynamic and kinetic) in which we are interested the general features should not depend on the particular type of interaction between the particles.

In order to discuss the behavior of a system of N (identical) hard spheres it is very convenient to introduce the so-called phase space, i.e., a $6N$-dimensional space where the Cartesian coordinates are the $3N$ components of the the N position vectors of the sphere centers x_i and the $3N$ components of the N velocities ξ_i. In this space, the state of the system, if known with absolute accuracy, is represented by a point whose coordinates are the $6N$ values of the components of the position vectors and velocities of the N particles. Let us denote by z the $6N$-dimensional position vector of this point in the phase space. If the state is not known with absolute accuracy, we must introduce a probability density $P(z,t)$, which gives the distribution of probability in phase space [whose precise meaning is given by Eq. (1.1)]. Given $P_0(z)$, its value at $t = 0$, we can compute $P(z,t)$ $(t > 0)$, provided we have an evolution equation for it. Here we have a difficulty because the hard sphere time evolution is discontinuous; in fact, when two particles collide, their velocities change instantaneously from incoming to outgoing. To deal with this difficulty, first we cancel the parts of the phase space corresponding to overlapping spheres and then we add suitable boundary conditions at the border of the remaining domain.

The evolution $z = T^t z_o$ of each phase point z_0 is then uniquely defined, provided that the phase points that lead to triple and higher collisions and those leading to infinitely many collisions in a finite time are neglected. The set of such points is of zero Lebesgue measure, as will be discussed in Chapter 4.

The probability that z will be found in a region $\mathbf{D}$ of phase space at time t is

$$\text{Prob}(z \in \mathbf{D}) = \int_{\mathbf{D}} P(z,t)dz \tag{1.1}$$

where dz denotes the Lebesgue measure in phase space. We remark that by writing Eq. (1.1) we are implicitly using the assumption that the measure defining the probability is absolutely continuous with respect to the Lebesgue measure.

The above probability is equal to the probability that the representative point was, at $t = 0$, in the region $\mathbf{D}_0$ consisting of the points z_0, which are the inverse images of the points $z \in \mathbf{D}$ in the mapping $z = T^t z_0$. In formulas, $\mathbf{D}_0 = \{z_0 \mid T^t z_0 \in \mathbf{D}\}$. Hence:

$$\int_{\mathbf{D}} P(z,t)dz = \int_{\mathbf{D}_0} P_0(z_0)dz_0. \tag{1.2}$$

We can now exploit the fact that the set of values of $z \in \mathbf{D}$ coincides with the set of points $z = T^t z_o$ with $z_0 \in \mathbf{D}_0$ and change the integration variable in the left-hand side from z to z_0. We obtain

$$\int_{\mathbf{D}} P(z,t)dz = \int_{\mathbf{D}_0} P(T^t z_o, t)J(z/z_0)dz_0 \tag{1.3}$$

where $J(z/z_o)$ is the Jacobian determinant of the old variables with respect to the new ones (which turns out to be positive; by continuity, if no collisions occur, but also in the presence of collisions, see below). Comparison of Eqs. (1.3) and (1.2) gives, due to the arbitrariness of $\mathbf{D}_0$,

$$P(T^t z_o, t)J(z/z_0) = P_0(z_0). \tag{1.4}$$

If we assume that no forces act on the particles, the Jacobian turns out to be unity. In fact, between two collisions each particle evolves independently, with $x_i = x_{0i} + \xi_{0i}t$, $\xi_i = \xi_{0i}$, were x_{0i} and ξ_{0i} are the initial values of the position and velocity vectors of the ith particle. It is clear that the Jacobian is unity (see Problems 1 and 2). We have to examine what happens at a collision.

When two spheres collide, conservation of momentum and energy must hold. Thus the velocities of the two particles after the impact (ξ_1, ξ_2) and before the impact (ξ_1', ξ_2') satisfy:

$$\xi_1 + \xi_2 = \xi_1' + \xi_2'$$

$$|\xi_1|^2 + |\xi_2|^2 = |\xi_1'|^2 + |\xi_2'|^2. \tag{1.5}$$

Let us introduce a unit vector n directed along $\xi_1 - \xi_1'$; this direction bisects the directions of $V = \xi_1 - \xi_2$ and $-V' = -(\xi_1' - \xi_2)$ (therefore, see Fig. 2, $n = (x_1 - x_2)/|x_1 - x_2|$ is the unit vector directed along the line joining the centers of the spheres, since the change of momentum at the moment of the impact between two smooth spheres must be directed along such a line). It is not hard to see (Problem 3) that Eqs. (1.5) imply

$$\xi_1' = \xi_1 - n[n \cdot (\xi_1 - \xi_2)]$$

$$\xi_2' = \xi_2 + n[n \cdot (\xi_1 - \xi_2)]. \tag{1.6}$$

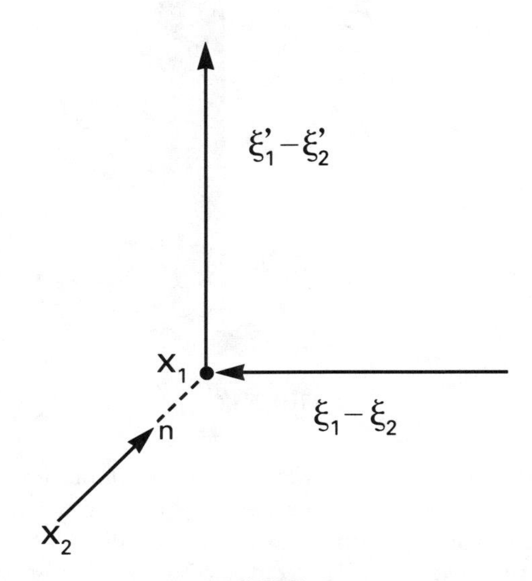

FIGURE 2.

We remark that the relative velocity

$$V = \xi_1 - \xi_2 \tag{1.7}$$

satisfies

$$V' = V - 2n(n \cdot V) \tag{1.8}$$

i.e., undergoes a specular reflection at the impact. This means that if we split V at the point of impact in a normal component V_n, directed along n and a tangential component V_t (in the plane normal to n), then V_n changes sign and V_t remains unchanged in a collision (Problem 4).

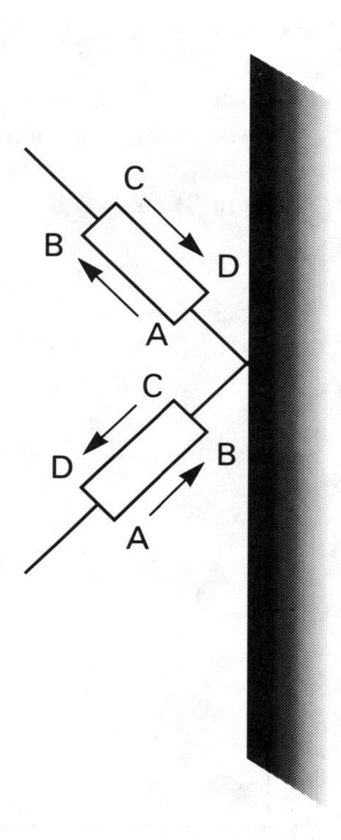

FIGURE 3.

It is now easy to compute the Jacobian of the velocities after the impact with respect to those before. The easiest way is to first transform each set of variables to the corresponding variables V (the relative velocity) and $\overline{\xi} = \frac{1}{2}(\xi_1 + \xi_2)$ (the velocity of the center of mass). These transformations are easily seen to have unit Jacobian (Problem 5). The Jacobian matrix of the transformation from $(\overline{\xi}, V)$ to $(\overline{\xi}', V')$ is now diagonal if we adopt normal and tangential components; it differs from the unit matrix, because one entry (corresponding to the normal component) is -1 rather than 1. Hence the Jacobian is -1.

We now need the Jacobian of the position variables after the impact with respect to those before the impact. It is clear that this Jacobian is -1, because the volume elements change their relative orientation upon impact (see Fig. 3).

The Jacobian J of the transformation occurring in phase space when a collision occurs is clearly the product of the Jacobians corresponding to the transformations undergone by space and velocity variables, respectively. Hence $J = (-1) \cdot (-1) = 1$.

We remark that the hard sphere dynamics can be obtained as a limiting case of the dynamics in which the particles interact via a repulsive potential that becomes infinity for $r \leq \sigma$ and zero for $r > \sigma$; this is easy to check for the two-body problem, although is a delicate question for an N-body problem. In the case of this potential one can easily prove the Liouville theorem on the invariance of the volume in phase space during the evolution, because then $\dot{z} = Z(z)$, where Z is a solenoidal field. Then $J(z/z_0) = 1$ and, by taking the limit, the same result holds for hard spheres. This is a different proof of the relation that we need.

We conclude that, in the absence of forces acting on the particles during their movement between two subsequent collisions, Eq. (1.4) simply becomes

$$P(T^t z_o, t) = P_0(z_0). \tag{1.9}$$

Hence P is constant along the trajectory of z in phase space.

P is defined in the set $\Omega^N \times \Re^{3N}$, where Ω is a subset of $\Re^3$ where the N particles move; it is, however, 0 at the points of this set that satisfy:

$$\exists i, j \in \{1, 2, \dots, N\} \quad (i \neq j) \ : \ |x_i - x_j| < \sigma \tag{1.10}$$

where σ is the sphere diameter. In fact, if z is a point in the set defined by (1.10), the ith and jth molecule would overlap, which is impossible, since they are assumed to be hard spheres. It is, accordingly, convenient to consider the set Λ, obtained by deleting from $\Omega^N \times \Re^{3N}$ the points satisfying (1.10).

If P is a differentiable function of the variables z, t in Λ, Eq. (1.9) implies that:

$$\frac{\partial P}{\partial t} + \sum_{i=1}^{N} \xi_i \cdot \frac{\partial P}{\partial x_i} = 0 \qquad (z \in \Lambda). \tag{1.11}$$

In fact, since $z = T^t z_0$ describes a rectilinear motion of all the particles inside Λ , it is obviously true that

$$dz/dt = (\xi_1, \xi_2, \dots, \xi_N, 0, 0, \dots, 0) \tag{1.12}$$

and Eq. (1.11) follows from Eq. (1.9) by differentiation. Eq. (1.11) is called the *Liouville equation* for the system under consideration.

Eq. (1.11) is a partial differential equation, and as such, must be accompanied by suitable initial and boundary conditions. The initial condition simply assigns the value of P at $t = 0$. As for the boundary conditions, they are present even if the gas is free to move without bounds in space (i.e., if $\Omega = \Re^3$). In fact, we had to introduce boundaries in order to define Λ (where (1.11) holds); these are the boundaries with the regions where the spheres would overlap. At these boundary points we must impose the condition dictated by Eq. (1.9); since P is always constant along the trajectory in Λ (boundary included), but the velocity variables undergo a discontinuous transformation there, we must impose that P is the same at z and z', where z and z' indicate points of the boundary of Λ that are transformed one into the other by the transformation associated with an impact:

$$P(z,t) = P(z',t) \qquad (z \in \partial\Lambda), \tag{1.13}$$

or, in more detail:

$$\begin{gathered} P(x_1, \xi_1, \ldots, x_i, \xi_i, \ldots, x_j, \xi_j, \ldots, x_N, \xi_N, t) = \\ P(x_1, \xi_1, \ldots x_i, \xi_i - n_{ij}(n_{ij} \cdot V_{ij}), \ldots, x_j, \xi_j + n_{ij}(n_{ij} \cdot V_{ij}), \ldots, x_N, \xi_N, t) \\ \text{if} \quad |x_i - x_j| = \sigma \quad (i \neq j) \end{gathered} \tag{1.14}$$

where $V_{ij} = \xi_i - \xi_j$ and n_{ij} is the unit vector directed as $x_i - x_j$.

If Ω does not coincide with the entire space $\Re^3$, then there are additional boundary points corresponding to those z for which at least one x_i is on $\partial\Lambda$. A suitable boundary condition must be assigned at these points as well. Frequently one assumes specular reflection ($\xi_i' = \xi_i - n_i(n_i \cdot \xi_i)$, where n_i is the normal at x_i), but other boundary conditions are used in practice (see Chapter 8). If Ω is a box, periodicity conditions are very popular; in that case one can avoid mentioning the boundaries and talk about a flat torus (after identification of opposite faces).

An important point to be mentioned is the circumstance that the initial values that we shall allow are symmetric upon interchange of any two particles (since the particles are identical):

$$\begin{gathered} P_0(x_1, \xi_1, \ldots, x_i, \xi_i, \ldots, x_j, \xi_j, \ldots, x_N, \xi_N) = \\ P_0(x_1, \xi_1, \ldots x_j, \xi_j, \ldots, x_i, \xi_i, \ldots, x_N, \xi_N). \qquad \forall (i,j). \end{gathered} \tag{1.15}$$

Since the time evolution is consistent with this symmetry (see Problems 9 and 10), Eq. (1.10) shows that the same symmetry is preserved for $t > 0$.

Problems

1. Show that if a particle's motion is uniform and rectilinear, then the Jacobian of the position and velocity components at time t with respect to their initial data is unity.
2. Show that in a system of N noninteracting particles the Jacobian of the position and velocity components at time t with respect to their initial data is the product of the Jacobians corresponding to the single particles.
3. Show that Eqs. (1.6) hold. (Remark: by definition we have $\xi_1 = \xi_1' - nC$, where C is a scalar to be determined...; in an equivalent way, project Eq. (1.5) along n and in a plane perpendicular to n...)
4. Check that if we split the relative velocity V at the point of impact into a normal component V_n directed along n and a tangential component V_t (in the plane normal to n), then V_n changes sign and V_t remains unchanged in a collision.
5. Show that if we transform from the variables ξ_1, ξ_2 to the variables V (the relative velocity) and $\bar{\xi} = \frac{1}{2}(\xi_1 + \xi_2)$ (the velocity of the center of mass), the transformation has unit Jacobian.
6. Check, by a direct calculation that the Jacobian of the transformation (1.7) is -1, if the collision occurs in a plane (i.e., ξ_1, ξ_2, ξ_1', and ξ_2' have just two components, while the components of n can be written $(\cos\theta, \sin\theta)$, where θ is a suitable angle).
7. Check Eq. (1.13).
8. Check Eq. (1.12).
9. Check that if the time evolution is dictated by $x_i = x_{0i} + \xi_{0i}t$, $\xi_i = \xi_{0i}$ $(i = 1, 2, \ldots, N)$, then if we interchange the initial data of two particles, say the first and the second, the solution changes only by the same interchange, i.e., the values of x_1 and ξ_1 at time t become those previously taken by x_2 and ξ_2 and the other way around.
10. Check that Eq. (1.12) remains unchanged if two particles are interchanged.

2.2 Boltzmann's Argument in a Modern Perspective

The Liouville equation discussed in the previous section is a useful conceptual tool, but it cannot in any way be used in practical calculations because of the large number of real variables on which the unknown depends (of the order of 10^{20}). This was realized by Maxwell and Boltzmann when they started to work with the one-particle probability density, or distribution function $P^{(1)}(x, \xi, t)$. The latter, at variance with the function $P(z, t)$ used in the previous section, depends on just seven real variables, i.e. the components of the two vectors x and ξ and time t. In particular, Boltzmann wrote an evolution equation for $P^{(1)}$ by means of a heuristic argument, which we shall try to present in such a way as to show where extra assumptions are introduced. One should realize that, as we shall see, one can obtain an exact equation for $P^{(1)}$ from the Liouville equation, but this equation contains $P^{(2)}$; a closed equation for $P^{(1)}$ is an extremely important step in the treatment of the problem. The equation written by Boltzmann and bearing his name can be justified in terms of statistical independence, as we shall see later.

Let us first consider the meaning of $P^{(1)}(x, \xi, t)$; it gives the probability density of finding one fixed particle (say, the one labeled 1) at a certain point (x, ξ) of the six-dimensional reduced phase space associated with the position and velocity of that particle. It is thus clear that there is a simple relation between $P^{(1)}$ and P; in fact

(2.1)
$$P^{(1)}(x_1, \xi_1, t) = \int_{\Omega \times \Re^{3N-3}} P(x_1, \xi_1, x_2, \xi_2, \ldots, x_N, \xi_N, t) dx_2 d\xi_2, \ldots dx_N d\xi_N$$

since $P^{(1)}$ is the probability of finding the first particle in a certain state no matter what the states of the particles labeled $2, \ldots, N$ are (in Eq. (2.1), of course, P is set equal to zero outside $\Lambda \times \Re^{3N}$) . Thus, in principle, the evolution of $P^{(1)}$ is contained in the Liouville equation; this remark will be useful later, but will presently be disregarded. In this section we shall try to write an equation for $P^{(1)}$ on the basis of its physical significance.

Let us remark that, in the absence of collisions, $P^{(1)}$ would satisfy the same equation as P [except that we should take $N = 1$ in Eq. (1.11)]. Accordingly we must evaluate the effects of collisions on the time evolution of $P^{(1)}$. We remark that the probability of occurrence of a collision will be related to the probability of finding another molecule with a center exactly one diameter from the center of the first one, whose distribution function is $P^{(1)}$. Thus, generally speaking, in order to write the evolution equation for $P^{(1)}$ we shall need another function, $P^{(2)}$, which gives the probability density of finding, at time t, the first molecule at x_1 with velocity ξ_1 and the second at x_2 with velocity ξ_2; obviously $P^{(2)} = P^{(2)}(x_1, \xi_1, x_2, \xi_2, t)$. Generally speaking we shall have

$$\frac{\partial P^{(1)}}{\partial t} + \xi_1 \cdot \frac{\partial P^{(1)}}{\partial x_1} = G - L. \tag{2.2}$$

Here $L dx_1 d\xi_1 dt$ gives the expected number of particles with position between x_1 and $x_1 + dx_1$ and velocity between ξ_1 and $\xi_1 + d\xi_1$ that disappear from these ranges of values because of a collision in the time interval between t and $t + dt$ and $G dx_1 d\xi_1 dt$ gives the analogous number of particles entering the same range in the same time interval. The count of these numbers is analogous to the one made in Chapter 1 to compute the transfer of momentum from the molecules to a wall, provided we use the trick of imagining particle 1 as a sphere at rest and endowed with twice the actual diameter σ and the other particles to be point masses with velocity $(\xi_i - \xi_1) = V_i$. In fact, each collision will send particle 1 out of the above range and the number of the collisions of particle 1 will be the number of expected collisions of any other particle with that sphere. Since there are exactly $(N-1)$ identical point masses and multiple collisions are disregarded, $G = (N-1)g$ and $L = (N-1)l$, where the lowercase letters indicate the contribution of a fixed particle, say particle 2. We shall then compute the effect of the collisions of particle 2 with 1.

Let x_2 be a point of the sphere such that the vector joining the center of the sphere with x_2 is σn, where n is a unit vector. A cylinder with height $|V_2 \cdot n|dt$ and base area $dS = \sigma^2 dn$ (where dn is the area of a surface element of the unit sphere about n) will contain the particles with velocity ξ_2 hitting the base dS in the time interval $(t, t+dt)$; its volume is $\sigma^2 dn|V_2 \cdot n|\mathrm{dt}$. Thus the probability of a collision of particle 2 with particle 1 in the ranges $(x_1, x_1+dx_1), (\xi_1, \xi_1+d\xi_1), (x_2, x_2+dx_2), (\xi_2, \xi_2+d\xi_2), (t, t+dt)$ occurring at points of dS is $P^{(2)}(x_1, x_2, \xi_1, \xi_2, t)dx_1 d\xi_1 d\xi_2 \times \sigma^2 dn|V_2 \cdot n|dt$. If we want the probability of collisions of particles 1 and 2, when the range of the former is fixed but the latter may have any velocity ξ_2 and any position x_2 on the sphere (i.e. any n), we integrate over the sphere and all the possible velocities of particle 2 to obtain:

(2.3)
$$l dx_1 d\xi_1 dt = dx_1 d\xi_1 dt \int_{\Re^3}\int_{S_-} P^{(2)}(x_1, \xi_1, x_1 + \sigma n, \xi_2, t)|V_2 \cdot n|\sigma^2 dn d\xi_2$$

where S_- is the hemisphere corresponding to $V_2 \cdot n < 0$ (the particles are moving one toward the other before the collision). Thus we have the following result:

$$(2.4)\quad L = (N-1)\sigma^2 \int_{\Re^3}\int_{S_-} P^{(2)}(x_1, \xi_1, x_1 + \sigma n, \xi_2, t)|(\xi_2 - \xi_1) \cdot n| d\xi_2 dn.$$

The calculation of the gain term G is exactly the same as the one for L, except that we are looking at particles that have velocities ξ_1 and ξ_2 after collision, and hence we have to integrate over the hemisphere S_+, defined by $V_2 \cdot > 0$ (the particles are moving away one from the other after the collision). Thus we have:

$$(2.5)\quad G = (N-1)\sigma^2 \int_{\Re^3}\int_{S_+} P^{(2)}(x_1, \xi_1, x_1 + \sigma n, \xi_2, t)|(\xi_2 - \xi_1) \cdot n| d\xi_2 dn.$$

We thus could write the right-hand side of Eq. (2.2) as a single expression:

$$(2.6)\ G-L=(N-1)\sigma^2\int_{\Re^3}\int_{S^2}P^{(2)}(x_1,\xi_1,x_1+\sigma n,\xi_2,t)(\xi_2-\xi_1)\cdot n d\xi_2 dn$$

where now S^2 is the entire unit sphere and we have abolished the bars of absolute value in the right-hand side.

Although our derivation of Eqs. (2.4) to (2.6) has been a little cavalier, the results can (and will) be justified with full rigor.

Eq. (2.6), although absolutely correct, is not very useful in this form. It turns out that it is much more convenient to keep the gain and loss terms separated. Only in this way, in fact, can we insert in Eq. (2.2) the information that the probability density $P^{(2)}$ is continuous at a collision, (Eq. 1.14). This, in turn, as we shall see, will permit us to use the essential circumstance that particles that are about to collide are statistically independent, while those that have just collided are not. In order to use Eq. (1.14), we remark that if we write for $i=1$, $j=2$ and integrate with respect to the positions and velocities of the remaining $N-2$ particles, we have:

$$(2.7)\ P^{(2)}(x_1,\xi_1,x_2,\xi_2,t)=P^{(2)}(x_1,\xi_1-n(n\cdot V),x_2,\xi_2+n(n\cdot V),t)$$

if $|x_1-x_2|=\sigma$ where we have written V for $V_{12}=\xi_1-\xi_2$ and n for $-n_{12}$ (in agreement with the notation used earlier). In order to shorten, we write [in agreement with Eq. (1.6)]:

$$(2.8)\qquad \xi_1'=\xi_1-n(n\cdot V)\qquad \xi_2'=\xi_2+n(n\cdot V).$$

Inserting Eq. (2.8) in Eq. (2.5) we thus obtain:

$$(2.9)\ G=(N-1)\sigma^2\int_{\Re^3}\int_{S_+}P^{(2)}(x_1,\xi_1',x_1+\sigma n,\xi_2',t)|(\xi_2-\xi_1)\cdot n|d\xi_2 dn$$

which is a frequently used form. Sometimes n is changed into $-n$ in order to have the same integration range as in L; the only change (in addition to the change in the range) is in the third argument of $P^{(2)}$, which becomes $x_1-\sigma n$.

At this point we are ready to understand Boltzmann's argument. In a rarefied gas N is a very large number and σ (expressed in common units, such as centimeters) is very small; to fix the ideas, let us say that we have a box whose volume is 1 cm^3 at room temperature and atmospheric pressure.Then $N\cong 10^{20}$ and $\sigma=10^{-8}$cm. Then $(N-1)\sigma^2\cong N\sigma^2=10^4\ \text{cm}^2=1\ \text{m}^2$ is a sizable quantity, while we can neglect the difference between x_1 and $x_1+\sigma n$. This means that the equation to be written can be rigorously valid only in the so-called Boltzmann–Grad limit, when $N\to\infty$, $\sigma\to 0$ with $N\sigma^2$ finite.

In addition, since the volume occupied by the particles is about $N\sigma^3 \cong 10^{-4}\text{cm}^3$, the collisions between two preselected particles is a rather rare event. Thus two spheres that happen to collide can be thought of as two randomly chosen particles, and it makes sense to assume that the probability density of finding the first molecule at x_1 with velocity ξ_1 and the second at x_2 with velocity ξ_2 is the product of the probability density of finding the first molecule at x_1 with velocity ξ_1 times the probability density of finding the second molecule at x_2 with velocity ξ_2. If we accept this we can write (assumption of molecular chaos):

$$P^{(2)}(x_1, \xi_1, x_2, \xi_2, t) = P^{(1)}(x_1, \xi_1, t)P^{(1)}(x_2, \xi_2, t) \tag{2.10}$$

for two particles that are about to collide, or:

$$P^{(2)}(x_1, \xi_1, x_1 + \sigma n, \xi_2, t) = P^{(1)}(x_1, \xi_1, t)P^{(1)}(x_1 + \sigma n, \xi_2, t) \quad \text{for} \quad (\xi_2 - \xi_1) \cdot n < 0. \tag{2.11}$$

Thus we can apply this recipe to the loss term (2.2) but not to the gain term in the form (2.3). It is possible, however, to apply Eq. (2.11) (with ξ_1', ξ_2' in place of ξ_1, ξ_2) to the form (2.9) of the gain term, because the transformation (2.8) maps the hemisphere S^+ onto the hemisphere S^-.

If we accept all the simplifying assumptions made by Boltzmann, we obtain the following form for the gain and loss terms:

$$G = N\sigma^2 \int_{\Re^3} \int_{S_+} P^{(1)}(x_1, \xi_1', t)P^{(1)}(x_1, \xi_2', t)|(\xi_2 - \xi_1) \cdot n| d\xi_2 dn, \tag{2.12}$$

$$L = N\sigma^2 \int_{\Re^3} \int_{S_-} P^{(1)}(x_1, \xi_1, t)P^{(1)}(x_1, \xi_2, t) \mid (\xi_2 - \xi_1) \cdot n \mid d\xi_2 dn. \tag{2.13}$$

By inserting these expressions in Eq. (2.2) we can write the Boltzmann equation in the following form:

$$\frac{\partial P^{(1)}}{\partial t} + \xi_1 \cdot \frac{\partial P^{(1)}}{\partial x_1} \tag{2.14}$$

$$= N\sigma^2 \int_{\Re^3} \int_{S_-} \left[P^{(1)}(x_1, \xi_1', t)P^{(1)}(x_1, \xi_2', t) - P^{(1)}(x_1, \xi_1, t)P^{(1)}(x_1, \xi_2, t)\right] \mid (\xi_2 - \xi_1) \cdot n \mid d\xi_2 dn.$$

The Boltzmann equation is an evolution equation for $P^{(1)}$ without any reference to $P^{(2)}$ or P. This is its main advantage. It has been obtained, however, at the price of several assumptions; the chaos assumption present in Eqs. (2.10) and (2.11) is particularly strong and will be discussed in the next section.

Problem

1. Check that the transformation (2.8) actually maps the hemisphere S_+ onto S_-.

2.3 Molecular Chaos. Critique and Justification

The molecular chaos, assumed in Eqs. (2.10) to (2.11), is clearly a property of randomness. Intuitively, one feels that collisions exert a randomizing influence, but it would be completely wrong to argue that the statistical independence described by Eq. (2.10) is a consequence of the dynamics. It is quite clear that we cannot expect every choice of the initial value for P to give a $P^{(1)}$ that agrees with the solution of the Boltzmann equation in the Boltzmann-Grad limit. In other words, molecular chaos must be present initially and we can only ask whether it is preserved by the time evolution of the system of hard spheres.

It is evident that the chaos property (2.10), if initially present, is almost immediately destroyed if we insist that it should be valid everywhere. In fact, if it were strictly valid everywhere, the gain and loss terms in the Boltzmann-Grad limit would be exactly equal and there would be no effect of the collisions on the time evolution of $P^{(1)}$. The essential point is that we need the chaos property only for particles that are about to collide, i.e. in the precise form stated in Eq. (2.11). It is clear, then, that even if $P^{(1)}$, as predicted by the Liouville equation, converges nicely to a solution of the Boltzmann equation, $P^{(2)}$ may converge to a product, as stated in Eq. (2.11), only in a way that is, in a certain sense, very singular. In fact, it is not enough to show that the convergence is almost everywhere, because we need to use the chaos property in a zero measure set. On the other hand we cannot try to show convergence everywhere, because this would be false; in fact, we have just remarked that Eq. (2.11) is simply not true for molecules that have just collided.

How can we approach the question of justifying the Boltzmann equation without invoking the molecular chaos assumption as an a priori hypothesis? Obviously, since $P^{(2)}$ appears in the evolution equation for $P^{(1)}$, we must investigate the time evolution for $P^{(2)}$; now, as is clear and as will be illustrated in the next section, the evolution equation for $P^{(2)}$ contains another function, $P^{(3)}$, which depends on time and the coordinates of three particles and gives the probability density of finding, at time t, the first molecule at x_1 with velocity ξ_1, the second at x_2 with velocity ξ_2, and the third at x_3 with velocity ξ_3. In general if we introduce a function

$P^{(s)} = P^{(s)}(x_1, x_2 \ldots, x_s, \xi_1, \xi_2 \ldots, \xi_s, t)$, the so-called s-particle distribution function, which gives the probability density of finding, at time t, the first molecule at x_1 with velocity ξ_1, the second at x_2 with velocity $\xi_2, \ldots$, and the sth at x_s with velocity ξ_s, we find the evolution equation of $P^{(s)}$ contains the next function $P^{(s+1)}$, until we reach $s = N$; in fact $P^{(N)}$ is nothing other than P, and it satisfies the Liouville equation. It is thus clear that we cannot proceed unless we handle all the $P^{(s)}$ at the same time and attempt to prove a generalized form of molecular chaos, i.e.

$$(3.1) \qquad P^{(s)}(x_1, \xi_1, , x_2, \xi_2, \ldots, x_s, \xi_s, t) = \prod_{j=1}^{s} P^{(1)}(x_s, \xi_s, t)$$

The task then becomes to show that if true at $t = 0$, this property remains preserved (for any fixed s) in the Boltzmann-Grad limit. This will be discussed in more detail in the next few sections.

There remains the problem of justifying the *initial chaos assumption*, according to which Eq. (3.1) is satisfied at $t = 0$. One can give two justifications, one of them being physical in nature and the second mathematical; essentially, they say the same thing, i.e., that it is hard to prepare an initial state for which Eq. (3.1) does not hold. The physical reason for this is that, in general, we cannot handle the single molecules but rather act on the gas as a whole at a macroscopic level, usually starting from an equilibrium state (for which Eq. (3.1) holds). The mathematical argument indicates that if we choose the initial data for the molecules at random there is an overwhelming probability[1,3] that Eq. (3.1) is satisfied for $t = 0$ (see Problem 1).

A word should be said about boundary conditions. When proving that chaos is preserved in the limit, it is absolutely necessary to have a boundary condition compatible (at least in the limit) with Eq. (3.1). If the boundary conditions are those of periodicity or specular reflection, no problems arise. In general, it is sufficient that the particles are scattered without losses from the boundary in a way that does not depend on the state of the other particles of the gas[1].

Problems

1. Give a reasonable definition of probability for the initial data in terms of P and show that it attains a constrained maximum (the constraint being that $P^{(1)}$ is assigned) when $P = P^{(N)}$ is chaotic, i.e. it satisfies Eq. (3.1) (with $s = N$ and $t = 0$). (see Ref. 1).
2. What happens if in the previous problem we add the constraint that P is zero outside $\Omega^N \times \Re^{3N}$?

2.4 The BBGKY Hierarchy

In this section we shall deal with the equations satisfied by the s-particle distribution functions $P^{(s)}$ as a consequence of the Liouville equation (1.11), which we rewrite here for the convenience of the reader:

$$\frac{\partial P}{\partial t} + \sum_{i=1}^{N} \xi_i \cdot \frac{\partial P}{\partial x_i} = 0 \qquad (z \in \Lambda). \tag{4.1}$$

A rigorous derivation of these equations involves some subtleties, which will be discussed in Chapter 4. Here we shall assume that P is a smooth function, so that all the steps to be performed are justified.

We first state the relation between $P^{(s)}$ and P, which follows from their definition and is similar to Eq. (2.1):

$$\begin{aligned} &P^{(s)}(x_1, \xi_1, x_2, \xi_2, \ldots, x_s \xi_s, t) \\ &= \int_{\Omega^s \times \Re^{3s}} P(x_1, \xi_1, x_2, \xi_2, \ldots, x_N, \xi_N, t) \prod_{j=s+1}^{N} dx_j d\xi_j. \end{aligned} \tag{4.2}$$

The first step to be performed in order to derive an evolution equation for $P^{(s)}$ is now rather obvious: we integrate Eq. (4.1) with respect to the variables x_j $(s+1 \le j \le N)$ over $\Omega^s \times \Re^{3s}$. It is convenient to keep the terms in the sum appearing in Eq. (4.1) with $i \le s$ from those with $i >$s.

$$\frac{\partial P^{(s)}}{\partial t} + \int \sum_{i=1}^{s} \xi_i \cdot \frac{\partial P}{\partial x_i} \prod_{j=s+1}^{N} dx_j d\xi_j + \sum_{k=s+1}^{N} \int \xi_k \cdot \frac{\partial P}{\partial x_k} \prod_{j=s+1}^{N} dx_j d\xi_j = 0 \tag{4.3}$$

where integration with respect to the velocity variables extends to the entire $\Re^{3N-3s}$, while it extends to Ω^{N-s} deprived of the spheres $|x_i - x_j| < \sigma$ $(i = 1, \ldots, N, i \neq j)$ with respect to the position variables. It is also expedient to call k the dummy suffix in the second sum rather than i.

A typical term in the first sum in Eq. (4.3) contains the integral of a derivative with respect to a variable, x_i, over which one does not integrate; it is not possible, however, to simply change the order of integration and differentiation to obtain a derivative of $P^{(s)}$, even if the function P is assumed to be smooth, because the domain has boundaries $(|x_i - x_j| = \sigma)$ depending upon x_i. To obtain the correct result, a boundary term has to be added:

$$\int \xi_i \cdot \frac{\partial P}{\partial x_i} \prod_{j=s+1}^{N} dx_j d\xi_j = \xi_i \cdot \frac{\partial P^{(s)}}{\partial x_i} - \sum_{k=s+1}^{N} \int P^{(s+1)} \xi_i \cdot n_{ik} d\sigma_{ik} d\xi_k \tag{4.4}$$

where n_{ik} is the outer normal to the sphere $|x_i - x_k| = \sigma$ (with center at x_k), $d\sigma_{ik}$ the surface element on the same sphere, and $P^{(s+1)}$ the $(s+1)$-particle distribution function with arguments (x_j, ξ_j) $(j = 1, 2, \ldots, s, k)$ (see Problem 1).

A typical term in the second sum in Eq. (4.3) can be immediately integrated by means of the Gauss theorem, since it involves the integration of a derivative taken with respect to one of the integration variables. We find:

$$\int \xi_k \cdot \frac{\partial P}{\partial x_k} \prod_{j=s+1}^{N} dx_j d\xi_j \tag{4.5}$$

$$= \sum_{i=1}^{s} \int P^{(s+1)} \xi_k \cdot n_{ik} d\sigma_{ik} d\xi_k$$

$$+ \sum_{\substack{i=s+1 \\ i \neq k}}^{N} \int P^{(s+2)} \xi_k \cdot n_{ik} d\sigma_{ik} d\xi_k dx_i d\xi_i + \int P^{(s+1)} \xi_k \cdot n_k dS_k d\xi_k$$

where dS_k is the surface element of the boundary of Ω in the three-dimensional subspace described by x_k, and n_k is the unit vector normal to such a surface element and pointing into the gas. The last term in Eq. (4.5) is the contribution from the solid boundary of Ω; if the boundary conditions are of the form described at the end of the last section (in particular if there are specular reflection or periodicity boundary conditions) the term under consideration is zero; henceforth it will be omitted.

Inserting Eqs. (4.4) and (4.5) into Eq. (4.3) we find:

$$\frac{\partial P^{(s)}}{\partial t} + \sum_{i=1}^{s} \xi_i \cdot \frac{\partial P^{(s)}}{\partial x_i} = \sum_{i=1}^{s} \sum_{k=s+1}^{N} \int P^{(s+1)} V_{ik} \cdot n_{ik} d\sigma_{ik} d\xi_k \tag{4.6}$$

$$+ \frac{1}{2} \sum_{\substack{i=s+1 \\ i \neq k}}^{N} \int P^{(s+2)} V_{ki} \cdot n_{ik} d\sigma_{ik} d\xi_k dx_i d\xi_i$$

where $V_{ik} = \xi_i - \xi_k$ is the relative velocity of the ith particle with respect to the kth particle and we have taken into account that $\xi_k \cdot n_{ik}$ can be replaced by $\frac{1}{2} V_{ki} \cdot n_{ik}$ in the second sum, because $n_{ik} = -n_{ki}$. The last integral is now easily seen to be zero. In fact, the integral over the sphere described by n_{ik} can be split into two parts; one refers to the hemisphere $V_{ki} \cdot n_{ik} > 0$, the other to the hemisphere $V_{ki} \cdot n_{ik} < 0$. Now Eq. (1.15) implies:

$$P^{(s)}(x_1, \xi_1, \ldots x_i, \xi_i, \ldots, x_j, \xi_j, \ldots, x_s, \xi_s, t) \tag{4.7}$$

$$= P^{(s)}(x_1, \xi_1, \ldots, x_i, \xi_i - n_{ij}(n_{ij} \cdot V_{ij}), \ldots, x_j, \xi_j + n_{ij}(n_{ij} \cdot V_{ij}), \ldots, x_s, \xi_s, t)$$

$$\text{if} \quad |x_i - x_j| = \sigma \quad (i, j = 1, 2, \ldots, s;\ i \neq j;\ s = 1, 2, \ldots, N).$$

Thus in the last integral of Eq. (4.6) any point of one hemisphere is mapped by a measure preserving transformation of the type shown in Eq. (2.8) into a point of the other hemisphere where $P^{(s+2)}$ takes the same value (Eq. (4.7) with $s+2$ in place of s). Since the factor $V_{ki} \cdot n_{ik}$ takes opposite values at these two points, the integral under consideration vanishes. Further, the first integral in Eq. (4.6) is the same no matter what the value of the dummy index k is; thus we can abolish this index and write x_*, ξ_* in place of x_k, ξ_k. Summarizing, we have:

$$\frac{\partial P^{(s)}}{\partial t} + \sum_{i=1}^{s} \xi_i \cdot \frac{\partial P^{(s)}}{\partial x_i} = (N-s)\sum_{i=1}^{s} \int P^{(s+1)} V_i \cdot n_i d\sigma_i d\xi_* \tag{4.8}$$

where $V_i = \xi_i - \xi_*, n_i = (x_i - x_*)/\sigma$ and the arguments of $P^{(s+1)}$ are $(x_1, \xi_1, \ldots, x_s, \xi_s, x_*, \xi_*, t)$.

It should be clear that the streaming operator in the left-hand side of Eq. (4.8) should be complemented with the boundary conditions on the boundary of Λ^s. This operator is the generator of the free motion of s particles. The physical meaning of Eq. (4.8) should be transparent: the s-particle distribution function evolves in time according to the s-particle dynamics, corrected by the effect of the interaction with the remaining $(N-s)$ particles. The effect of this interaction is described by the right-hand side of Eq. (4.8).

We remark that for $s = 2$ Eq. (4.8) reduces to Eq. (2.1) when the right-hand side $G - L$ is written in the form (2.5). That expression is thus rigorously justified for functions P that are smooth enough. However, as we remarked in Section 2, the form (2.5) is not the most convenient for the right-hand side of Eq. (2.1). It is better to keep the contributions from the two hemispheres $\pm V \cdot n > 0$ separate. For the same reason, here we separate in Eq. (4.8) the contributions from the two hemispheres ${S_+}^i$ and ${S_-}^i$, defined by $V_i \cdot n_i > 0$ and $V_i \cdot n_i < 0$, respectively. In addition, we remark that $d\sigma_i = \sigma^2 dn_i$(where dn_i is the surface element on the unit sphere described by n_i) and write:

$$\frac{\partial P^{(s)}}{\partial t} + \sum_{i=1}^{s} \xi_i \cdot \frac{\partial P^{(s)}}{\partial x_i} = (N-s)\sigma^2 \tag{4.9}$$

$$\sum_{i=1}^{s} \left(\int_{\Re^3} \int_{{S_+}^i} P^{(s+1)} |V_i \cdot n_i| dn_i d\xi_* - \int_{\Re^3} \int_{S_-^i} P^{(s+1)} |V_i \cdot n_i| dn_i d\xi_* \right)$$

Now, as we did in Section 2 for the particular case where $s = 2$, we use the laws of elastic impact and the continuity of the distribution functions embodied in Eq. (4.7) (with $s+1$ in place of s) to obtain

$$\frac{\partial P^{(s)}}{\partial t} + \sum_{i=1}^{s} \xi_i \cdot \frac{\partial P^{(s)}}{\partial x_i} = (N-s)\sigma^2 \tag{4.10}$$

$$\sum_{i=1}^{s}\left(\int_{\Re^3}\int_{S_+} P^{(s+1)'}|V_i\cdot n_i|dn_i d\xi_* - \int_{\Re^3}\int_{S_-} P^{(s+1)}|V_i\cdot n_i|dn_i d\xi_*\right)$$

where $P^{(s+1)'}$ means that in $P^{(s+1)}$ we replace the arguments ξ_i and ξ_* with ξ_i' and ξ_*', given by:

$$\xi_i' = \xi_i - n_i(n_i \cdot V_i) \qquad \xi_*' = \xi_* + n_i(n_i \cdot V_i). \tag{4.11}$$

We may transform the two integrals extended to S_+^i and S_-^i into a single integral by changing, e.g., n_i into $-n_i$ in the second integral; we may even abolish the index i in n_i, provided the argument x_* in the second integral of the ith term is replaced by

$$x_* = x_i - n\sigma \tag{4.12}$$

(x_* is replaced by $x_i + n\sigma$ in the first integral, of course). Thus we have:

$$\frac{\partial P^{(s)}}{\partial t} + \sum_{i=1}^{s} \xi_i \cdot \frac{\partial P^{(s)}}{\partial x_i} \tag{4.13}$$

$$= (N-s)\sigma^2 \sum_{i=1}^{s} \int_{\Re^3}\int_{S^2} (P^{(s+1)'}|V_i\cdot n| - P^{(s+1)}|V_i\cdot n|)dnd\xi_*.$$

This system of equations is usually called the BBGKY hierarchy for a hard sphere gas.

Problems

1. Show that Eq. (4.4) holds.
2. Show that the last integral in Eq. (4.4) is zero if the boundary conditions are such that the change of state of a particle at the boundary is independent of the state of the other particles (see Ref. 2).

2.5 The Boltzmann Hierarchy and Its Relation to the Boltzmann Equation

Let us consider the Boltzmann-Grad limit ($N \to \infty$ and $\sigma \to 0$ in such a way that $N\sigma^2$ remains finite). Then we obtain (for each fixed s) that if each $P^{(s)}$ tends to a limit (which we denote by the same symbol) and this limit is sufficiently smooth, the finite hierarchy of Eqs. (4.13) becomes in the limit:

$$\frac{\partial P^{(s)}}{\partial t} + \sum_{i=1}^{s} \xi_i \cdot \frac{\partial P^{(s)}}{\partial x_i} = N\sigma^2 \sum_{i=1}^{s} \int_{\Re^3} \int_{S_+} (P^{(s+1)'} - P^{(s+1)})|V_i \cdot n| dn d\xi_* \tag{5.1}$$

where the arguments of $P^{(s+1)'}$ and $P^{(s+1)}$ are the same as above, except that $x'_* = x_* = x_i$ in agreement with Eq. (4.12) for $\sigma \to 0$. Eqs. (5.1) give a complete description of the time evolution of a Boltzmann gas (i.e. the ideal gas obtained in the Boltzmann-Grad limit), provided the initial value problem is well posed for this infinite system of equations, which appears to have been first written in Ref. 3 and is usually called *the Boltzmann hierarchy.*

As we already know, Eq. (5.1) is not equivalent to the Boltzmann equation, unless a special assumption on the initial data is made. Indeed, as discussed by Spohn[4], the solutions of the Boltzmann hierarchy describe the evolution of a Boltzmann gas, when the chaos assumption given by Eq. (3.1) is not satisfied by the initial data at $t = 0$. The solutions of the Boltzmann hierarchy in the case when the factorization property is not fulfilled for $t = 0$ will be given in Section 7 of Chapter 4. Here we shall assume that the data satisfy Eq. (3.1), which we rewrite here for $t = 0$:

$$P^{(s)}(x_1, \xi_1, x_2, \xi_2, \ldots, x_s, \xi_s, 0) = \prod_{j=1}^{s} P^{(1)}(x_s, \xi_s, 0). \tag{5.2}$$

It is now a simple remark, made in Ref. 3, that if Eq. (5.2) is satisfied and the Boltzmann equation, given by Eq. (2.14) or, shortly, ($V = \xi - \xi_*$) by:

$$\frac{\partial P^{(1)}}{\partial t} + \xi \cdot \frac{\partial P^{(1)}}{\partial x} = N\sigma^2 \int_{\Re^3} \int_{S_+} (P^{(1)'} P_*^{(1)'} - P^{(1)} P_*^{(1)})|V \cdot n| d\xi_* dn \tag{5.3}$$

admits a solution $P^{(1)}(x_s, \xi_s, t)$ for given initial data $P^{(1)}(x_s, \xi_s, 0)$, then the Boltzmann hierarchy Eq. (5.1) has at least a solution, given by

$$P^{(s)}(x_1, \xi_1, x_2, \xi_2, \ldots, x_s, \xi_s, t) = \prod_{j=1}^{s} P^{(1)}(x_s, \xi_s, t). \tag{5.4}$$

Therefore the chaos assumption, Eq. (5.4), is not inconsistent with the dynamics of rigid spheres in the Boltzmann-Grad limit; actually, if the Boltzmann hierarchy has a unique solution for data that satisfy Eq. (5.2), then Eq. (5.4) necessarily holds at any time if it holds at $t = 0$. Then the Boltzmann equation is justified.

We stress, however, the fact that we made several assumptions (existence of limits, their smoothness, an existence theorem for the Boltzmann equation, a uniqueness theorem for the Boltzmann hierarchy) that might not be satisfied. A few cases in which these properties have been shown to hold and thus the Boltzmann equation has been shown to be valid, will be discussed later (see Chapter 4).

We end this chapter with a few remarks on the Boltzmann equation, Eq. (5.3). First, we can omit the superscript $^{(1)}$, which is no longer needed, and we rewrite Eq. (5.3) as follows

$$\frac{\partial P}{\partial t} + \xi \cdot \frac{\partial P}{\partial x} = N\sigma^2 \int_{\Re^3} \int_{S_+} (P'P'_* - PP_*)|V \cdot n| d\xi_* dn. \tag{5.5}$$

Then it should be clear that the arguments of P are x, ξ, t,those of P_* x, ξ_*, t, those of P' x, ξ', t and those of P'_* x, ξ'_*, t, where

$$\xi' = \xi - n(n \cdot V) \qquad \xi'_* = \xi_* + n(n \cdot V), \qquad V = \xi - \xi_*. \tag{5.6}$$

Finally we observe that the integral in Eq. (5.5) is extended to the hemisphere S_+, but could be equivalently extended to the entire sphere S^2 provided a factor 1/2 is inserted in front of the integral itself. In fact changing n into $-n$ does not change the integrand.

The considerations of this and the previous sections could be extended to the case when an external force per unit mass X acts on the molecules; the only difference would be to add a term $X \cdot \partial P/\partial \xi$ in the left-hand side of Eq. (5.5). Since we shall usually consider cases when the external action on the gas, if any, is exerted through solid boundaries (surface forces), we shall not usually write the abovementioned term; it should be kept in mind, however, that such simplification implies neglecting, *inter alia*, gravity.

Extensions of the Boltzmann equation to molecular models different from the hard spheres are possible. This line started with Boltzmann himself, who, following previous calculations made by Maxwell, considered molecules modeled as point masses that repel each other with a central force. It is not hard to write a Boltzmann equation for this case (see Refs. 1–3) but, since the rigorous theory for these molecular models is in a very preliminary stage, we shall not consider it any longer in this book.

We finally mention that it is possible to retain some of the effects of the finite size of the molecules that disappear in the Boltzmann-Grad limit, as shown by Enskog in 1921. The relation between the Enskog equation and the Liouville equation is unclear from a rigorous standpoint. Once accepted, however,the Enskog equation lends itself to interesting mathematical investigations.

References

1. C. Cercignani, *The Boltzmann Equation and Its Applications,* Springer, New York (1988).

2. C. Cercignani, "On the Boltzmann equation for rigid spheres," *Transp. Theory Stat. Phys.*, **2**, 211–225 (1972).
3. H. Grad, "The many faces of entropy," *Comm. Pure Appl. Math.* **14**, 323–354 (1961).
4. H. Spohn, "Boltzmann Hierarchy and Boltzmann Equation," in *Kinetic Theories and the Boltzmann Equation*, C. Cercignani, ed., **LNM 1048**, Springer, Berlin (1984).

3
Elementary Properties of the Solutions

3.1 Collision Invariants

In this chapter we shall devote ourselves to a study of the main properties of the solutions of the Boltzmann equation. We assume that our solutions are as smooth as required. It will be the purpose of the remaining part of the book to show that sufficiently smooth solutions exist for which the manipulations presented here make sense.

Before embarking in the study of the properties of the Boltzmann equation we remark that the unknown of the latter is not always chosen to be a probability density as we have done so far; it may be multiplied by a suitable factor and transformed into an (expected) number density or an (expected) mass density (in phase space, of course). The only thing that changes is the factor in front of Eq. (II.5.5), which is no longer $N\sigma^2$. In order to avoid any commitment to a special choice of that factor we replace $N\sigma^2$ by a constant α and the unknown P by another letter, f (which is also the most commonly used letter to denote the one-particle distribution function, no matter what its normalization is). In some physically interesting situations in which the gas domain is the entire $\Re^3$ and the total mass (or total number of particles) is only locally finite (i.e. $\int_{\Lambda\times\Re^3} f(x,\xi)dxd\xi < +\infty$, for any bounded $\Lambda \in \Re^3$) the distribution function cannot even be normalized (i.e. $\int_{\Re^6} f(x,\xi)dxd\xi = +\infty$). Let us then rewrite Eq. (II.5.5) in the following form:

$$\frac{\partial f}{\partial t} + \xi \cdot \frac{\partial f}{\partial x} = \alpha \int_{\Re^3} \int_{S_+} (f'f'_* - ff_*)|V \cdot n| d\xi_* dn \tag{1.1}$$

The right-hand side contains a quadratic expression $Q(f,f)$, given by:

$$Q(f,f) = \int_{\Re^3}\int_{S_+} (f'f'_* - ff_*)|V\cdot n|d\xi_* dn. \tag{1.2}$$

This expression is called the collision integral or simply the collision term, and the quadratic operator Q goes under the name of collision operator. In this section we study some elementary properties of Q. Actually it turns out that it is more convenient to study the slightly more general bilinear expression associated with $Q(f,f)$, i.e.:

$$Q(f,g) = \frac{1}{2}\int_{\Re^3}\int_{S_+} (f'g'_* + g'f'_* + fg_* - gf_*)|V\cdot n|d\xi_* dn. \tag{1.3}$$

It is clear that when $g=f$, Eq. (1.3) reduces to Eq. (1.2) and

$$Q(f,g) = Q(g,f). \tag{1.4}$$

Our first aim is to study the eightfold integral:

$$\int_{\Re^3} Q(f,g)\phi(\xi)d\xi = \frac{1}{2}\int_{\Re^3}\int_{\Re^3}\int_{S_+} (f'g'_* + g'f'_* - fg_* - gf_*)\phi(\xi))|V\cdot n|d\xi_* dn \tag{1.5}$$

where f and ϕ are functions such that the indicated integrals exist and the order of integration does not matter. A simple interchange of the starred and unstarred variables [with a glance to Eq. (II.5.6)] shows that

$$\int_{\Re^3} Q(f,g)\phi(\xi)d\xi \tag{1.6}$$
$$= \frac{1}{2}\int_{\Re^3}\int_{\Re^3}\int_{S_+} (f'g'_* + g'f'_* - fg_* - gf_*)\phi(\xi_*)|V\cdot n|d\xi d\xi_* dn.$$

Next, we consider another transformation of variables, the exchange of primed and unprimed variables (which is possible because the transformation in Eq. (II.5.6) is its own inverse). This gives

$$\int_{\Re^3} Q(f,g)\phi(\xi)d\xi \tag{1.7}$$
$$= \frac{1}{2}\int_{\Re^3}\int_{\Re^3}\int_{S_+} (fg_* + gf_* - f'g'_* - g'f'_*)\phi(\xi')|V\cdot n|d\xi' d\xi'_* dn.$$

(Actually since $V'\cdot n = -V\cdot n$, we should write S_- in place of S_+; changing n into $-n$, however, gives exactly the expression written here.)

The absolute value of the Jacobian from ξ, ξ_* to ξ', ξ'_* is unity; thus we can write $d\xi d\xi_*$ in place of $d\xi' d\xi'_*$ and Eq. (1.7) becomes:

(1.8)
$$\int_{\Re^3} Q(f,g)\phi(\xi)d\xi = \frac{1}{2}\int_{\Re^3}\int_{\Re^3}\int_{S_+}(fg_*+gf_*-f'g'_*-g'f'_*)\phi(\xi')|V\cdot n|d\xi d\xi_* dn$$

Finally we can interchange the starred and unstarred variables in Eq. (1.8) to find:

$$\int_{\Re^3} Q(f,g)\phi(\xi)d\xi \tag{1.9}$$

$$= \frac{1}{2}\int_{\Re^3}\int_{\Re^3}\int_{S_+}(fg_* + gf_* - f'g'_* - g'f'_*)\phi(\xi'_*)|V\cdot n|d\xi d\xi_* dn.$$

Eqs.(1.6), (1.8), and (1.9) differ from Eq. (1.5) because the factor $\phi(\xi)$ is replaced by $\phi(\xi_*)$, $-\phi(\xi')$, and $-\phi(\xi'_*)$ respectively. We can now obtain more expressions for the integral in the left hand side by taking linear combinations of the four different expressions available. Among them, the most interesting one is the expression obtained by taking the sum of Eqs. (1.5), (1.6), (1.8), and (1.9) and dividing by four. The result is:

$$\int_{\Re^3} Q(f,g)\phi(\xi)d\xi \tag{1.10}$$

$$= \frac{1}{8}\int_{\Re^3}\int_{\Re^3}\int_{S_+}(f'g'_* + g'f'_* - fg_* - gf_*)(\phi + \phi* - \phi' - \phi*')|V\cdot n|d\xi d\xi_* dn.$$

This relation expresses a basic property of the collision term, which is frequently used. In particular, when $g = f$, Eq. (1.10) reads

$$\int_{\Re^3} Q(f,f)\phi(\xi)d\xi \tag{1.11}$$

$$= \frac{1}{4}\int_{\Re^3}\int_{\Re^3}\int_{S_+}(f'f'_* - ff_*)(\phi + \phi* - \phi' - \phi*')|V\cdot n|d\xi d\xi_* dn.$$

We remark that the following form also holds:

(1.12)
$$\int_{\Re^3} Q(f,f)\phi(\xi)d\xi = \frac{1}{2}\int_{\Re^3}\int_{\Re^3}\int_{S_+} ff_*(\phi' + \phi*' - \phi - \phi*)|V\cdot n|d\xi d\xi_* dn.$$

In fact, the integral in Eq. (1.11) can be split into the difference of two integrals (one containing $f'f'_*$, the other ff_*); the two integrals are just the opposite of each other, as an exchange between primed and unprimed variables shows, and Eq. (1.12) holds.

We now observe that the integral in Eq. (1.10) is zero independent of the particular functions f and g, if

$$\phi + \phi_* = \phi' + \phi'_* \tag{1.13}$$

is valid almost everywhere in velocity space. Since the integral appearing in the left-hand side of Eq. (1.11) is the rate of change of the average value of the function ϕ due to collisions, the functions satisfying Eq. (1.13) are called "collision invariants."

The first discussion of Eq. (1.13) is due to Boltzmann[6,7], who assumed ϕ to be differentiable twice and arrived at the result that the most general solution of Eq. (1.13) is given by

$$\phi(\xi) = A + B \cdot \xi + C|\xi|^2. \tag{1.14}$$

After Boltzmann, the matter of finding the solutions of Eq. (1.13) was investigated by Gronwall[14,15] (who was the first to reduce the problem to Cauchy's functional equation for linear functions), Carleman[9], and Grad[13]. All these authors assumed ϕ to be continuous and proved that it must be of the form given in Eq. (1.14). Slightly different versions of Carleman's proof are given in Refs. 11 and 22. In the latter monograph[22] the authors prove that the solution is of the form (1.14), even if the function ϕ is assumed to be measurable rather than continuous. In fact, they use a result on the solutions of Cauchy's equation:

$$f(u+v) = f(u) + f(v) \qquad (u, v \in \Re^n \text{or } \Re_+) \tag{1.15}$$

valid for measurable functions. When passing from continuous to (possibly) discontinuous functions, however, one should insist on the fact that Eq. (1.14) is satisfied almost everywhere and not everywhere in $\Re^3 \times \Re^3 \times S^2$, as assumed in Ref. 22. It should be possible, although this was never attempted, to transform the proof in Ref. 22 into a proof that the collision invariants are the classical ones under the assumption that Eq. (1.14) holds almost everywhere.

The problem of solving Eq. (1.13) was tackled by Cercignani[10] with the aim of proving that Eq. (1.14) gives the most general solution of Eq. (1.13), when the latter is satisfied almost everywhere in $\Re^3 \times \Re^3 \times S^2$, under the assumption that ϕ is in the Hilbert space H_ω of the square integrable functions with respect to a Maxwellian weight $\omega(|\xi|) = (\beta/\pi)^{3/2}\exp(-\beta|\xi|^2)$, $\beta > 0$. The first step was to show that the linear manifold of the solutions had a polynomial basis. After that it was enough to look for smooth solutions. The existence of these can be made very simple if we look for C^2 solutions.

A completely different proof of the same result (under the assumption that $\phi \in L^1_{\text{loc}}$) was contained in a paper by Arkeryd[2], but remained largely ignored in the literature. As shown in a paper by Arkeryd and Cercignani[3], Arkeryd's argument, when combined with the proof for C^2 functions of Ref. 10, allows a very simple proof of the fact that (1.14) is the most general solution when $\phi \in L^1_{\text{loc}}$ and Eq. (1.13) is satisfied almost everywhere.

Alternatively, it is possible to prove which the continuous solutions are, in such a way that the case of L^1_{loc} solutions follows by the continuous proof "with a. e. added at suitable places." We shall deal with Eq. (1.15) in the set $\mathbf{M} = \{u, v \in \Re^3; u \cdot v = 0\}$, because solving this equation is equivalent to solving Eq. (1.14) (see Problem 10).

In fact, the proof we shall now present can be directly used under the weaker assumption that f is measurable and finite a. e. and that Eq. (1.15) holds for a. e. $(u, v) \in \mathbf{M}$. Following Carleman[9] we split f into an even part $k(u) = f(u) + f(-u)$ and an odd part $h(u) = f(u) - f(-u)$, which separately satisfy Eq. (1.15). Carleman's study of k is simple and holds also in the measurable case "with a. e. added at suitable places." As for h, his construction uses in an essential way a set of measure zero, not easily adaptable to the measurable case. Here we will use a different strategy, which was recently proposed by Arkeryd and Cercignani[3]. Following the latter paper, we shall prove the following.

(3.1.1) Theorem. *If $f : \Re^3 \to \Re$ is continuous and satisfies (1.15) for $(u, v) \in \mathbf{M}$, then for some $B \in \Re^3, C \in \Re$, it holds that*

$$f(u) = B \cdot u + C \mid u \mid^2 . \tag{1.16}$$

If f is measurable, finite a. e., and satisfies (1.11) for a. e. $(u, v) \in \mathbf{M}$, then (1.16) holds for a. e. $u \in \Re^3$.

The proof in the continuous case uses Cauchy's result (Problem 1) that any continuous function χ satisfying

$$\chi(x) + \chi(y) = \chi(x + y), \qquad x, y \in \Re (\text{or } \Re_+) \tag{1.17}$$

is of the form $\chi(x) = \beta x$ for some $\beta \in \Re$.

A generalization of this result[3] can be used to prove the proposition in the measurable case.

(3.1.2) Lemma. *If χ is a measurable function from $\Re$ (or $\Re_+$), finite a. e., and satisfying (1.17) for a. e. $(x, y) \in \Re^2$ (or $\Re^2{}_+$), then there is $\beta \in \Re$ such that $\chi(x) = \beta x$ for a. e. $x \in \Re$ (or $\Re_+$).*

Proof. The idea is to show that $\chi \in L^\infty_{\text{loc}}$ and then make a study of $\int_0^1 \chi(xt) dt$ as in Arkeryd's proof[2]. We first let the domain of χ be $\Re$. Given an interval $\mathbf{I} = (-a/2, a/2)$, by Lusin's theorem there is a continuous function F on $\Re$ such that $\chi(x) = F(x)$ for all $x \in \mathbf{I}$ outside of a measurable set of measure less than $a/3$. For some $\delta > 0, \mid F(x+h) - F(x) \mid < 1$ if $\mid h \mid < \delta, x \in \mathbf{I}$. Take $\delta < a/3$ and notice that for each h with $\mid h \mid < \delta, \chi(x+h) = F(x+h)$ for all $x \in \mathbf{I}$ outside of a measurable set of measure less than $a/3 + \delta < 2a/3$.

Thus, given h with $\mid h \mid < \delta$, there is a subset $\Omega_h \subset \mathbf{I}$ of a measure larger than $a/3 - \delta > 0$, with $\mid \chi(x) - \chi(x+h) \mid < 1$ for $x \in \Omega_h$. But for a. a. $(x, h) \in \mathbf{I} \times (-\delta, \delta)$:

$$\chi(x+h) - \chi(x) = \chi(h). \tag{1.18}$$

In particular for a. a. $h \in (-\delta, \delta)$ there is an $x_0 \in \Omega_h$ such that

$$1 > \mid \chi(x_0+h) - \chi(x_0) \mid = \mid \chi(h) \mid = \mid \chi(x+h) - \chi(x) \mid \quad \text{for a. e. } x \in \mathbf{I}. \tag{1.19}$$

Hence by Fubini's theorem it holds for a. e. $x \in \mathbf{I}$ that

$$\mid \chi(x+h) - \chi(x) \mid < 1 \text{ for a. a. } h \text{ with } \mid h \mid < \delta. \tag{1.20}$$

It follows that $\chi \in L^\infty(\mathbf{I})$ and, since $\mathbf{I}$ is arbitrary, that $\chi \in L^1_{\text{loc}}$. Thus for $x \neq 0$

$$g(x) = \int_0^1 \chi(tx)dt = \int_0^x \chi(s)ds/x \tag{1.21}$$

is well defined and continuous. With $g(0) = 0$ it satisfies

$$g(x) + g(y) = g(x+y) \text{ for } (x, y) \in \Re^2. \tag{1.22}$$

We use now the elementary result (see Problem 1) that if g is continuous and satisfies Eq. (1.22) then $g(x) = \beta x$; hence from Eq. (1.21), $\chi(x) = 2\beta x$ a. e. □

We are now ready to prove Theorem 3.1.1.

Proof. For the even continuous solution k of (1.15) Carleman[9] noted that

$$k(u) + k(v) = f(u \pm v) + f(-(u \pm v)), (u, v) \in \mathbf{M}. \tag{1.23}$$

In particular for $p_1, p_2 \in \Re^3$ with $\mid p_1 \mid = \mid p_2 \mid = r$, and $u = (p_1 + p_2)/2, v = (p_1 - p_2)/2$, this gives:

$$(1.24) \qquad k(p_1) = k(u+v) = k(u-v) = k(p_2).$$

So there is a function Φ with $k(p) = \Phi(r^2)$. Finally, we obtain

$$(1.25) \qquad \Phi(|\, p_1 \,|^2) + \Phi(|\, p_2 \,|^2) = \Phi(|\, p_1 \,|^2 + |\, p_2 \,|^2)$$

and by Cauchy's result we get

$$(1.26) \qquad k(u) = \Phi(|\, u \,|^2) = 2C \,|\, u \,|^2,$$

where we replaced β by 2C.

In the measurable case, starting from (1.15) for k and a. e. $(u, v) \in \mathbf{M}$, we can argue in the same way and by the lemma conclude that (1.26) holds for a. e. $u \in \Re^3$.

For the odd solution h of Eq. (1.15), in the continuous case we let e_1, e_2, e_3 be an arbitrary orthonormal basis in $\Re^3$ and notice that (1.15) holds for h and this basis. For (u, v) in $\mathbf{M}$ set $u = \sum u_j e_j$ and $v = \sum v_j e_j$. By (1.15)

$$(1.27) \qquad \sum(h(u_j e_j) + h(u_j e_j)) = h(\sum u_j e_j) + h(\sum v_j e_j)$$

$$= h(u) + h(v) = h(u+v) = h(\sum((u_j + v_j)e_j)) = \sum h((u_j + v_j)e_j).$$

And so:

(1.28)

$$h(u_1 e_1) + h(v_1 e_1) - h((u_1 + v_1)e_1) = -\sum_2^3 (h(u_i e_i) + h(v_i e_i) + h((u_i + v_i)e_i)).$$

Since h is odd this gives

$$(1.29) \qquad h(u_i e_i) + h(v_i e_i) - h((u_i + v_i)e_i) = 0 \quad (i = 1).$$

An analogous result holds for $i = 2, 3$. So by Cauchy's result, for some $B_i \in \Re$

$$(1.30) \qquad h(u_i e_i) = 2B_i u_i$$

and with $B = \sum B_i e_i$:

$$(1.31) \qquad h(u) = 2B \cdot u.$$

By the discussion of Eq. (1.15), in the measurable case there is an orthonormal basis e_1, e_2, e_3 such that Eq. (1.31) holds for h and a. e. $u \in \Re^3$. Using this basis, the discussion holds for a. e. $(u, v) \in \mathbf{M}$, in particular Eq. (1.13) holds for almost everywhere $u \in \Re^3$. Finally Eq. (1.16) follows (for a. e. $u \in \Re^3$) by adding Eqs. (1.26) and (1.31). □

Problems

1. Show that if x is a vector in an n-dimensional vector space E_n and $f(x)$ a function continuous in at least one point and satisfying $f(x) + f(y) = f(x+y)$ for any $x, y \in E_n$, then $f(x) = A \cdot x$, where A is a constant vector. (Hint: show that f is actually continuous everywhere and satisfies $f(rx) = rf(x)$ for any integer r; extend this property to any rational and then to any real r; then use a basis in E_n.)
2. Show that the even part of a function ϕ satisfying Eq. (1.13) is a function of $|\xi|^2$ alone. (Hint: $\phi + \phi_*$ is constant if and only if $\xi + \xi_*$ and$|\xi|^2 + |\xi_*|^2$ are constant, and $\xi + \xi_*$ vanishes for $\xi_* = -\xi$.)
3. Show that the even part of a continuous function satisfying Eq. (1.13) has the form $a + c|\xi|^2$, where a and c are constants. (Hint: let $a = \phi(0)$ and use the results of the two previous problems.)
4. Show that if ξ and ξ_* are orthogonal then the even part of a collision invariant ϕ satisfies $\phi(\xi) + \phi(\xi_*) = \phi(\xi + \xi_*)$.
5. Extend the result of the previous problem to a pair of vectors ξ and ξ_*, not necessarily orthogonal. (Hint: consider another vector ρ orthogonal to both of them with magnitude $|\xi \cdot \xi_*|^{1/2}$ and consider the vectors $\xi + \rho$, $\xi_* \pm \rho$, to which the result of the previous problem applies.)
6. Apply the results of Problems 1 and 5 to show that the odd part of a collision invariant, if continuous in ξ, must have the form $b \cdot \xi$ where b is a constant vector, so that, because of the result of Problem 3 a collision invariant must have the form shown in Eq. (1.14).
7. Extend the results of the previous problems to measurable functions using the fact that the result of Problem 1 is valid if f is assumed to be measurable as discussed in the main text (see also Ref. 3).
8. Extend the result of Problem 6 to the case of a function ϕ in H_ω, the Hilbert space of functions, which are square integrable with respect to the Maxwellian weight $\omega(|\xi|) = (\beta/\pi)^{3/2} \exp(-\beta|\xi|^2), \beta > 0$, when Eq. (1.13) is satisfied almost everywhere. (Hint: define the operator K in the following way: $K\psi = \frac{1}{4\pi} \int_{\Re^3 \times S^2} \omega(|\xi|)(\psi(\xi_*) + \psi(\xi') - \psi(\xi_*))dnd\xi_*$ and show that K is a bounded self-adjoint operator in H_ω. Then prove that K transforms polynomials of the mth degree into polynomials of degree not larger than m. Then, noting that ψ is a collision invariant iff it is an eigenfunction of K corresponding to the unit eigenvalue, prove that the collision invariants are polynomials in ξ. Then apply the result of Problem 6; see Ref. 10.)
9. Show that Eq. (1.13) can be written as follows
$$\phi(\xi + u + v) + \phi(\xi) = \phi(\xi + u) + \phi(\xi + v)$$
provided u and v are two vectors such that

$$u \cdot v = 0.$$

10. Show that in order to solve Eq. (1.13) it is enough to solve
$$f(u+v) = f(u) + f(v) \qquad (u \cdot v = 0).$$
(Hint: set $f(u) = \psi(\xi + u) - \psi(\xi)$ in the equation of Problem 9.)
11. Introduce an orthonormal basis $e_i (i = 1,2,3)$ in $\Re^3$ and write $u = \sum_i u_i e_i$, so that if f satisfies $f(u+v) = f(u) + f(v) \quad (u \cdot v = 0)$, then $f(u) = \sum_i f(u_i e_i)$ if f is continuous. Show that if f is measurable and the mentioned equation holds a. e. $u, v \in \mathbf{M} = \{u, v \in \Re^3; u \cdot v = 0\}$, then we can pick an orthonormal basis e_1, e_2, e_3 so that $f(u) = \sum_i f(u_i e_i)$ for a. e. $u = \sum u_j e_j \in \Re^3$ (see Ref. 3).
12. If $f \in L^1_{\text{loc}}$ satisfies $f(u+v) = f(u) + f(v)(u \cdot v = 0)$ show that $f(tu)$, $(t \in [0,1])$ is L^1_{loc} in t for a. a. $u \in \Re^3$ and if we define $g(u) = \int_0^1 f(tu)dt$, g turns out to be C^0 and satisfies
$$g(u+v) = g(u) + g(v) \qquad (u \cdot v = 0).$$
(Hint: by means of an orthonormal basis $e_i (i = 1,2,3)$ in $\Re^3$, write $u = \sum_i u_i e_i$, so that $f(u) = \sum_i f(u_i e_i)$. Next show that
$$\int_0^{u_1} dv_1 \int_0^{u_2} \int_0^{u_3} dv_2 dv_3 f(v)$$
exists and equals $u_1 u_2 u_3 g(u)$. Then everything is easily proved for $u, v \neq 0$. The latter restriction can also be eliminated; see Refs. 2 and 3.)
13. Prove that if f is continuous, $n-1$ times differentiable $(n \geq 1)$, and satisfies $f(u+v) = f(u) + f(v)(u \cdot v = 0)$, then $g = \int_0^1 ftu)dt$ is n times differentiable and $u \cdot \frac{\partial g}{\partial u} + g = f$. (Hint: proceeding as in the previous problem, first prove that $g(u) = \sum_i \frac{1}{u_i} \int_0^{u_i} dv_i f(v_i e_i)$ for $u_i \neq 0$; then see Ref. 3.)
14. Let f be a measurable solution of $f(u+v) = f(u) + f(v)(u \cdot v = 0$. Prove that $\phi = f$ is a solution of Eq. (1.13) even if the equations are satisfied a. e. in $\mathbf{M} = \{u, v \in \Re^3; u \cdot v = 0\}$ and in $\Re^3 \times \Re^3 \times S^2$, respectively. (Hint: let u, v, and t be three vectors with $u \cdot v = 0$ and decompose t as $t = t_u + t_v + t_o$, where $t_u$and t_v are directed as u and v, respectively, while t_o is orthogonal to both; see Ref. 3.)
15. Prove that if f satisfies $f(u+v) = f(u) + f(v) \quad (u \cdot v = 0)$, in a. e. sense, then if u and v are generic vectors (with $u \cdot v \neq 0$, in general), $f(u) + f(v)$ is a function of $u+v$ and $|u|^2 + |v|^2$ in a. e. sense. (Hint: use the previous lemma to prove that $f(t) + f(w) = f(t+u) + f(w-u)$ provided that t and w are arbitrary vectors and u such that $|t+u|^2 + |w-u|^2 = |t|^2 + |w|^2$. Then remark that $t' = t + u, w' = w - u$ with u satisfying the latter constraint is the most general transformation leaving both $t + w$ and $|t|^2 + |w|^2$ invariant; see Ref. 3.)

16. Prove that if $f \in C^2$, then the most general solution of $f(u+v) = f(u) + f(v)$ $(u \cdot v = 0)$ is given by $f(u) = B \cdot u + C \mid u \mid^2$. (Hint: according to the previous problem, we have $f(u) + f(v) = F(x, y)$, where $x = u + v; y = \frac{1}{2}(\mid u \mid^2 + \mid v \mid^2)$; differentiate this relation with respect to u and subtract from the result the analogous derivative with respect to v; eliminate the derivatives of F to obtain a relation between the derivatives of f; then differentiate with respect to a generic component of u; a further differentiation with respect to a generic component of v gives relations that straightforwardly imply the result; see Refs. 3 and 10.)
17. By means of the results proved in the previous problems (in particular Problems 13 and 16) prove that if $f : \Re^3 \to \Re$ is in L^1_{loc} and satisfies $f(u+v) = f(u)+f(v)$ $(u \cdot v = 0)$, then for some $B \in \Re^3, C \in \Re$, $f(u) = B \cdot u + C \mid u \mid^2$. (Hint: use the fact that the set of functions having this shape is invariant with respect to the transformation from f to g and its inverse, defined in Problem 13; see Ref. 3.)

3.2 The Boltzmann Inequality and the Maxwell Distributions

In this section we investigate the existence of positive functions f that give a vanishing collision integral:

$$Q(f, f) = \int_{\Re^3} \int_{S_+} (f' f'_* - f f_*) |V \cdot n| d\xi_* dn = 0. \tag{2.1}$$

In order to solve this equation, we prove a preliminary result that plays an important role in the theory of the Boltzmann equation: if f is a non-negative function such that $\log f Q(f, f)$ is integrable and the manipulations of the previous section hold when $\phi = \log f$, then the *Boltzmann inequality*:

$$\int_{\Re^3} \log f Q(f, f) d\xi \leq 0 \tag{2.2}$$

holds; further, the equality sign applies if and only if $\log f$ is a collision invariant, or equivalently:

$$f = \exp(a + b \cdot \xi + c \mid \xi \mid^2). \tag{2.3}$$

To prove Eq. (2.2) it is enough to use Eq. (1.11) with $\phi = \log f$:

$$\int_{\Re^3} \log f Q(f,f) d\xi = \frac{1}{4} \int_{\Re^3} \int_{\Re^3} \int_{S_+} \log(f f_* / f' f'_*)(f' f'_* - f f_*) |V \cdot n| d\xi_* dn \tag{2.4}$$

and Eq. (2.2) follows thanks to the elementary inequality

$$(z - y)\log(y/z) \leq 0 \quad (y, z \in \Re^+). \tag{2.5}$$

Eq. (2.5) becomes an equality if and only if $y = z$; thus the equality sign holds in Eq. (2.2) if and only if:

$$f' f'_* = f f_* \tag{2.6}$$

applies almost everywhere. But taking the logarithms of both sides of Eq. (2.6), we find that $\phi = \log f$ satisfies Eq. (1.13) and is thus given by Eq. (1.14). $f = \exp(\phi)$ is then given by Eq. (2.3).

We remark that in the latter equation c must be negative, since $f \in L^1(R^3)$. If we let $c = -\beta, b = 2\beta v$ (where v is another constant vector), Eq. (2.3) can be rewritten as follows:

$$f = A \exp(-\beta \mid \xi - v \mid^2) \tag{2.7}$$

where A is a positive constant related to $a, c, \mid b \mid^2$ (β, v, A constitute a new set of constants). The function appearing in Eq. (2.7) is the so–called Maxwell distribution or Maxwellian. Frequently one considers Maxwellians with $v = 0$ (nondrifting Maxwellians), which can be obtained from drifting Maxwellians by a change of the origin in velocity space.

Let us return now to the problem of solving Eq. (2.1). Multiplying both sides by $\log f$ gives Eq. (2.2) with the equality sign. This implies that f is a Maxwellian, by the result just shown. Suppose now that f is a Maxwellian; then $f = \exp(\phi)$ where ϕ is a collision invariant and Eq. (2.6) holds; then Eq.(2.1) also holds. Thus there are functions that satisfy Eq. (2.1), and they are all Maxwellians, Eq. (2.7).

Problem

1. Prove (2.5).

3.3 The Macroscopic Balance Equations

In this section we compare the microscopic description supplied by kinetic theory with the macroscopic description supplied by continuum gas dynamics. For definiteness, in this section f will be assumed to be an expected mass density in phase space. In order to obtain a density, $\rho = \rho(x, t)$, in ordinary space, we must integrate f with respect to ξ:

$$\rho = \int_{\Re^3} f d\xi. \tag{3.1}$$

The bulk velocity v of the gas (e.g., the velocity of a wind) is the average of the molecular velocities ξ at a certain point x and time instant t; since f is proportional to the probability for a molecule to have a given velocity, v is given by

$$v = \left(\int_{\Re^3} \xi f d\xi\right) / \left(\int_{\Re^3} f d\xi\right) \tag{3.2}$$

(the denominator is required even if f is taken to be a probability density in phase space, because we are considering a conditional probability, referring to the position x). Eq. (3.2) can also be written as follows:

$$\rho v = \int_{\Re^3} \xi f d\xi \tag{3.3}$$

or, using components:

$$\rho v_i = \int_{\Re^3} \xi_i f d\xi \qquad (i = 1, 2, 3). \tag{3.4}$$

The bulk velocity v is what we can directly perceive of the molecular motion by means of macroscopic observations; it is zero for a gas in equilibrium in a box at rest. Each molecule has its own velocity ξ, which can be decomposed into the sum of v and another velocity

$$c = \xi - v \tag{3.5}$$

called the random or peculiar velocity; c is clearly due to the deviations of ξ from v. It is clear that the average of c is zero (Problem 1).

The quantity ρv_i that appears in Eq. (3.4) is the ith component of the mass flow or of the momentum density of the gas. Other quantities of similar nature are: the momentum flow

$$m_{ij} = \int_{\Re^3} \xi_i \xi_j f d\xi \qquad (i, j = 1, 2, 3); \tag{3.6}$$

the energy density per unit volume:

$$w = \frac{1}{2} \int_{\Re^3} |\xi|^2 f d\xi; \tag{3.7}$$

and the energy flow:

$$(3.8) \qquad r_i = \frac{1}{2}\int_{\Re^3} \xi_i |\xi|^2 f d\xi.$$

Eq. (3.8) shows that the momentum flow is described by the components of a symmetric tensor of second order, because we need to describe the flow in the ith direction of the momentum in the jth direction. It is to be expected that in a macroscopic description only a part of this tensor will be identified as a bulk momentum flow, because in general, m_{ij} will be different from zero even in the absence of a macroscopic motion ($v = 0$). It is thus convenient to reexpress m_{ij} in terms of c and v. Then we have:

$$(3.9) \qquad m_{ij} = \rho v_i v_j + p_{ij}$$

where:

$$(3.10) \qquad p_{ij} = \int_{\Re^3} c_i c_j f d\xi \qquad (i, j = 1, 2, 3)$$

plays the role of stress tensor (because the microscopic momentum flow associated with it is equivalent to forces distributed on the boundary of any region of gas, according to the macroscopic description).

Similarly, one has:

$$(3.11) \qquad w = \frac{1}{2}\rho |v|^2 + \rho e,$$

where e is the internal energy per unit mass (associated with random motions) defined by:

$$(3.12) \qquad \rho e = \frac{1}{2}\int_{\Re^3} | c |^2 f d\xi;$$

and

$$(3.13) \qquad r_i = \rho v_i (\frac{1}{2}|v|^2 + e) + \sum_{j=1}^{3} v_j p_{ij} + q_i \qquad (i = 1, 2, 3),$$

where q_i are the components of the so-called heat-flow vector:

$$(3.14) \qquad q_i = \frac{1}{2}\int_{\Re^3} c_i | c |^2 f d\xi.$$

The decomposition in Eq. (3.13) shows that the microscopic energy flow is a sum of a macroscopic flow of energy (both kinetic and internal), of the work (per unit area and unit time) done by stresses, and of the heat flow.

In order to complete the connection, as a simple mathematical consequence of the Boltzmann equation, one can derive five differential relations satisfied by the macroscopic quantities introduced above; these relations describe the balance of mass, momentum, and energy and have the same form as in continuum mechanics. To this end let us consider the Boltzmann equation

$$\frac{\partial f}{\partial t} + \xi \cdot \frac{\partial f}{\partial x} = \alpha Q(f, f). \tag{3.15}$$

If we multiply both sides by one of the elementary collision invariants ψ_ν ($\nu = 0, 1, 2, 3, 4$) defined in Section 1 and integrate with respect to ξ, we have, thanks to Eq.(1.15) with $g = f$ and $\phi = \psi_\nu$:

$$\int_{\Re^3} \psi_\nu(\xi) Q(f, f) d\xi = 0, \tag{3.16}$$

and hence, if it is permitted to change the order by which we differentiate with respect to t and integrate with respect to ξ:

$$\frac{\partial}{\partial t} \int \psi_\nu f d\xi + \sum_{i=1}^{3} \frac{\partial}{\partial x_i} \int \xi_i \psi_\nu f d\xi = 0 \qquad (\nu = 0, 1, 2, 3, 4). \tag{3.17}$$

If we take successively $\nu = 0, 1, 2, 3, 4$ and use the definitions introduced above, we obtain

$$\frac{\partial \rho}{\partial t} + \sum_{i=1}^{3} \frac{\partial}{\partial x_i} (\rho v_i) = 0, \tag{3.18}$$

$$\frac{\partial}{\partial t} (\rho v_j) + \sum_{i=1}^{3} \frac{\partial}{\partial x_i} (\rho v_i v_j + p_{ij}) = 0, \qquad (j = 1, 2, 3) \tag{3.19}$$

$$\frac{\partial}{\partial t} (\frac{1}{2} \rho \mid v \mid^2 + \rho e) + \sum_{i=1}^{3} \frac{\partial}{\partial x_i} [\rho v_i (\frac{1}{2} |v|^2 + e) + \sum_{j=1}^{3} v_j p_{ij} + q_i] = 0. \tag{3.20}$$

These equations have the so-called conservation form because they express the circumstance that a certain quantity (whose density appears differentiated with respect to time) is created or destroyed in a certain region Ω because something is flowing through the boundary $\partial\Omega$. In fact, when integrating both sides of the equations with respect to x over Ω, the terms differentiated with respect to x can be replaced by surface integrals over

$\partial\Omega$, thanks to the divergence theorem. If these surface integrals turn out to be zero then we obtain that the total mass,

$$M = \int_\Omega \rho dx, \tag{3.21}$$

the total momentum,

$$Q = \int_\Omega \rho v dx, \tag{3.22}$$

and the total energy,

$$E = \int_\Omega (\frac{1}{2}\rho \mid v \mid^2 + \rho e) dx, \tag{3.23}$$

are conserved in Ω. Typical cases when this occurs are: a) Ω is $\Re^3$ and suitable conditions at infinity ensure that the fluxes of the mass, momentum and energy flow vectors through a large sphere vanish when the radius of the sphere tends to infinity; b) Ω is a box with periodicity conditions (flat torus), because essentially there are no boundaries. When Ω is a compact domain with the condition of specular reflection on Ω then the boundary terms on $\partial\Omega$ disappear in the mass and energy equations but not in the momentum equation; thus only M and E are conserved.

We also remark that in the so-called space-homogeneous case, the various quantities do not depend on x; all the space derivatives then disappear from Eqs. (3.18–3.20) and the densities $\rho, \rho v$, and $\frac{1}{2}\rho \mid v \mid^2 + \rho e$ are conserved, i.e. do not change with time.

The considerations of this section apply to all solutions of the Boltzmann equation. The definitions, however, can be applied to any positive function for which they make sense. In particular if we take f to be a Maxwellian in the form (2.7), we find that the constant vector v appearing there is actually the bulk velocity as defined in Eq. (3.2) while β and A are related to the internal energy e and the density ρ in the following way:

$$\beta = 3/(4e), \qquad A = \rho(4\pi e/3)^{-3/2}. \tag{3.24}$$

Furthermore the stress tensor turns out to be diagonal ($p_{ij} = (\frac{2}{3}\rho e)\delta_{ij}$, where δ_{ij} is the so-called Kronecker delta ($=1$ if $i = j; = 0$ if $i \neq j$)), while the heat-flow vector is zero.

We end this section with the definition of pressure p in terms of $f; p$ is nothing other than 1/3 of the spur or trace (i.e. the sum of the three diagonal terms) of p_{ij} and is thus given by:

$$p = \frac{1}{3} \int_{\Re^3} | c |^2 f dc. \tag{3.25}$$

If we compare this with the definition of the specific internal energy e, given in Eq. (3.12), we obtain the relation:

$$p = \frac{2}{3} \rho e. \tag{3.26}$$

This is the state equation that was already obtained in Chapter 1 through an elementary argument. In the case of a Maxwellian distribution, as we have seen, the stress tensor is diagonal; the common value of the three nonzero components coincides with the pressure.

It is not worthless to mention, at this point, that Eqs. (3.18–3.20) are *not* fluid-dynamic equations. Actually they cannot even be solved without first solving the Boltzmann equation to determine p_{ij} and q_i. There are situations, however, where the distribution function can be shown to be very close to a Maxwellian so that q_i and the anisotropic part of p_{ij} are negligible, and, by taking

$$q_i = 0, \qquad p_{ij} = p\delta_{ij}, \tag{3.27}$$

we can describe the gas by means of the Euler equations. How to pass from the kinetic regime (described by the Boltzmann equation) to the hydrodynamic regime (described by the Euler equations) will be described in Section 8, and some rigorous results regarding this transition are given in Chapter 11.

Problems

1. Prove that $\int_{\Re^3} c f d\xi = 0$, where c is the random velocity given by Eq. (3.5).
2. Prove Eq. (3.9).
3. Prove Eq. (3.11).
4. Prove Eq. (3.12).
5. Check Eqs. (3.18)–(3.20).
6. Check that the flows of mass and energy vanish at a boundary where the molecules are specularly reflected.
7. Prove that Eqs. (3.24) hold for a Maxwellian.
8. Prove that the heat-flow vector vanishes and the stress is diagonal if f is a Maxwellian.
9. In Chapter 1 we proved the state equation, Eq. (3.26), using the additional assumption that the averages of the squares of the three components of ξ were equal. Here Eq. (3.26) was obtained for a quite general situation. Explain why we needed the symmetry assumption in Chapter 1.

3.4 The H-Theorem

Let us consider a further application of the properties of the collision term $Q(f,f)$ of the Boltzmann equation:

(4.1) $$\frac{\partial f}{\partial t} + \xi \cdot \frac{\partial f}{\partial x} = \alpha Q(f,f).$$

If we multiply both sides of this equation by $\log f$ and integrate with respect to ξ, we obtain:

(4.2) $$\frac{\partial \mathcal{H}}{\partial t} + \frac{\partial}{\partial x} \cdot \mathcal{J} = \mathcal{S}$$

where

(4.3) $$\mathcal{H} = \int_{\Re^3} f \log f d\xi$$

(4.4) $$\mathcal{J} = \int_{\Re^3} \xi f \log f d\xi$$

(4.5) $$\mathcal{S} = \alpha \int_{\Re^3} \log\, f Q(f,f) d\xi.$$

Eq. (4.2) differs from the balance equations considered in the previous section because the right side, generally speaking, does not vanish. We know, however, that the Boltzmann inequality, Eq. (2.2), implies:

(4.6) $$\mathcal{S} \leq 0 \quad \text{and} \quad \mathcal{S} = 0 \quad \text{iff} \quad f \quad \text{is a Maxwellian.}$$

Because of this inequality, Eq. (4.2) plays an important role in the theory of the Boltzmann equation. We illustrate the role of Eq. (4.6) in the case of space homogeneous solutions. In this case the various quantities do not depend on x, and Eq. (4.2) reduces to

(4.7) $$\frac{\partial \mathcal{H}}{\partial t} = \mathcal{S} \leq 0.$$

This implies the so-called H-theorem (for the space homogeneous case) : $\mathcal{H}$ is a decreasing quantity, unless f is a Maxwellian (in which case the time derivative of $\mathcal{H}$ is zero). Remember now that in this case the densities ρ, ρv, and ρe are constant in time; we can thus build a Maxwellian M that has, at any time, the same ρ, v, and e as any solution f corresponding to given initial data. Since $\mathcal{H}$ decreases unless f is a Maxwellian (i.e. $f = M$), it is tempting to conclude that f tends to M when $t \to \infty$. The temptation is strengthened when we realize that $\mathcal{H}$ is bounded from below by $\mathcal{H}_M$ (Problem 1), the value taken by the functional $\mathcal{H}$ when $f = M$. In fact $\mathcal{H}$ is decreasing , its derivative is nonpositive unless it takes the value $\mathcal{H}_M$; one feels that $\mathcal{H}$ tends to $\mathcal{H}_M$! This conclusion is, however, unwarranted without a more detailed consideration of the source term $\mathcal{S}$ in Eq. (4.7). This is deferred to Chapter 6, when the existence and properties of f will be proved. Here we only remark that if $\mathcal{H}$ tends to $\mathcal{H}_M$, then it is easy to conclude that f tends to M, thanks to the inequality (see Problem 2):

$$f \log f - f \log M + M - f \geq cg\left(\frac{| f - M |}{M}\right) | f - M | \tag{4.8}$$

where c is a constant (independent of f) and

$$g(z) = \left\{ \begin{matrix} z & \text{if} \quad 0 \leq z \leq 1 \\ 1 & \text{if} \quad z \geq 1 \end{matrix} \right\}. \tag{4.9}$$

Integrating both sides of Eq. (4.8) gives

$$\mathcal{H} - \mathcal{H}_M \geq c\left[\int_{L_t} |f - M| d\xi + \int_{S_t} |f - M|^2 M^{-1} d\xi\right] \tag{4.10}$$

where L_t and S_t denote the sets (depending on t) where $| f - M |$ is larger (resp. smaller) than M. Since $\mathcal{H}$ is assumed to tend to $\mathcal{H}_M$, it follows that both integrals tend to zero when $t \to \infty$. The fact that the second integral tends to zero implies, by Schwarz's inequality, that

$$\int_{S_t} |f - M| d\xi \leq \left[\int_{S_t} |f - M|^2 M^{-1} d\xi\right]^{1/2} \left[\int_{S_t} M d\xi\right]^{1/2} \to 0. \tag{4.11}$$

Then

$$\int_{\Re^3} |f - M| d\xi = \int_{L_t} |f - M| d\xi + \int_{S_t} |f - M| d\xi \tag{4.12}$$

also tends to zero and f tends strongly to M in L^1.

If the state of the gas is not space homogeneous, the situation becomes more complicated. In this case it is convenient to introduce the quantity

$$H = \int_{\Omega} \mathcal{H} dx \tag{4.13}$$

where Ω is the space domain occupied by the gas (assumed here to be time-independent). Then Eq. (4.2) implies

$$\frac{dH}{dt} \leq \int_{\partial\Omega} \mathcal{J} \cdot n d\sigma \tag{4.14}$$

where n is the inward normal and $d\sigma$ the measure on $\partial\Omega$. Clearly, several situations may arise. Among the most typical ones, we quote:

1. Ω is a box with periodicity boundary conditions (flat torus). Then there is no boundary, $dH/dt \leq 0$ and one can repeat about H what was said about $\mathcal{H}$ in the space homogeneous case. In particular, there is a natural (space homogeneous) Maxwellian associated with the total mass, momentum, and energy (which are conserved as was remarked in the previous section).
2. Ω is a compact domain with specular reflection. In this case the boundary term also disappears because the integrand of $\mathcal{J} \cdot n$ is odd on $\partial\Omega$ and the situation is similar to that in case 1. There might seem to be a difficulty for the choice of the natural Maxwellian because momentum is not conserved, but a simple argument shows that the total momentum must vanish when $t \to \infty$. Thus M is a nondrifting Maxwellian.
3. Ω is the entire space. Then the asymptotic behavior of the initial values at ∞ is of paramount importance. If the gas is initially more concentrated at finite distances from the origin, one physically expects and can mathematically prove (with arguments akin to those of Theorem 9.5.1) that the gas escapes through infinity and the asymptotic state is vacuum.
4. Ω is a compact domain but the boundary conditions on $\partial\Omega$ are different from specular reflection. Then the asymptotic state may be completely different from a Maxwellian. This case will be described in more detail in Chapters 8 and 9.

Problems

1. Show that if $\mathcal{H}(f)$ is the functional defined in Eq. (4.3) and M is the Maxwellian with the same density, velocity, and internal energy as f, then $\mathcal{H}(f) \geq \mathcal{H}(M)$. (Hint: use the inequality $z\log z - z\log y + y - z \geq 0$, valid for non-negative y and z and the fact that $\log M$ is a collision invariant so that $\int_{\Re^3} \log M(f-M)d\xi = 0$. Another possibility is to find for what functions f the first variation $\delta \int f \log f d\xi$ vanishes under the constraint that $\int \psi f d\xi$ is given for any collision invariant ψ.
2. Show that inequality (4.8) (which is an improvement of the inequality used in the previous problem) is true (Hint: study the function $h(x)$ of the real variable x, defined by $h(x) = x\log x + 1 - x - cg(|\, x-1\,|)\,|\, x-1\,|$, with a sufficiently small constant c and then let $x = f/M$).

3.5 Loschmidt's Paradox

Boltzmann's H-theorem is of basic importance because it shows that his equation has a basic feature of irreversibility: the quantities $\mathcal{H}$ (in the space homogeneous case) and H (in other cases where the gas does not exchange mass and energy with a solid boundary) always decrease in time. This result seems to be in conflict with the fact that the molecules constituting the gas follow the laws of classical mechanics, which are time reversible. Accordingly, given a motion at $t = t_0$ with velocities $\xi^{(1)}, \xi^{(2)}, \ldots, \xi^{(N)}$, we can always consider the motion with velocities $-\xi^{(1)}, -\xi^{(2)}, \ldots, -\xi^{(N)}$ (and the same positions as before) at $t = t_0$; the backward evolution of the latter state will be equal to the forward evolution of the original one. Therefore if $dH/dt < 0$ in the first case, we shall have $dH/d(-t) < 0$ or $dH/dt > 0$ in the second case, which contradicts Boltzmann's H-theorem.

This paradox is mentioned by Thomson in a short paper[21], which is seldom quoted. This paper appeared in 1874 and contains a substantial part of the physical aspects of the modern interpretation of irreversibility not only for gases, but also for more general systems made up of molecules. Thomson notes that "the instantaneous reversal of the motion of every moving particle of a system causes the system to move backwards, each particle of it along its old path, and at the same speed as before, when again in the same position. That is to say, in mathematical language, any solution remains a solution when t is changed into $-t$... If, then, the motion of every particle of matter in the universe were precisely reversed at any instant, the course of nature would be simply reversed for ever after. The bursting bubble of foam at the foot of a waterfall would reunite and descend into the water; the thermal motions would reconcentrate their energy, and throw the mass up the fall in drops re-forming into a close column of ascending water. Heat generated by the friction of solids and dissipated by conduction, and radiation with absorption, would come again to the place of contact, and throw the moving body back against the force to which it had previously yielded. Boulders would recover from the mud the materials required to rebuild them into their previous jagged forms, and would become reunited to the mountain peak from which they had formerly broken away. And if the materialistic hypothesis of life were true, living creatures would grow backwards, with conscious knowledge of the future, but no memory of the past, and would become again unborn." He also remarks: "If no selective influence, such as that of the ideal 'demon,' guides individual molecules, the average result of their free motions and collisions must be to equalize the distribution of energy among them in the gross..." In other words, the impossibility of observing macroscopic phenomena that run backwards with respect to those actually observed is, in the last analysis, due to the large number of molecules present even in macroscopically small volumes.

Josef Loschmidt, to whom the paradox is usually attributed, mentioned it briefly in the first[17] of four articles devoted to the thermal equilibrium of a system of bodies subject to gravitational forces. His intention was to show that the heat death of the universe (which seems to follow from the second principle of thermodynamics) is not inevitable. In a passage of the paper he says that the entire course of events in the universe would be retraced if at some instant the velocities of all its parts were reversed. In spite of the obscure arguments of Loschmidt, Boltzmann quickly got the point and gave a thorough discussion of the paradox[8], ending with a conclusion similar to that of Thomson.

Nowadays we are more prepared to discuss this kind of question. In fact, we remark that when giving a justification of the Boltzmann equation in the previous chapter, we used the laws of elastic collisions and the continuity of the probability density at the impact to express the distribution functions corresponding to an after-collision state in terms of the distribution functions corresponding to the state before the collision, rather than the latter in terms of the former. It is obvious that the first way is the right one to follow if the equations are to be used to predict the future from the past and *not vice versa*; it is clear, however, that this choice introduced a connection with the everyday concepts of past and future, which are extraneous to molecular dynamics and are based on our macroscopic experience. When we took the Boltzmann–Grad limit we obtained equations that describe the statistical behavior of the gas molecules: a striking consequence of our choice is that the Boltzmann equation describes motions for which the quantity H (or $\mathcal{H}$) has a tendency to decrease, while the opposite choice would have led to an equation having a negative sign in front of the collision term, and hence describing only motions with increasing H. We must remark that, in order to derive the Boltzmann equation, we took special (although highly probable) initial data; thus certain special data were excluded. As the discussion in the previous chapter (Section 3) shows, these excluded data correspond to a state in which the molecular velocities of the molecules that are about to collide show an unusual correlation. This situation can be simulated by studying the dynamics of many interacting particles on a computer and leads to an evolution in which there is an increasing H, as expected, while "randomly" chosen initial data invariably lead to an evolution with decreasing H[1,4]. In other words, the fact that H decreases is not an intrinsic property of the dynamical system but a property of the level of description.

It is not the place here to discuss the relation of the H-theorem with the notions of past and future[11,12]. We only comment on an amusing example of Miller and Shinbrot[18], which shows the pitfalls of the subject. They defined a system on $\Re^2$ with the evolution $T_t(x, y) = (xe^t, ye^{-t})$ or, in differential form:

$$\frac{dx}{dt} = x; \qquad \frac{dy}{dt} = -y. \tag{5.1}$$

This system is clearly time-reversible because $t \Rightarrow -t$, $x \Rightarrow y$, $y \Rightarrow x$

changes the system into itself. The corresponding "Liouville equation" is:

$$\frac{\partial P}{\partial t} + x\frac{\partial P}{\partial x} - y\frac{\partial P}{\partial y} = 0. \tag{5.2}$$

Miller and Shinbrot[18] define a reduced description based on

$$P^{(1)} = \int_{\Re} P(x, y)dy; \tag{5.3}$$

$P^{(1)}$ satisfies:

$$\frac{\partial P^{(1)}}{\partial t} + x\frac{\partial P^{(1)}}{\partial x} + P^{(1)} = 0. \tag{5.4}$$

Hence, if we define:

$$H = \int_{\Re} P^{(1)} \mathrm{log} \mathrm{P}^{(1)} dx, \tag{5.5}$$

we obtain:

$$\frac{dH}{dt} = -\int_{\Re} P^{(1)} dx \leq 0, \tag{5.6}$$

which shows a formal resemblance to the H-theorem. It would seem that we have obtained a cheap example of how to obtain irreversibility from a time-reversible model, without the subtleties related to the Boltzmann-Grad limit. The pitfall lies in the fact that when we perform the transformation $t \Rightarrow -t$, $x \Rightarrow y$, $y \Rightarrow x$, the functional H, defined by Eq. (5.5) does not transform into itself; Boltzmann's H, on the other hand, transforms into itself when all the molecular velocities are changed into their opposites.

We remark that the term "time-reversible" has different meanings in the mathematical literature; a detailed discussion of the various definitions is given in a paper by R. Illner and H. Neunzert[16].

It is perhaps not out of place to comment on a statement that is frequently made, to the effect that no kind of irreversibility can follow by correct mathematics from the analytical dynamics of a conservative system and hence some assumption of kinetic theory must contradict analytical dynamics. It should be clear that it is not a new assumption that is introduced, but the fact that we study asymptotic properties of a conservative system in the Boltzmann-Grad limit, under the assumption that the initial probability distributions are factorized in the way indicated in Eq. (5.2).

Problems

1. Show that Eq. (5.2) is the appropriate Liouville equation for the dynamical system defined by Eq. (5.1).
2. Show that Eq. (5.4) follows from Eq. (5.2).
3. Show that Eq. (5.6) holds.
4. Obtain Eq. (5.6) from the explicit dynamics of the Miller-Shinbrot model, i.e. $T_t(x, y) = (xe^t, ye^{-t})$.

3.6 Poincaré's Recurrence and Zermelo's Paradox

There is another objection that can be raised against the H-theorem when presented as a rigorous consequence of the laws of dynamics. The starting point is a theorem of Poincaré[20] (the so-called recurrence theorem), which says that any conservative system, whose possible states form a compact set in phase space, will return arbitrarily close to its initial state, for almost any choice of the latter, provided we wait long enough. This applies to a gas of hard sphere molecules, enclosed in a specularly reflecting box, because the set of the possible states S with a given energy is compact and has a finite measure $\mu(S)$ (induced by the Lebesgue measure). If A is a subset of S with measure $\mu(A)$, which evolves into a set A_t at time t (according to the dynamics of the system), then $\mu(A_t) = \mu(A)$. To prove Poincaré's theorem, let us assume that there is a subset A whose points will never come back to A. We choose A small enough and τ large enough so that A_τ and A do not overlap (if this is impossible, the theorem is trivially true); then none of the sets $A_{2\tau}$, $A_{3\tau}, \ldots$ overlap, because, if $A_{n\tau}$ and $A_{(n+k)\tau}$ had points in common, then by tracing the motion backwards and using the uniqueness of the motion through any given phase space point, it would follow that A and $A_{k\tau}$ must have common points, and this would contradict the definition of A. If A, A_τ, $A_{2\tau}, \ldots$ do not overlap, then since $\mu(A) = \mu(A_\tau) = \mu(A_{2\tau}) = \ldots$, the total measure of the union of these disjoint sets would be infinite (which is impossible because $\mu(S) < \infty$), unless $\mu(A) = 0$, and Poincaré's recurrence theorem is proved.

This theorem implies that our molecules can have, after a "recurrence time," positions and velocities so close to the initial ones that the one particle distribution function f would be practically the same; therefore H should also be practically the same, and if it decreased initially, then it must have increased at some later time. This paradox goes under the name of Zermelo, who stated it in 1896[23], but it was actually mentioned before in a short paper by Poincaré[19]. The traditional answer to Zermelo's paradox was given by Boltzmann himself[5]: the recurrence time is so large that, practically speaking, one would never observe a significant portion of the recurrence cycle. In fact, according to an estimate made by Boltzmann[5], the recurrence time for a typical amount of gas is a huge number even if the estimated age of the universe is taken as the time unit.

In view of the fact that we claim validity for the Boltzmann equation in the Boltzmann-Grad limit only, we do not have to worry about the recurrence paradox; in fact, the set S is no longer compact when $N \to \infty$ and the recurrence time is expected to go to infinity with N (at a much faster rate).

3.7 Equilibrium States and Maxwellian Distributions

The trend toward a Maxwellian distribution expressed by the H-theorem indicates that this particular distribution is a good candidate to describe a gas in a (statistical) equilibrium state. In order to prove that a Maxwellian describes the equilibrium states of a gas, however, we must give a definition of equilibrium. Intuitively, a gas is in equilibrium if, in a situation where it does not exchange mass and energy with other bodies, its state does not change with time. Thus for the moment we define an equilibrium state to be one of a gas in a steady situation in a box with periodic or specular reflection boundary conditions. It is then clear that the distribution function must be a Maxwellian; in fact, Eq. (4.2) implies (when $\mathcal{H}$ does not depend on time):

$$(7.1) \qquad -\int_{\partial\Omega} \mathcal{J} \cdot n d\sigma = \int_{\Omega} S dx \leq 0$$

where n is the inward normal and equality holds if and only if f is Maxwellian. But $\mathcal{J} \cdot n$ is zero for the situation under consideration and the only possibility is that f be a Maxwellian. We must now impose the condition that this Maxwellian must be a steady solution of the Boltzmann equation i.e. it must satisfy

$$(7.2) \qquad \xi \cdot \frac{\partial f}{\partial x} = \alpha Q(f, f).$$

This readily implies that both the right- and the left-hand sides of the Boltzmann equation must vanish; as a consequence, the parameters A, β, and v appearing in the Maxwellian

$$(7.3) \qquad f = A \exp(-\beta \mid \xi - v \mid^2)$$

must be of the form $v = v_0 + \omega \wedge x, A = A_0 \exp[| \ \omega \ |^2 | \ x \ |^2 \ -(\omega \cdot x)^2]$, β =constant (where v_0 and ω are constant vectors and A_0 is a constant scalar; $\wedge$ denotes the vector product). If the periodicity condition or the specular reflection boundary condition is imposed, it turns out that A and v must also be constant (and not space-dependent). Thus a Maxwellian with constant parameters is the most general equilibrium solution of the Boltzmann equation.

The question immediately arises, whether there are solutions of the Boltzmann equation that are Maxwellians with parameters depending on x and t. Since the right-hand side of the Boltzmann equation vanishes identically if f is a Maxwellian, it turns out that a Maxwellian, i.e., a function of the form specified in Eq. (7.3), can be a solution of the Boltzmann equation if and only if A, β, and v depend on t and x in such a way that f also satisfies:

$$\frac{\partial f}{\partial t} + \xi \cdot \frac{\partial f}{\partial x} = 0. \tag{7.4}$$

Since the general solution of this equation has the form $f = f(x - vt, x \wedge \xi, \xi)$, it turns out that there are several solutions of this form; they were investigated by Boltzmann[7] in 1876. Among them we quote the case met above in which $v = v_0 + \omega \wedge x$, $A = A_0 \exp[\mid \omega \mid^2 \mid x \mid^2 - (\omega \cdot x)^2]$, $\beta =$ constant (with v_0, A_0 and ω constants) and the case in which $A =$ constant, $\beta = \beta_0(1 + t/t_0)^2, v = (t + t_0)^{-1}x$. The latter solution describes a compression if t_0 is negative (but the solution ceases to exist for $t > \mid t_0 \mid$), and an expansion if $t_0 > 0$ (in which case the solution exists globally).

Problems

1. Prove the statement that in Eq. (7.3) the parameters must be constant in an equilibrium state.
2. Check that the general solution of Eq. (7.4) is of the form $f = f(x - vt, x \wedge \xi, \xi)$.
3. Find all the Maxwellians that are solutions of the Boltzmann equation (see Refs. 4, 6, and 7).

3.8 Hydrodynamical Limit and Other Scalings

A point of great relevance in the study of the Boltzmann equation is the analysis of the scaling properties: a large system, as we shall see, can be more conveniently described in terms of fluid-dynamic equations, when it is considered on a suitable space-time scale.

Let us consider a gas obeying the Boltzmann equation, confined to a large box Λ_ϵ of side ϵ^{-1}, ϵ being a parameter to be sent to zero. Let $f^\epsilon = f^\epsilon(x, \xi, t)$, $x \in \Lambda_\epsilon$, be the number density of the particles. We assume that the total number of particles is proportional to the volume of the box, i.e. we normalize f^ϵ as follows:

$$\int_{\Lambda^\epsilon \times \Re^3} f^\epsilon(x,\xi)dxd\xi = \epsilon^{-3}. \tag{8.1}$$

We also assume that the time evolution is given by the Boltzmann equation

$$\frac{\partial f^\epsilon}{\partial t} + \xi \cdot \frac{\partial f^\epsilon}{\partial x} = \alpha Q(f^\epsilon, f^\epsilon) \tag{8.2}$$

and look at the behavior of the system on the scale of the box; in this case we have to use appropriate space and time variables, because in terms of the variable x, the box is of size ϵ^{-1}, while we would like to regard it as being of order unity. Thus we introduce the new independent and dependent variables

$$r = \epsilon x, \quad \tau = \epsilon t; \qquad (r \in \Lambda) \tag{8.3}$$

$$\hat{f}(r,\xi,t) = f^\epsilon(x,\xi,t). \tag{8.4}$$

Clearly, $\hat{f}$ describes the gas on the scale of the box and is normalized to unity:

$$\int_{\Lambda \times \Re^3} \hat{f}(r,\xi)drd\xi = 1. \tag{8.5}$$

The picture of the (same) system in terms of the variables r and τ is called *macroscopic*, while the picture in terms of x and t is called *microscopic*. Note that on the macroscopic scale the typical length for the kinetic phenomena described by the Boltzmann equation, i.e. the mean free path, turns out to be of order ϵ (since it is of order unity on the scale described by x). Thus sending the size of the box to infinity like ϵ^{-1} or the mean free path to zero like ϵ are equivalent limiting processes.

In terms of the macroscopic variables, Eq. (8.2) reads as follows:

$$\frac{\partial \hat{f}}{\partial \tau} + \xi \cdot \frac{\partial \hat{f}}{\partial r} = \epsilon^{-1}\alpha Q(\hat{f}, \hat{f}). \tag{8.6}$$

Thus, on the scale of the box, the mean free path (inversely proportional to the factor in front of Q) is reduced by a factor ϵ. This means that the average number of collisions diverges when $\epsilon \to 0$ and the collisions become dominant. For Eq. (8.6) to hold, $Q(\hat{f}, \hat{f})$ must be small of order ϵ, so that $\hat{f}$

is expected to be close to a Maxwellian, whose parameters are, in general, space- and time-dependent. In this case the macroscopic balance equations (3.18–3.20) can be closed through Eqs. (3.26) and (3.27) t o obtain the Euler equations for a perfect compressible fluid. These considerations can be made rigorous and will be illustrated in detail in Chapter 11.

For now, let us mention other physical considerations concerning our scaling. To this end, let us consider a small portion of fluid in a neighborhood of a point $r \in \Lambda$ (Fig. 4): By the scaling transformation this portion is magnified into a large system of particles, which is seen to evolve on a long time scale. It will have a tendency to "thermalize" so that its distribution will quickly become a local Mawellian with parameters $A(\epsilon^{-1}r), \beta(\epsilon^{-1}r), v(\epsilon^{-1}r)$ suitably related to the fluid-dynamic fields ρ, e, v. These will evolve according to the Euler equations on a much slower scale of times. Thus we have illustrated two different time scales. The fast one, which we call *kinetic*, is of the order of the time necessary to reach a local equilibrium, a process described by the Boltzmann equation. The slow scale, which we call *fluid-dynamic*, describes the time evolution of the parameters of the local Maxwellian.

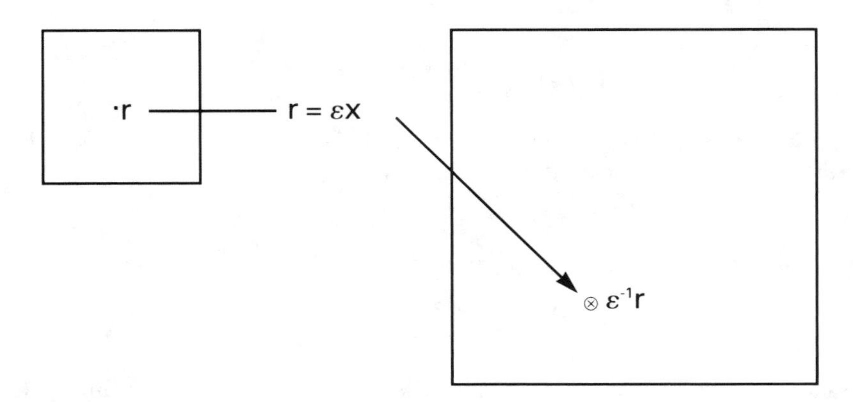

FIGURE 4.

We notice that the same considerations could apply to the Newton equations (or the corresponding hierarchy of equations for the s-particle distribution functions). Although one might expect that the Newton equations, under the above scaling, should yield the Euler equations, our ignorance of the long–time behavior of Hamiltonian systems is such that, at the moment, we are quite far from a rigorous derivation of the hydrodynamical equations starting from the basic laws of classical mechanics.

Let us now analyze another scaling, which clarifies the nature of the

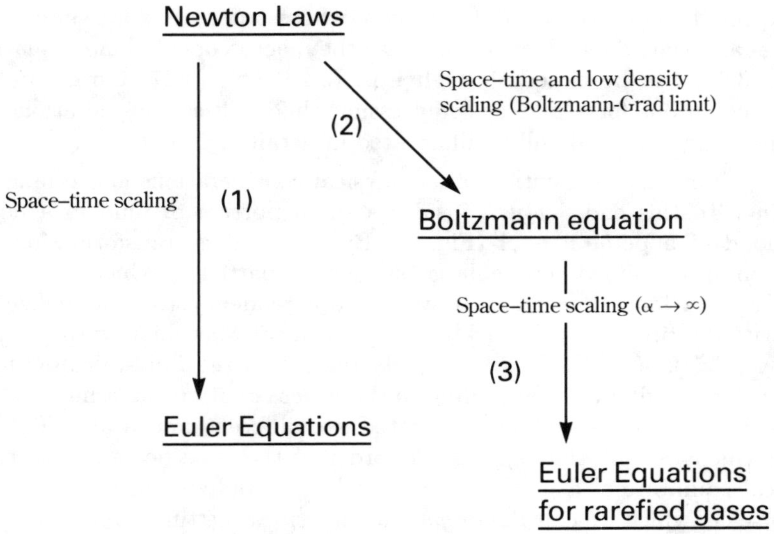

Newton Laws

Space–time and low density scaling (Boltzmann-Grad limit)

(2)

Space–time scaling

(1)

Boltzmann equation

Space–time scaling ($\alpha \to \infty$)

(3)

Euler Equations

Euler Equations for rarefied gases

FIGURE 5.

Boltzmann-Grad limit. We now require the number of particles in Λ_ϵ to be of the order of ϵ^{-2}, i.e., we replace Eq.(8.1) by

$$\int_{\Lambda^\epsilon \times \Re^3} f^\epsilon(x,\xi)\,dx\,d\xi = \epsilon^{-2}. \tag{8.7}$$

In order to keep the normalization to unity of $\hat{f}(r,\xi,t)$, expressed by Eq. (8.5) we change the scaling from Eq. (8.4)

$$\hat{f}(r,\xi,t) = \epsilon^{-1} f^\epsilon(x,\xi,t). \tag{8.8}$$

Then we obtain, in place of Eq. (8.6)

$$\frac{\partial \hat{f}}{\partial \tau} + \xi \cdot \frac{\partial \hat{f}}{\partial r} = \alpha Q(\hat{f},\hat{f}). \tag{8.9}$$

Hence the Boltzmann equation is invariant for the space-time scaling (4.3), provided that the particle number goes as the power 2/3 of the volume. This invariance property suggests that the Boltzmann equation can be derived from the BBGKY hierarchy via a space–time scaling with the total number of particles proportional to ϵ^{-2}; this is what can be checked at a formal level and is essentially what we did in Section 2.5, where ϵ, of course, was the molecular diameter σ. It is also clear why the Boltzmann-Grad limit is frequently called the *low–density* limit; in fact, in this limit, the particle number in a large box divided by the volume of the box goes to zero. The number of collisions per unit (macroscopic) time stays finite, while it diverges in the hydrodynamical limit, as we saw before.

We summarize the content of this section in Fig. 5.

As we said earlier, nothing is known at a rigorous level about the limit indicated by arrow (1); the limit corresponding to arrow (2) has been proved for short times and for an expanding rare cloud of gas, as we demonstrate in the next chapter; the limit (3) is quite well understood.

Notice that the Euler equations indicated by arrows 1 and 3 in Fig. 5 are in general not the same; the first describe the hydrodynamical behavior of a particle system, the second that of a rarefied gas evolving according to the Boltzmann equation. As a consequence, the state equation relating pressure and density in the first case is in general not that of a perfect gas, as in the second case. This important difference will be discussed in some detail in Chapter 11.

Problem

1. Check that the low–density limit based on Eqs. (8.7) and (8.8) is formally equivalent to the limiting procedure used in Section 2.5 to obtain Eq. (5.1) from Eq. (4.13).

References

1. B. J. Alder and T. E. Wainwright, "Studies in molecular dynamics II. Behavior of a small number of elastic spheres," *J. Chem. Phys.* **33**, 1439 (1960).
2. L. Arkeryd, "On the Boltzmann equation. Part II: The full initial value problem," *Arch. Rat. Mech. Anal.* **45**, 17–34 (1972).
3. L. Arkeryd and C. Cercignani, "On a functional equation arising in the kinetic theory of gases," *Rend. Mat. Acc. Lincei* **s.9**, **v.1**, 139–149 (1990).
4. A. Bellemans and J. Orban, "Velocity inversion and irreversibility in a dilute gas of hard disks," *Phys. Letters* **24a**, 620 (1967).
5. L. Boltzmann, "Entgegnung auf die wärmetheoretischen Betrachtungen des Hrn. E. Zermelo," Annalen der Physik **57**, 773–784 (1896).
6. L. Boltzmann, "Über das Wärmegleichgewicht von Gasen, auf welche äussere Kräfte wirken," *Sitzungsberichte der Akademie der Wissenschaften Wien* **72**, 427–457 (1875).
7. L. Boltzmann, "Über die Aufstellung und Integration der Gleichungen, welche die Molekularbewegungen in Gasen bestimmen," *Sitzungsberichte der Akademie der Wissenschaften Wien* **74**, 503–552 (1876).
8. L. Boltzmann, "Über die Beziehung eines allgemeinen mechanischen Satzes zum zweiten Hauptsatze der Wärmetheorie," *Sitzungsberichte Akad. Wiss.*, II,**75**, 67–73 (1877).
9. T. Carleman, *Problèmes Mathématiques dans la Théorie Cinétique des Gaz,* Almqvist & Wiksell, Uppsala (1957).
10. C. Cercignani, "Are there more than five linearly independent collision invariants for the Boltzmann equation?" *J. Statistical Phys.* **58**, 817–823 (1990).
11. C. Cercignani, *The Boltzmann Equation and Its Applications,* Springer, New York (1988).
12. C. Cercignani "Le radici fisiche e matematiche dell'irreversibilità temporale," (accompanied by a translation into English), *Alma Mater Studiorum* **2**, 37–52 (1989).
13. H. Grad, "On the kinetic theory of rarefied gases," *Communications on Pure and Applied Mathematics,* **2**, 331–407 (1949).

14. T. H. Gronwall, "A functional equation in the kinetic theory of gases," *Annals of Mathematics (2)* **17**, 1–4 (1915).
15. T. H. Gronwall, "Sur une équation fonctionelle dans la théorie cinétique des gaz," *Comptes Rendus de l'Académie des Sciences* (Paris) **162**, 415–418 (1916).
16. R. Illner and H. Neunzert, "The concept of irreversibility in the kinetic theory of gases," *Transport Th. and Stat. Phys.* **16 (1)**, 89–112 (1987).
17. J. Loschmidt, "Über den Zustand des Wärmegleichgewichtes eines Systems von Körpern mit Rücksicht auf die Schwerkraft," *Wien. Ber.* **73**, 128 (1876).
18. G. Miller and M. Shinbrot, "The H-theorem and reversibility," *Transport Th. and Stat. Phys.* **12**, 195 (1983).
19. H. Poincaré, "Le mécanisme et l'expérience," *Revue de Metaphysique et de Morale* **1**, 534–537 (1893).
20. H. Poincaré, "Sur le problème des trois corps et les équation de la dynamique," *Acta Mathematica* **13**, 1–270 (1890).
21. W. Thomson (Lord Kelvin), "The kinetic theory of the dissipation of energy," *Proc. Royal Soc. Edinburgh*, Vol. 8, 325–334 (1874).
22. C. Truesdell and R. G. Muncaster, *Fundamentals of Maxwell's Kinetic Theory of a Simple Monatomic Gas*, Academic Press, New York (1980).
23. E. Zermelo, "Über einen Satz der Dynamik und die mechanische Warmetheorie," *Annalen der Physik* **54**, 485–494 (1896).

4
Rigorous Validity of the Boltzmann Equation

4.1 Significance of the Problem

In Chapter 2 we gave a formal derivation of the Boltzmann equation from the basic laws of mechanics. In particular, we introduced the Liouville equation, the BBGKY hierarchy, the Boltzmann hierarchy, and the Boltzmann equation, and we discussed the assumptions that allowed us to make the transitions from each of those to the next. The objective of this chapter is to do all these steps rigorously, wherever possible. In particular, our discussion will lead to a rigorous validity and existence result for the Boltzmann equation, locally for a general situation and globally for a rare gas cloud in vacuum.

The importance of the problem is evident: We have to settle the fundamental question of whether the *irreversible* Boltzmann equation can be rigorously obtained from *reversible* mechanics. The answer to this query is yes, as we shall see here. In particular, there is no contradiction between the second law of thermodynamics and the reversibility of molecular dynamics, at least for the hard-sphere model of a rarefied gas.

We will start, as in Chapter 2, from hard-sphere dynamics and the Liouville equation. One difficulty we face is to give a rigorous derivation of the BBGKY hierarchy for hard spheres—the problem we must deal with is that the flow operators $(T^t)_{t\in\Re}$, introduced in Chapter 2, are not defined through multiple collisions (here, and in the sequel, "multiple collision" will always mean simultaneous contact of more than two hard spheres). Also, they are not globally defined on phase points leading to infinitely many collisions in finite time. Therefore, we first have to define the hard-sphere dynamics, at least almost everywhere, by deleting suitable sets (of measure zero) of phase points leading to configurations for which the flow cannot be continued any longer. Then we have to prove that the dynamics, in terms of probability distributions give rise to the BBGKY hierarchy introduced in Chapter 2.

The latter is not an obvious step. For example, as we anticipate that the hard-sphere flow T_t is only almost everywhere defined, the marginal distribution densities $P^{(s)}$ will only be L^1-functions, no matter how regular they were at time zero. However, the right-hand side of the BBGKY hierarchy involves restrictions of $P^{(s)}$ to sets of codimension one, a restriction that does not immediately make sense at this level of regularity. We therefore have to justify the BBGKY hierarchy first and explain in what sense it holds. This rigorous derivation is important, but unfortunately rather technical. As it is probably of marginal interest for most readers, we confine it to Appendix 4B together with certain other properties of the hard-sphere dynamics.

The next and still more difficult step is to take the Boltzmann–Grad limit and prove that the solution of the BBGKY hierarchy converges to a solution of the Boltzmann hierarchy. For short times, this was first done by O. Lanford in a classical paper [19]. Lanford's paper contains only a sketch of the proof; a more detailed treatment is given by H. Spohn in his recent monograph [23]. The approach we take here uses somewhat different estimates. We also present an extension [17,18], due to two of the authors, which gives a global validity result for a gas cloud in all space if the mean free path is large. This is the only global result known so far. In both cases, we also prove propagation of chaos and existence and uniqueness of solutions for the Boltzmann equation.

4.2 Hard-Sphere Dynamics

Consider a system of N hard spheres of equal mass and diameter $\sigma > 0$ in a domain $\Lambda \subset \Re^3$. Sometimes, we let $\Lambda = \Re^3$. If $\Lambda \neq \Re^3$, we assume that $\partial\Lambda$ is so smooth that a unit normal to $\partial\Lambda$ exists at every $x \in \partial\Lambda$. The state of the system is given by a phase point $z = (x_1, \ldots, x_N, \xi_1, \ldots, \xi_N) \in \Lambda \times \Re^{3N}$. The phase space is

$$\Gamma = \{z \in \Lambda^N \times \Re^{3N}; |x_i - x_j| \geq \sigma, i \neq j\}.$$

We say that two particles at x_i and $x_j = x_i + n\sigma$, $n \in S^2$, are in an ingoing collision configuration if their velocities ξ_i, ξ_j satisfy $n \cdot (\xi_i - \xi_j) > 0$ (in a grazing configuration of $n \cdot (\xi_i - \xi_j) = 0$, and in an outgoing configuration if $n \cdot (\xi_i - \xi_j) < 0$). Let the collision transformation J be defined by

$$J : (\xi_i, n, \xi_j) \longrightarrow (\xi_i', -n, \xi_j'), \tag{2.1}$$

where ξ_i' , ξ_j' are given by (2.1.6). J is easily seen to be an involution, i.e., $J^2 = id$; as proved in Section (2.1), it preserves Lebesgue measure on $\Re^3 \times S^2 \times \Re^3$ (this also follows from $J^2 = id$, see Problem 1), and it takes ingoing (outgoing) configurations into ingoing (outgoing) configurations.

We remind the reader of the laws of momentum and energy conservation (2.1.5). Between collisions, the spheres move on straight lines with their velocities unchanged (there are no outer forces or gravitational forces between the particles). If a particle hits $\partial\Lambda$ at a point x with velocity ξ_i, it gets reflected with velocity

$$\xi_i' = \xi_i - 2n(x)(n(x) \cdot \xi_i), \tag{2.2}$$

where $n(x)$ is the inner normal at x to $\partial\Lambda$. There are many other possible boundary conditions, but we only discuss (2.2) for simplicity. The condition (2.2) satisfies $\xi_i'^2 = \xi_i^2$, i.e., the particle neither gains nor loses energy.

Eqs. (2.1), (2.2), and the free flow determine completely the time evolution of all phase points z for which the particles experience (in backward or forward evolution) only pair collisions and hit the boundary $\partial\Lambda$ only in isolated collisions. If the time evolution leads a phase point into a triple or higher-multiple collision, or into a situation where two particles collide with each other and at the same time with $\partial\Lambda$, the flow through such situations is not determined (it is easy to see that momentum and energy conservation leave too many degrees of freedom). Fortunately, we have the following.

(4.2.1) Theorem. *The following sets are of Lebesgue measure zero in phase space:*

1. *the set of all phase points that are led into a multiple collision under forward or backward evolution,*
2. *the set of all phase points such that there is a cluster point of collision instants under forward or backward evolution,*
3. *the set of all phase points such that there is a cluster point of collision instants with the boundary under forward or backward evolution.*

We discuss Theorem 4.2.1 in Appendix 4.A. For the moment we observe that by deleting the null sets defined in the theorem from phase space, we arrive at a set $\Gamma_0 \subset \Lambda^N \times \Re^{3N}$ on which the time evolution of every phase point z is globally defined backward and forward.

For $z \in \Gamma_0$, let $T^t z$ be the state of the system t units of time later. The family $\{T^t; t \in \Re\}$ is a group: $T^0 = id$, $T^t \circ T^s = T^{t+s}$ for all $t, s \in \Re$.

We refer to this group as "the flow in phase space." It has the additional property of *mechanical reversibility*, which we formulate as follows.

Define an involution $S : \Gamma_0 \longrightarrow \Gamma_0$ by

$$S(x_1 \ldots x_N, \xi_1 \ldots \xi_N) = (x_1 \ldots x_N, -\xi_1 \ldots -\xi_N),$$

then

(2.3) $$T^t S T^t z = Sz$$

for all $z \in \Gamma_0$, $t \in \Re$.

Eq. (2.3) is a consequence of the involutive property of J. We can equivalently write $ST^t = T^{-t}S$ on Γ_0. Equation (2.3) must not be confused with *reversibility with time inversion* ($T^{-t} \circ T^t = id$) or *Poincaré–reversibility* if Λ is bounded (see Chapter 2), which are also true for $\{T^t\}_{t\in\Re}$.

We conclude this section by discussing some properties of the flow T^t that will turn out essential for the validity proof for the Boltzmann equation describing the evolution of a rare gas cloud in vacuum, and for the rigorous derivation of the BBGKY hierarchy. For convenience, we abbreviate

$$(x(t), \xi(t)) = (x_1(t) \dots x_N(t), \xi_1(t) \dots \xi_N(t)) = T^t z,$$

then $x_i(t), \xi_i(t) \in \Re^3$, $x(t), \xi(t) \in \Re^{3N}$, $z = (x(0), \xi(0))$.

Also, we set

$$x^0(t) = x(0) + t\xi(0), \ \ \xi^0(t) = \xi(0),$$

then the so defined group T_0^t describes the (fictitious) evolution of a system in which the particles do not interact.

We focus on the case where $\Lambda = \Re^3$. In this case, we have the following.

(4.2.2) Lemma.

1. $\|\xi(t)\| = \|\xi(0)\|$ *(this is also true if* $\Lambda \neq \Re^3$ *and if the boundary condition preserves energy)*
2. $\sum_{i=1}^N \xi_i(t) = \sum_{i=1}^N \xi_i(0)$.
3. $\sum_{i=1}^N x_i(t) = \sum_{i=1}^N x_i(0) + t\sum_{i=1}^N \xi_i(0)$.

Proof. 1 and 2 are immediate from the conservation of energy and momentum. 3 also follows from momentum conservation— the details are left as an exercise. □

Let $z \in \Gamma_0$, $t > 0$, and consider $T^t z$. Suppose that during [0,t], there are k collisions at $0 \leq t_1 \leq t_2 \leq \dots \leq t_k \leq t$, and that (η_i, n_i, η_i') is the ingoing collision configuration in the ith collision.

(4.2.3) Lemma.

$$\|x(t)\|^2 = \|x^0(t)\|^2 + 2\sigma \sum_{i=1}^k (t - t_i) n_i \cdot (\eta_i - \eta_i').$$

In particular, because $n_i \cdot (\eta_i - \eta_i') > 0$, $\|x(t)\| \geq \|x^0(t)\|$, *and* $\|x(t)\| \longrightarrow \infty$ *as* $t \longrightarrow \infty$ *provided that* $\xi(0) \neq 0$.

Proof. Let $I(T^t z) = \|x(t)\|^2$, $(T^{t_i} z)^- = \lim_{t \nearrow t_i} T^t z$, $(T^{t_i} z)^+ = \lim_{t \searrow t_i} T^t z$, then, by energy conservation,

$$
\begin{aligned}
I(T^t z) &= I(T_0^{t-t_k}(T^{t_k} z)^+) \\
&= I((T^{t_k} z)^+) + 2(t - t_k) \sum_{i=1}^{N} x_i(t_k) \cdot \xi_i(t_k)^+ + (t - t_k)^2 \|\xi(t_k)\|^2.
\end{aligned}
$$

For simplicity, suppose that only one collision happens at time t_k, with ingoing velocities η_k, η_k' and outgoing velocities ζ_k, ζ_k'. If the colliding particles are at y_k, y_k', we have

$$
\begin{aligned}
y_k \cdot \zeta_k + y_k' \cdot \zeta_k' &= y_k(\zeta_k + \zeta_k') + (y_k' - y_k)\zeta_k' \\
&= y_k(\eta_k + \eta_k') + (y_k' - y_k)\zeta_k' \\
&= y_k \cdot \eta_k + y_k' \cdot \eta_k' + (y_k' - y_k)(\zeta_k' - \eta_k') \\
&= y_k \cdot \eta_k + y_k' \cdot \eta_k' + \sigma n_k \cdot (\eta_k - \eta_k').
\end{aligned}
$$

It follows that

$$
I(T^t z) = I(T_0^{t-t_k}(T^{t_k} z)^-) + 2\sigma(t - t_k) n_k \cdot (\eta_k - \eta_k').
$$

Repeated application of this calculation leads to the assertion. □

Later we will need a small generalization of Lemma 4.2.3, which we formulate and prove now.

(4.2.4) Lemma. *$I(T_0^s \circ T^t z) \geq I(T_0^{t+s} z)$ whenever $s \geq 0$, $t \geq 0$ or $s \leq 0$, $t \leq 0$.*

Proof. For $s \geq 0$, $t \geq 0$ the proof of Lemma 4.2.3 applies without change. For $s \leq 0$, $t \leq 0$, let S again denote velocity inversion. Using the identities $I(Sz) = I(z)$, $ST^{-t} = T^t S$ and $ST_0^{-t} = T_0^t S$, we find

$$
\begin{aligned}
I(T_0^{-|s|} T^{-|t|} z) = I(S T_0^{|s|} S T^{-|t|} z) &= I(T_0^{|s|} T^{|t|} S z) \\
&\geq I(T_0^{|s|+|t|} S z) \\
&= I(T_0^{-|s|-|t|} z).
\end{aligned}
$$

□

Remark. Parts 2 and 3 of Lemma 4.2.2, Lemma 4.2.3, and Lemma 4.2.4 are in general wrong if $\Lambda \neq \Re^3$.

Finally, we list some crucial consequences of Theorem 4.2.1.

The flow T^t is only almost everywhere defined in Γ_0 with respect to Lebesgue measure. However, it is also well defined, again almost everywhere, with respect to a suitable measure, on certain surfaces of codimension one. Namely, consider the set $\mathcal{F}^+ \subset \Gamma_0$

$$\mathcal{F}^+ = \{z \in \Gamma_0; |x_i - x_j| = \sigma \text{ for some } i \neq j \text{ and } (\xi_i - \xi_j) \cdot n_{ij} > 0\}$$

where $n_{ij} = \frac{(x_i - x_j)}{\|x_i - x_j\|}$. In other words, $\mathcal{F}^+$ is the set of all phase points in which there are two particles at contact with outgoing velocities. On $\mathcal{F}^+$, consider the measure

$$d\sigma = dx_1 \dots dx_i \dots dx_{j-1} dx_{j+1} \dots dx_N d\xi_1 \dots d\xi_N dy_{ij} n_{ij} \cdot (\xi_i - \xi_j),$$

where dy_{ij} is the surface element over the sphere of radius σ centered at x_i. In the proof of Theorem 4.2.1 in Appendix 4.A, we also prove

(4.2.5) Lemma. *The flow T^t is also defined for σ- almost all $y \in \mathcal{F}^+$.*

This lemma, discussed in some detail in Appendix 4.A, is crucial for a rigorous validation of the BBGKY hierarchy. For the moment, however, we will only briefly reflect on the implications.

Consider all points $z \in \Gamma_0$ that experienced a collision in the past. Every such point z can be parametrized by a point $y \in \mathcal{F}^+$ and by a time $t < \alpha(y)$ such that $T^t y = z$; here, $\alpha(y)$ is the time of the first collision experienced by y in the future. As demonstrated in Appendix 4.A, the change of variables $z \longrightarrow (y, t)$ is such that the Lebesgue measure transforms like $dz \longrightarrow d\sigma(y)\, dt$. Therefore, if the flow T^t were undefined on a set A with $\sigma(A) > 0$, it would also be undefined on the "tube" based on A, i.e., on the set $\{(y,t); y \in A, 0 \leq t < \alpha(y)\}$. This set has positive Lebesgue measure, and we have a contradiction to Theorem 4.2.1.

Problems

1. Use $J^2 = id$ to show that $\left| det \frac{\partial(\xi_1', \xi_2')}{\partial(\xi_1, \xi_2)} \right| = 1$, and conclude that J preserves Lebesgue measure on $\Re^3 \times S^2 \times \Re^3$.
2. Construct an example of a boundary $\partial\Lambda$ and a velocity ξ such that a particle moving with ξ initially will have infinitely many collisions with $\partial\Lambda$ in finite time. Two dimensions are sufficient.
3. Prove 3 in Lemma 4.2.2.

4.3 Transition to L^1. The Liouville Equation and the BBGKY Hierarchy Revisited

The flow $\{T^t\}$ we introduced in Section 4.2 gives, from a strictly deterministic and mechanical point of view, a complete solution to the evolution problem for a hard-sphere system. For rarefied gas dynamics, however, this solution is unsatisfactory for the following reasons:

1. In realistic cases we are interested in huge particle numbers, like $N \approx 10^{23}$. It is then impossible to follow the flow in detail or to determine the initial state z exactly.
2. In addition, the information we could obtain by following a trajectory $\{T^t z\}_{t\in\Re}$ exactly might be of little relevance (atypical) from a physical point of view—there are, for example, some paths $\{T^t z\}$ along which very pathological behavior is displayed—in particular, a system of hard spheres can be arranged inside a rectangular box with reflecting boundary conditions such that all the particles will undergo periodic oscillations.
3. The Poincaré recurrence theorem applies to a system of N hard spheres in a bounded domain and predicts that almost all phase points in Γ_0 return to within any neighborhood of their initial state infinitely often.

No such behavior is ever observed in real gases. We always see an approach to some kind of equilibrium, and this suggests that we introduce a method that can in some way describe the evolution of the particle system from "less likely" to "more likely" states. The next logical step is to abandon the consideration of individual phase points altogether and consider instead a probability density function $P_0 \in L^1_+(\Gamma)$ whose time evolution is then given by the Liouville equation (2.1.11).

We now derive a version of the Liouville equation for which only minimal regularity of $P(z,t)$ is required. If $P(z,t)$ is the probability density of the system at time t, we must have

$$\int_{T^t A} P(z,t)\,dz = \int_A P_0(z)\,dz \tag{3.1}$$

for all Borel sets A in phase space. Because the Lebesgue measure dz is invariant under T^t (see Chapter 2),

$$\int_A P(T^t z,t)\,dz = \int_{T^t A} P(z,t)\,dz = \int_A P_0(z)\,dz \tag{3.2}$$

and so $P(T^t z,t) = P_0(z)$ for almost all z, or

$$\frac{d}{dt}[P(T^t z,t)] = 0. \tag{3.3}$$

We refer to (3.3) as the Liouville equation in mild formulation; the boundary condition (2.1.15) is implicitly contained in (3.3). Clearly $P(z,t) := P_0(T^{-t}z)$ solves (3.3).

Let us next discuss the BBGKY hierarchy arising from (3.3), introduced at a formal level in Chapter 2.

Let

$$P^{(s)}(z^s,t) = \int P(z^s, z^{N-s}, t)\,dz^{N-s}.$$

Here, $z^s = (x_1 \dots x_s, \xi_1 \dots \xi_s)$ is used as shorthand for the $6s$-dimensional phase point describing the state of the first s particles. We write $z = (z^s, z^{N-s})$, with the obvious meaning of the symbols.

Recall that

1. Since the particles are identical, $P_0(z)$ is symmetric with respect to all particles, i.e., for any permutation Π (reordering of the particles) $P_0(\Pi z) = P_0(z)$. Because $\Pi T^t z = T^t \Pi z$, it follows that $P(\cdot, t)$ also has this symmetry.

By Γ^s, $1 \leq s \leq N$, we denote the phase space of s particles (phase points leading to multiple collisions, etc., are deleted). The time evolution of the s particles of which $P^{(s)}(z^s, t)$ keeps track is influenced by the interactions of the s particles with the remaining $N - s$ particles. In order to be able to quantify these interactions, we need one further assumption on P_0:

2. We require that $t \longrightarrow P_0(T^t z)$ is continuous for almost all $z \in \Gamma$.

Assumption 2 is of tantamount importance in the sequel, such that it deserves a further comment. If z is an $N-$ particle precollisional phase point and z' is the corresponding post-collisional phase point, i.e.,

$$z = (x_1 \dots x_i \dots x_j \dots x_N, \xi_1 \dots \xi_i \dots \xi_j \dots \xi_N)$$

with $x_j = x_i + n\sigma$ and $n \cdot (\xi_i - \xi_j) > 0$,

$$z' = (x_1 \dots x_i \dots x_j \dots x_N, \xi_1 \dots \xi_i' \dots \xi_j' \dots \xi_N),$$

assumption 2 means that P_0 is continuous outside the contact points and

$$P_0(z) = P_0(z'). \tag{3.4}$$

In other words, "good" initial probability distributions are those that do not distinguish between precollisional and post-collisional configurations.

We next address the problem of giving rigorous meaning to the right-hand side of the BBGKY hierarchy.

Consider a phase point z belonging to the manifold $\mathcal{F}$ of codimension 1 defined by

$$(x_1 \dots x_i \dots x_i + n\sigma \dots x_N, \xi_1 \dots \xi_i \dots \xi_j \dots \xi_N) \text{ for some } i \neq j.$$

Then, according to Proposition 4.2.5,

$$P(z, t) = P_0(T^{-t} z) \tag{3.5}$$

is well defined for almost all $(x_1 \dots x_{j-1} x_{j+1} \dots x_N, \xi_1 \dots \xi_j \dots \xi_N, n)$.

For $j < s$, we integrate $P(z, t)$ over the last $N - (s+1)$ variables. We obtain that

$$P^{(s+1)}(x_1 \dots x_i \dots x_s, x_i - n\sigma, \xi_1 \dots \xi_i \dots \xi_{s+1}, t)$$

is well defined for almost all $(x_1 \dots x_s, \xi_1 \dots \xi_s, \xi_{s+1}, n)$ and all t. Therefore, the operator Q_{s+1}^σ, acting on the time-evolved marginal distributions via

$$Q^\sigma_{s+1}P^{(s+1)}(x_1 \ldots x_s, \xi_1 \ldots \xi_s, t)$$

$$= \sum_{j=1}^{s}(N-s)\sigma^2 \int_{S^2} dn \int d\xi_{s+1} n \cdot (\xi_j - \xi_{s+1}) \tag{3.6}$$

$$P^{(s+1)}(x_1 \ldots x_s x_j - n\sigma, \xi_1 \ldots \xi_{s+1}, t)$$

is also well defined for all t and almost all z^s. Moreover, we can prove the following.

(4.3.1) Theorem. *Under assumptions 1 and 2, $Q^\sigma_{s+1}P^{(s+1)}(T^t z^s, t)$ is continuous in t for almost all z^s, and the $P^s(\cdot, t)$, $1 \le s \le N$, satisfy the BBGKY hierarchy in the mild sense, i.e.,*

$$\frac{d}{dt}[P^{(s)}(T^t z^s, t)] = Q^\sigma_{s+1}P^{(s+1)}(T^t z^s, t) \tag{3.7}$$

for almost all z^s.

We trust that the ideas presented in Chapter 2 and the properties of the flow T^t discussed in the previous section will suffice to convince most readers of the validity of Eq. (3.7). We present a rigorous derivation of this equation in Appendix 4.B.

Other rigorous discussions regarding the BBGKY hierarchy can be found in the works of Uchiyama [24] and Petrina et al. [10,11,21].

Sometimes Eq. (3.7) is written in the following form:

$$\partial_t P^{(s)} = \left\{H^\sigma_s, P^{(s)}\right\}(z^s, t) + Q^\sigma_{s+1}P^{(s+1)} \tag{3.8}$$

where $\{H^\sigma_s, \cdot\}$ denotes the generator (Liouville operator) of the $s-$ particle dynamics.

It is easy to see that Eqs. (3.7) and (3.8) are equivalent from a formal point of view. The latter is reminiscent of the usual BBGKY hierarchy for systems interacting via smooth two-body potentials. However, due to the singularity of the hard–sphere interaction, the Liouville operator $\{H^\sigma_s, \cdot\}$ is rather degenerate in our situation (remember that it involves collisions in s-particle dynamics).

We summarize the meaning of Eqs. (3.8) or (3.7). The changes in $P^{(s)}$ are due to the $s-$ particle dynamics (expressed above in $\{H^\sigma_s, \cdot\}$) and due to the interaction of the tagged group of s particles with coordinates z^s with the rest. This interaction, which is clearly the dominant part in the Boltzmann–Grad limit $\sigma \longrightarrow 0$, $N \longrightarrow \infty$, $N\sigma^2 \longrightarrow$ *const.*, is given by the collision operator Q^σ_{s+1}. This operator acts only on $P^{(s+1)}$ (and not on P) because of the symmetry of P and the binary character of the interactions.

Finally, we mention that our consideration of hard–sphere systems has advantages and disadvantages with respect to the Boltzmann–Grad limit.

The main disadvantage is that we had to be careful even with the definition of $N-$ particle dynamics because of the singular character of the collisions. Therefore, the derivation of the BBGKY hierarchy is more subtle for this case than for interactions via smooth potentials. The main advantage is that once we have the BBGKY hierarchy in the above form, it is so close in structure to the Boltzmann hierarchy that an investigation of its behavior in the Boltzmann–Grad limit seems natural and convenient.

In contrast, the collision operator of the BBGKY hierarchy for smooth potentials involves derivatives with respect to velocity; this requires extra algebraic work and makes the derivation less transparent. Moreover, if the potential is really long range, the possibility of finding a suitable scaling under which a Boltzmann equation can even formally be derived is not clear from either a mathematical nor a physical point of view.

4.4 Rigorous Validity of the Boltzmann Equation

We saw in Chapter 2.5 how the Boltzmann equation arises in the Boltzmann– Grad limit from the BBGKY hierarchy via the Boltzmann hierarchy. We will now show how this transition can be done rigorously.

First, however, we discuss a few aspects of the informal derivation that should convince the reader that the Boltzmann–Grad limit is truly subtle and requires a rigorous analysis.

Consider the BBGKY hierarchy (3.7)

$$\frac{d}{dt}P^{(s)}(T_\sigma^t z^s, t) = \left(Q_{s+1}^\sigma P^{(s+1)}\right)(T_\sigma^t z^s, t) \tag{4.1}$$

(notice that we have here added an index σ to the flow operators, as a reminder that these operators do change as $\sigma \longrightarrow 0$), and let us analyze what happens to the collision operator in the Boltzmann–Grad limit.

We begin with expression (3.6) for the collision operator and split the integration over S^2 into integrations over two hemispheres. On the hemisphere given by $(\xi_j - \xi_{s+1}) \cdot n \leq 0$ (incoming configurations), we leave the argument of $P^{(s+1)}$ untouched, but we make here the coordinate change $n \longrightarrow -n$ and therefore change this part of the integral to an integral over the hemisphere $(\xi_j - \xi_{s+1}) \cdot n \geq 0$.

On the hemisphere given originally by this latter inequality (outgoing configurations), we leave n untouched, but we take advantage of the continuity of $P^{(s+1)}$ through collisions to replace the velocities ξ_j and ξ_{s+1} by their precollisional counterparts ξ_j' and ξ_{s+1}'. The result of these operations is

$$
\begin{aligned}
&Q_{s+1}^{\sigma} P^{(s+1)}(x_1 \ldots x_s, \xi_1 \ldots \xi_s) \\
&= \sum_{j=1}^{s} (N-s)\sigma^2 \int d\xi_{s+1} \int_{\{n\cdot(\xi_j-\xi_{s+1})\geq 0} dn\, n\cdot(\xi_j - \xi_{s+1}) \\
&\Big\{ P^{(s+1}(x_1 \ldots x_j \ldots x_j - \sigma n, \xi_1 \ldots \xi_j' \ldots \xi_{s+1}') \\
&- P^{(s+1}(x_1 \ldots x_j \ldots x_j + \sigma n, \xi_1 \ldots \xi_j \ldots \xi_{s+1}) \Big\}.
\end{aligned}
\tag{4.2}
$$

In the Boltzmann–Grad limit, formally $Q^{\sigma} \longrightarrow Q$, where

$$
\begin{aligned}
&Q_{s+1} f^{(s+1)}(x_1 \ldots x_s, \xi_1 \ldots \xi_s) \\
&= \alpha \sum_{j=1}^{s} \int d\xi_{s+1} \int_{n\cdot(\xi_j-\xi_{s+1})\geq 0} dn\, n\cdot(\xi_j - \xi_{s+1}) \\
&\Big\{ f^{(s+1}(x_1 \ldots x_j \ldots x_j, \xi_1 \ldots \xi_j' \ldots \xi_{s+1}') \\
&- f^{(s+1}(x_1 \ldots x_j \ldots x_j, \xi_1 \ldots \xi_j \ldots \xi_{s+1}) \Big\}.
\end{aligned}
\tag{4.3}
$$

As for fixed s $T_{\sigma}^{t} z^s \longrightarrow T_0^t z^s$ for almost all z^s in the limit $\sigma \longrightarrow 0$ (T_0^t denotes collisionless flow), we expect that the $P^{(s)}$ will converge to a sequence of functions $f^{(s)}$ solving the Boltzmann hierarchy:

$$
\frac{d}{dt} f^{(s)}(T_0^t z^s, t) = Q_{s+1} f^{(s+1)}(T_0^t z^s, t). \tag{4.4}
$$

The relationship of the Boltzmann hierarchy (4.4) to the Boltzmann equation is as already described in Chapter 2; If $f(\cdot, t)$ solves the Boltzmann equation, then

$$
f^{(s)}(x_1 \ldots x_s, \xi_1 \ldots \xi_s; t) = \prod_{j=1}^{s} f(x_j, \xi_j; t)
$$

solves the Boltzmann hierarchy. Thus, if the $f^{(s)}$ in (4.4) factorize initially and *if* the factorization is preserved in time (the second *if*, usually referred to as propagation of chaos, must be proved), the Boltzmann hierarchy and the Boltzmann equation are equivalent.

By proving the convergence $P^{(s)} \longrightarrow f^{(s)}$ and propagation of chaos, we will complete our objective. First, however, we make some observations that will clarify what we can expect to achieve.

In the derivation of the operator Q, we chose to represent collision phase points in terms of ingoing configurations. Given the assumed continuity along trajectories, we could use the representation in terms of outgoing configurations, and that would lead formally to a limit that is Eq. (4.4) with a minus sign in front of the collision operator. Also, if we take the formal limit in the right-hand side of Eq. (3.7) without first splitting into gain and loss terms, we obtain zero, because the integrations over the two hemispheres compensate each other. We are thus compelled to ask whether the representation in terms of ingoing configurations is the right one, i.e., physically meaningful. As we shall see later in a more careful analysis of the validity problem, the representation in terms of ingoing configurations follows automatically from hard–sphere dynamics and is, indeed, not a matter of an a priori choice.

We now consider Eq. (4.1) for $s = 1$ and discuss the propagation of chaos. Assuming representation in terms of ingoing configurations, the right-hand side reads

$$(N-1)\sigma^2 \int_{\Re^3} d\xi_2 \int_{n\cdot(\xi_1-\xi_2)\geq 0} dn\; n\cdot(\xi_1-\xi_2)$$

$$\begin{aligned}\{P^{(2)}(x_1+t\xi_1, x_1+t\xi_1-n\sigma, \xi_1', \xi_2'; t)\\ -P^{(2)}(x_1+t\xi_1, x_1+t\xi_1+n\sigma, \xi_1, \xi_2; t)\}.\end{aligned} \tag{4.5}$$

In order to get the right-hand side of the Boltzmann equation, we have to make the crucial assumption that in the limit $N \longrightarrow \infty$, $\sigma \longrightarrow 0$ and $N\sigma^2 \longrightarrow \alpha > 0$ there is a function $f = f(x, \xi, t)$ such that

$$\begin{aligned}\lim_{N\to\infty} P^{(1)}(x,\xi,t) &= f(x,\xi,t)\\ \lim_{N\to\infty} P^{(2)}(x, x+n\sigma, \xi_1, \xi_2; t) &= f(x,\xi_1,t)f(x,\xi_2,t)\end{aligned} \tag{4.6}$$

provided that the configuration is *ingoing*. If this convergence holds, we get the Boltzmann equation in the mild formulation

$$\frac{d}{dt}f(x+t\xi,\xi,t) = \alpha \int d\xi_* \int_{S^2} dn\, n\cdot(\xi-\xi_*)\,[f'f_*' - ff_*]\,(x+t\xi,t).$$

Our convention here is that $f_*(x+t\xi,t) = f(x+t\xi,\xi_*,t)$, etc.

It turns out that (4.6) is a statement stronger than what we need. Worse, (4.6) can be violated even at time $t = 0$ for quite reasonable $P^{(2)}$s, as we will show by an example. The form of propagation of chaos we will be able to prove is that

$$P^{(2)}(x_1, x_2, \xi_1, \xi_2; t) \longrightarrow f(x_1,\xi_1,t)f(x_2,\xi_2,t)$$

for almost all x_1, x_2, ξ_1, ξ_2 [and not on a manifold of codimension one, as in (4.6)].

The following is an example for which (4.6) fails. Suppose that f_0 is given and smooth. Let

$$P^{(N)}(z) = C\prod_{i=1}^{N} f_0(x_i,\xi_i)\prod_{i<j}\chi_{ij}(x_i,x_j),$$

where $\chi_{ij}(x_i,x_j)$ is in $C_+^\infty(\Re^3\times\Re^3)$ for all $i<j$ and

$$\chi_{ij}(x_i,x_j)=\begin{cases}1 \text{ if } |x_i-x_j|>\sigma+\epsilon\\ 0 \text{ if } |x_i-x_j|\le\sigma.\end{cases}$$

The constant $C=C(\epsilon)$ is there to normalize $P^{(N)}$. If $\epsilon \searrow 0$ as $\sigma \longrightarrow 0$ and $N \longrightarrow \infty$, we certainly have

$$\lim_{N\to\infty} P^{(s)}(x_1\ldots x_s,\xi_1\ldots\xi_s)=\prod_{i=1}^{s} f_0(x_i,\xi_i),$$

and assumptions 1 and 2 hold. However, (4.6) is violated by construction.

The observations we have made so far show how dangerous it is to argue with generators in this derivation, because we clearly *cannot* have

$$\frac{d}{dt}\left\{P^{(1)}(T^t_\sigma(x,\xi),t)\right\}\Big|_{t=0} \longrightarrow \frac{d}{dt}\left\{f(x+t\xi,\xi,t)\right\}|_{t=0}$$

in the Boltzmann–Grad limit.

In Appendix 4.C we discuss an example of a discrete velocity model for which the derivation done in the present section fails completely.

Given that convergence of the derivatives is not to be expected, what we are going to do is look at the solution of the BBGKY hierarchy and the Boltzmann hierarchy as a whole (and not at their derivatives) and show that the first converge a.e. to the second. The solution concept that allows us to do this is a series solution concept we now introduce.

After integration from 0 to t, we get from (4.1) that

$$P^{(s)}(T^t_\sigma z^s,t)=P_0^{(s)}(z^s)+\int_0^t dt_1\left(Q^\sigma_{s+1}P^{(s+1)}\right)(T^{t_1}_\sigma z^s,t_1)$$

or

$$P^{(s)}(z^s,t)=\left(S_\sigma(t)P_0^{(s)}\right)(z^s)+\int_0^t dt_1\, S_\sigma(t-t_1)Q^\sigma_{s+1}P^{(s+1)}(z^s,t_1)$$

where $S_\sigma(t)f(z^s)=f\left(T_\sigma^{-t}z^s\right)$.

By iterating the last equation $N-s$ times and using the convention that $P_0^{(s)}=0$ for $s>N$, we can express $P^{(s)}(z^s,t)$ as a finite sum of multiple integrals involving only the functions $P_0^{(r)}$ for $r\ge s$:

$$\begin{aligned}P^{(s)}(z^s,t)=&\sum_{n=0}^{\infty}\int_0^t dt_1\int_0^{t_1}dt_2\ldots\int_0^{t_{n-1}}dt_n\\ &S_\sigma(t-t_1)Q^\sigma_{s+1}S_\sigma(t_1-t_2)\ldots Q^\sigma_{s+n}S_\sigma(t_n)P_0(z^s)\end{aligned}\tag{4.7}$$

Note that the sum is actually finite because of our convention. For $s=N$ (4.7) reduces to $P^{(N)}(z,t)=S_\sigma(t)P_0^{(N)}(z)$, which is the solution of the Liouville equation.

Equation (4.7) is an equality that holds for almost all z^s. We hope that the discussion given in the previous section has convinced the reader that the right-hand side of Eq. (4.7) is well defined (that it actually is follows rigorously from our rigorous derivation of the BBGKY hierarchy given in Appendix 4.B).

Similarly, a formal series solution can be written down for the Boltzmann hierarchy (4.4). It is

$$f^{(s)}(\cdot,t) = \sum_{n=0}^{\infty} \int_0^t dt_1 \int_0^{t_1} dt_2 \dots \int_0^{t_{n-1}} dt_n \quad S_0(t-t_1)Q_{s+1}S_0(t_1-t_2)\dots Q_{s+n}S_0(t_n)f_0^{(s+n)} \tag{4.8}$$

where

$$S_0(t)f(z^s) = f(T_0^{-t}z^s)$$

(the reader should not confuse the summation index n with the unit vector n occurring in (4.2) and elsewhere).

In contrast to (4.7), (4.8) is an infinite series, and the question of convergence becomes critical.

There is an obvious trace problem in the definition of Q on L^1-functions (see Definition (4.2)); the series solution concept we adopt nicely avoids this problem, because each term of the series (4.8) involves $f_0^{(s)}$ evaluated at phase points that are computed by repeated adjoinment of collision partners (this is what Q does) and backward free streaming (this is what S_0 does), such that each term on the right-hand side of (4.8) makes sense if $f_0^{(s)}$ is assumed to be sufficiently smooth.

We emphasize at this point that (4.7) and (4.8) are profoundly different, in spite of their formal similarity. From a physical point of view, (4.7) describes a Hamiltonian (reversible) dynamical system, while (4.8) describes a dissipative evolution that is compatible with the H–theorem. Also, technically speaking, Q is more singular than Q^σ: Indeed, Q_{s+1} involves the trace of $f^{(s+1)}$ on a manifold of codimension three, while Q_{s+1}^σ needs the trace of $P^{(s+1)}$ on a manifold of codimension 1 (the sphere has shrunk to a point).

On the other hand, the presence of the flow T_σ^t in (4.7) makes it hard to interpret the BBGKY hierarchy from a pure PDE point of view. The hierarchy in Eq. (4.1) is a family of equations that can only be established once the flow T_σ^t is defined and the properties of this flow are well understood. We did this in the previous section.

After this long introduction we are finally able to formulate our rigorous validity result.

Consider an N-particle system in a region Λ. We shall assume either $\Lambda = \Re^3$ or $\Lambda \subset \Re^3$ bounded, with a smooth boundary. T_σ^t and T_0^t will refer, as always, to the N-particle dynamics and the free flow, both with reflecting boundary conditions on $\partial\Lambda$. The case in which Λ is a rectangle with periodic boundary conditions can also be considered.

Suppose that $f_0^{(s)}(z^s) = \prod_{i=1}^s f_0(x_i, \xi_i)$ is the factorizing initial value for the sth equation in the Boltzmann hierarchy (molecular chaos or statistical independence is taken for granted at $t = 0$), and that $P_0^{(s)}(z^s)$ is the initial value for the sth equation in the BBGKY hierarchy. We send $N \longrightarrow \infty$, $\sigma \longrightarrow 0$ in such a way that $N\sigma^2 = \alpha$ and assume that

i) if $\left(\Lambda \times \Re^3\right)_{\neq}^{s,\sigma} = \left\{z^s \in \Lambda^s \times \Re^3;\ |x_i - x_j| > \sigma,\ i \neq j,\ \sigma \geq 0\right\}$, then the $P_0^{(s)}$ are continuous on $\left(\Lambda \times \Re^3\right)_{\neq}^{s,\sigma}$ and at the collision points (continuity along trajectories). f_0 is continuous and $\lim_{N \to \infty} P_0^{(s)} = f_0^{(s)}$ uniformly on compact subsets of $\left(\Lambda \times \Re^3\right)_{\neq}^{s,\sigma}$ for all $s = 1, 2, \ldots$

ii) there are positive constants β, C, and b such that

$$\sup_{z^s} P_0^{(s)}(z^s) \exp\left\{\beta \sum_{i=1}^s \xi_i^2\right\} \leq C \cdot b^s \tag{4.9}$$

for all s.

(4.4.1) Theorem. *Suppose that i) and ii) hold. Then on a sufficiently small interval $[0, t_0]$, the series solution $P^{(s)}(\cdot, t)$ of the BBGKY hierarchy converges in the Boltzmann–Grad limit almost everywhere to the series solution of the Boltzmann hierarchy $f^{(s)}(\cdot, t)$. This solution exists, is unique, and is of the form*

$$f^{(s)}(z^s, t) = \prod_{i=1}^s f(x_i, \xi_i, t),$$

where f is a mild solution of the Boltzmann equation to the initial value f_0.

Proof. The proof of this theorem is done in four steps.

In step 1, we show that the convergence of the series (4.7) to (4.8) holds term by term. This step is technically rather straightforward, but, as we shall see, conceptually deep.

Step 2 is completely straightforward. We simply observe that the convergence proof will follow from step 1 if there is a non-negative converging series whose terms bound simultaneously those of (4.7) and (4.8).

We construct such a series in step 3. This is the technical part of the proof and the only part where we use the smallness assumption on the time interval.

Finally, in step 4, we prove uniqueness and the factorization property.

Step 1. We want to prove that by virtue of assumption i) in the Boltzmann–Grad limit $N \longrightarrow \infty$, $\sigma \longrightarrow 0$, $N\sigma^2 = \alpha$

$$(4.10)\quad \begin{aligned} &\int_0^t dt_1 \int_0^{t_1} dt_2 \ldots \int_0^{t_{n-1}} dt_n \\ &S_\sigma(t-t_1)Q^\sigma_{s+1}S_\sigma(t_1-t_2)\ldots Q^\sigma_{s+n}S_\sigma(t_n)P_0^{(s+n)}(z^s) \\ &\longrightarrow \int_0^t dt_1 \int_0^{t_1} dt_2 \ldots \int_0^{t_{n-1}} dt_n \\ &S_0(t-t_1)Q_{s+1}S_0(t_1-t_2)\ldots Q_{s+n}S_\sigma(t_n)f_0^{(s+n)}(z^s) \end{aligned}$$

almost everywhere.

We first discuss the simplest case $s = n = 1$, where we have to prove that

$$(4.11)\quad \begin{aligned} &\int_0^t dt_1\, S_\sigma(t-t_1)Q_2^{\sigma\pm}S_\sigma(t_1)P_0^{(2)}(x,\xi) \\ &\longrightarrow \int_0^t dt_1\, S_0(t-t_1)Q_2^{\pm}S_0(t_1)f_0^{(2)}(x,\xi) \end{aligned}$$

with

$$Q^\sigma = Q^{\sigma+} - Q^{\sigma-}$$

and

$$Q = Q^+ - Q^-.$$

The last two identities are the decomposition, based on the ingoing representations of collision configurations, of the collision operator in gain and loss part [see (4.2) and (4.3)]. We first prove (4.11) for Q^-, in which case (4.11) reads explicitly as

$$(4.12)\quad \begin{aligned} &\int_0^t dt_1 \int d\xi_2 \int_{n\cdot(\xi_1-\xi_2)\geq 0} dn\, n\cdot(\xi_1-\xi_2) \\ &S_\sigma(t_1)P_0^{(2)}(x_1-\xi_1(t-t_1), x_1-\xi_1(t-t_1)+n\sigma, \xi_1, \xi_2) \\ &\longrightarrow \int_0^t dt_1 \int d\xi_2 \int_{n\cdot(\xi_1-\xi_2)\geq 0} dn\, n\cdot(\xi_1-\xi_2) \\ &S_0(t_1)f_0^{(2)}(x_1-\xi_1(t-t_1), x_1-\xi_1(t-t_1), \xi_1, \xi_2). \end{aligned}$$

Next notice that as $\sigma \longrightarrow 0$

$$(4.13a)\quad \begin{aligned} &T_\sigma^{-t_1}(x_1-\xi_1(t-t_1), x_1-\xi_1(t-t_1)+n\sigma, \xi_1, \xi_2) \\ \longrightarrow &T_0^{-t_1}(x_1-\xi_1(t-t_1), x_1-\xi_1(t-t_1), \xi_1, \xi_2) \\ = &(x_1-\xi_1 t, x_1-\xi_1(t-t_1)-\xi_2 t_1) \end{aligned}$$

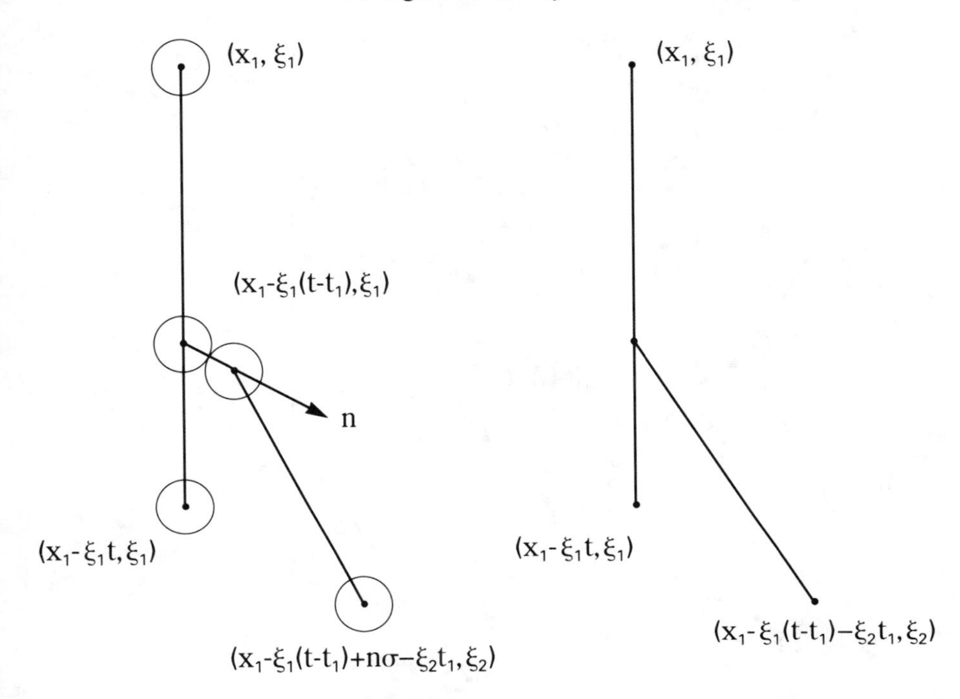

FIGURE 6.

as follows by direct inspection. We use the graphical representation in Fig. 6 to visualize this convergence for the reader.

At the instant $t-t_1$, we adjoin a second particle in ingoing configuration to the original particle at (x_1, ξ_1). As we move further back in time, the free motion of the original particle is unaffected.

If we adjoin this second particle such that the triple (n, ξ_1, ξ_2) is in outgoing configuration, we are in the situation that arises in the gain term $Q^{\sigma+}$(because (n, ξ_1', ξ_2') is then ingoing). Proceeding as above for $Q^{\sigma+}$, we have to study the limit of the expression

$$T_\sigma^{-t_1}(x_1 - \xi_1(t-t_1), x_1 - \xi_1(t-t_1) - n\sigma, \xi_1, \xi_2) \tag{4.13b}$$

as $\sigma \searrow 0$. Consider the graphical representation in Fig. 7.

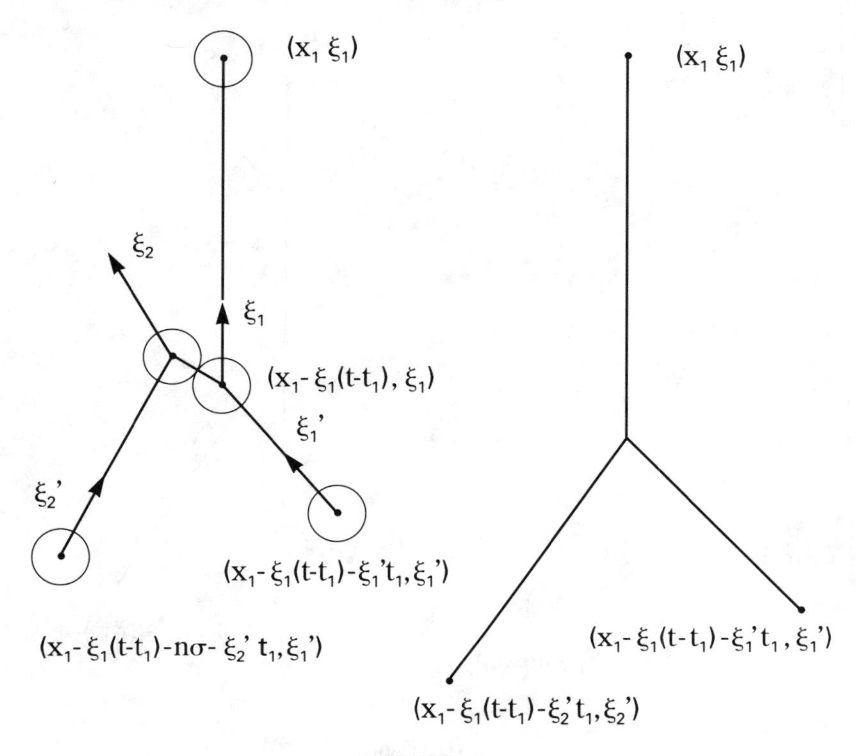

FIGURE 7.

As $(-n, \xi_1, \xi_2)$ is outgoing if (n, ξ_1, ξ_2) is ingoing, we have to make the transformation from outgoing to ingoing representations as we go back in time, and the expression (4.13b) therefore converges to

$$(x_1 - \xi_1(t - t_1) - \xi_2' t_1, x_1 - \xi_1(t - t_1) - \xi_1' t_1, \xi_1', \xi_2').$$

Summarizing, we write

$$\begin{aligned} &\int_0^t dt_1 S_\sigma(t - t_1) Q_2 S_\sigma(t_1) P_0^{(2)}(x, \xi) \\ &= \int_0^t dt_1 \int d\xi_2 \int dn\ n \cdot (\xi_1 - \xi_2) P_0^{(2)} \\ &\qquad \left(T_\sigma^{-t_1}\left(T_\sigma^{-(t-t_1)}(x, \xi) \cup (y - n\sigma, \xi_2)\right)\right) \end{aligned}$$

where $y = x - \xi_1(t - t_1)$, and we have used the symbol $\cup$ for the adjoinment of the particle. As pointed out in the earlier discussion, the argument of $P_0^{(2)}$ converges to $(x_1 - \xi_1 t, x_1 - \xi_1(t - t_1) - \xi_2 t_1, \xi_1, \xi_2)$ or to $(x_1 - \xi_1(t -$

$t_1) - \xi_1' t_1, x_1 - \xi_1(t - t_1) - \xi_2' t_1, \xi_1', \xi_2')$, depending on whether the triple (n, ξ_1, ξ_2) is in ingoing configuration or not.

Notice that we do not have a choice of the representation of the collision point in terms of ingoing or outgoing velocities; a representation just arises automatically, and the correct expression of the limit collision terms follows from the calculations. By assumption i), we have convergence (pointwise, and not just almost everywhere for the case $n = s = 1$) of the integrands in (4.12). By assumption ii) and the dominated convergence theorem the convergence in (4.12) follows.

We next consider the general case, where n and s are larger than 1. Unfortunately, as the reader may check, an induction proof is not feasible. We have no choice but to directly investigate the complicated configuration arising after repeated application of the operators T_σ and Q_σ. This investigation leads to obvious notational problems, and the temptation is there to just say that the argument for $n = s = 1$ extends to arbitrary n and s. However, as we show for a counterexample of a discrete velocity model in Appendix 4.C, this reasoning would be extremely dangerous! It is in fact important to verify the convergence spelled out in (4.10) explicitly, because it actually fails for the discrete velocity case.

First, observe that as $N \longrightarrow \infty$, $\sigma \longrightarrow 0$ with $N\sigma^2 = \alpha$,

$$(N - s)(N - s - 1)\dots(N - s - n)\sigma^{2n} \longrightarrow \alpha^n.$$

Also,

$$T_\sigma^{-\tau} z^s \longrightarrow T_0^{-\tau} z^s$$

except on a set of measure zero. The exceptional set is formed by those phase points for which, in the free–flow evolution, two point particles occupy the same spot at the same time, a set that is clearly of measure zero.

Therefore, as we adjoin a particle with coordinates $z_{s,1}^{k_1}$ (the index s here reminds us that we start with the phase point z^s, the index 1 that the particle is the first of n to be adjoined, and the index $k_1 \in \{1, \dots, s\}$ tells us who is the collision partner), for almost all choices of ξ_{s+1}, the point

$$T_\sigma^{-\tau_1} \left(T^{-(t_1 - t)} z^s \cup z_{s,1}^{k_1} \right),$$

$\tau_1 < t_1$, will converge to

$$T_0^{-\tau_1} \left(T_0^{(t_1 - t)} z^s \cup z_{s,1}^{k_1} \right)$$

or to

$$T_0^{-\tau_1} J \left(T_0^{(t_1 - t)} z^s \cup z_{s,1}^{k_1} \right).$$

Here, J is the collision transformation changing ξ, ξ_{s+1} into ξ', ξ_{s+1}', ξ being the velocity of the k_1th particle in the configuration $T_0^{(t_1-t)} z^s$. Which of the two limits will apply depends, as before, on whether the particle $z_{s,1}^{k_1}$ was added in ingoing or outgoing configuration.

The typical phase point occurring in the argument of $P_0^{(s+n)}$ on the left-hand side of (4.10) is of the form

$$Y^{s+n} = T_\sigma^{-t_n}$$

$$\left(z_{s,n}^{k_n} \cup T_\sigma^{-(t_{n-1}-t_n)} \left(z_{s,n-1}^{k_{n-1}} \cup T_\sigma^{-(t_{n-2}-t_{n-1})} \dots T_\sigma^{-(t-t_1)} z^s\right) \dots\right),$$

which converges to

$$Y_0^{n+s} = T_0^{-t_n} J^{\gamma_n}$$

$$\left(z_{s,n}^{k_n} \cup T_0^{-(t_{n-1}-t_n)} J^{\gamma_{n-1}} \left(z_{s,n-1}^{k_{n-1}} \cup T_0^{-(t_{n-2}-t_{n-1})} J^{\gamma_{n-2}} \dots T_0^{-(t-t_1)} z^s\right)\right)$$

for almost all choices of $\xi_{s+1}, \dots, \xi_{s+n}$. Here, γ_r is 0 or 1, depending on whether $z_{s,r}^{k_r}$ is adjoined in ingoing configuration or not. Note that $z_{s,r}^{k_r}$ depends on σ.

Consider now a typical integrand in (4.10). If we delete for the moment the indices in the collision operators indicating the number of particles but keep track of whether a collision configuration is ingoing or outgoing, we get terms like

$$S_\sigma(t-t_1) Q^{\sigma,\gamma_1} S_\sigma(t_1 - t_2) \dots Q^{\sigma,\gamma_n} S_\sigma(t_n) P_\sigma^s(z^s; 0) \tag{4.14}$$

where $\gamma_i = +$ or $-$. To prove the convergence of (4.14) to

$$S_0(t-t_1) Q_0^{\gamma_1} S_0(t_1 - t_2) \dots Q^{\gamma_n} S_0(t_n) f^s(z^s; 0), \tag{4.15}$$

we observe that (4.14) and (4.15) contain integrals with respect to velocities and impact parameters of $P_\sigma^{s+n}(Y^{s+n}; 0)$ and $f^{s+n}(Y_0^{s+n}; 0)$. As the convergence of $P_\sigma^{s+n}(Y^{s+n}; 0)$ to $f^{s+n}(Y_0^{s+n}; 0)$ a.e. is already proved, the convergence of (4.14) to (4.15) will follow from the dominated convergence theorem if we can find a suitable upper bound on the P_σ^{s+n} (uniform in σ resp. N). This bound should decay fast enough at infinity for large velocities in order to compensate for the unbounded cross section. The construction of this bound is what we do in step 3.

Remark and Warning. The reasoning from step 1 is not applicable to discrete velocity models because we have only a finite set of velocities to choose from; therefore, none of them can be excluded by a measure-zero argument. We demonstrate in Appendix 4.C that the convergence proof fails.

The preceding discussion also shows why the convergence of $P_\sigma(\cdot, t)$ to $f(\cdot, t)$ is only almost everywhere at every level. Consider the first term in either series: We have

$$\lim_{\sigma \to 0} \left(S_\sigma(t) P_0^{(s)} - S_0(t) f_0^{(s)}\right)(z^s) = 0$$

only if

$$P_0^{(s)}\left(T_s^{-t} z^s\right) - f_0^{(s)}\left(T_{0,s}^{-t} z^s\right) \longrightarrow 0,$$

and this is, for $t > 0$, in general violated on the set of phase points z^s for which there is a $\tau \in (0, t)$ such that two particles in $T_{0,s}^{-\tau} z^s$ occupy the same spot; for such phase points, $T_s^{-t} z^s$ does simply not converge to $T_{0,s}^{-t} z^s$ as $\sigma \longrightarrow 0$.

Step 2. In step 1, we have established that the series solution to the BBGGY hierarchy, say $\sum a_n^\sigma$, and the (formal) series solution to the Boltzmann hierarchy (say $\sum a_n$), satisfy $a_n^\sigma \longrightarrow a_n$ as $\sigma \longrightarrow 0$ almost everywhere for each n. We will show that for short times $\sum |a_n|$ converges and that

$$\sum a_n^\sigma \longrightarrow \sum a_n$$

as $\sigma \longrightarrow 0$. This is certainly true if we can find a sequence $\{b_n\}_{n \in N}$ such that $0 \le |a_n^\sigma| \le b_n$, $0 \le |a_n| \le b_n$, and $\sum b_n < \infty$ (see Problem 2).

Step 3. Finding the sequence $\{b_n\}$ is the key technical step in the proof. We introduce the following norm:

$$\|P^{(s)}\|_\beta = \sup_{x_1 \dots x_s, \xi_1 \dots \xi_s} \exp e^{\beta E_s} |P^{(s)}(x_1 \dots x_s, \xi_1 \dots \xi_s)|$$

where

$$E_s = \frac{1}{2} \sum_{j=1}^{s} \xi_j^2$$

is the energy of the configuration. Then

$$|Q_{s+1}^\sigma P^{(s+1)}(x_1 \dots x_s, \xi_1 \dots \xi_s)| \le C\alpha \sum_{j=1}^{s} \int d\xi_{s+1} \left(|\xi_j| + |\xi_{s+1}|\right)$$
$$\times \ \|P^{(s+1)}\|_\beta e^{-\beta E_{s+1}}.$$

Therefore, for $\beta' < \beta$

$$\|Q_{s+1}^\sigma P^{(s+1)}\|_{\beta'} \le C\alpha \|P^{(s+1)}\|_\beta \int d\xi_{s+1} e^{-\frac{\beta}{4}\xi_{s+1}^2}$$
$$\sum_{j=1}^{s} \left(|\xi_j| + |\xi_{s+1}|\right) e^{-\frac{\beta}{4}\xi_{s+1}^2} e^{-(\beta-\beta')E_s}.$$

Using the estimates

$$\sum_{j=1}^{s} |\xi_j| e^{-(\beta-\beta')E_s} \le C \frac{\sqrt{s}}{\sqrt{\beta - \beta'}}$$

(see Problem 3) and

$$\sum_{j=1}^{s} |\xi_{s+1}| e^{-\frac{\beta}{4}\xi_{s+1}^2} \le \frac{C}{\sqrt{\beta}} s,$$

we conclude that

$$\|Q^{\sigma}_{s+1}P^{(s+1)}\|_{\beta'} \leq \frac{C\alpha}{\beta^{\frac{3}{2}}}\|P^{(s+1)}\|_{\beta}\left(\frac{\sqrt{s}}{\sqrt{\beta-\beta'}}+\frac{s}{\sqrt{\beta}}\right). \tag{4.16}$$

On the other hand, by energy conservation

$$\|S_{\sigma}(t)P^{(s)}\|_{\beta} \leq \|P^{(s)}\|_{s},$$

such that

$$\begin{aligned}\|S_{\sigma}(t-t_1)Q^{\sigma}S_{\sigma}(t_1-t_2)Q^{\sigma}\dots S_{\sigma}(t_n)P^{(s+n)}\|_{\beta'}\\ \leq (s+n)^n C^n A(\beta,\beta')^n\|P^{(s+n)}\|_{\beta},\end{aligned} \tag{4.17}$$

where

$$A(\beta,\beta') = \max\left(\beta'^{-2}, \frac{1}{\sqrt{\beta-\beta'}}\cdot\frac{1}{(\beta')^{\frac{3}{2}}}\right).$$

In order to derive estimate (4.17), we have applied (4.16) to each of the Q^{σ}s appearing in the left-hand side of (4.17), with $\beta-\beta'$ replaced by $\frac{\beta-\beta'}{n}$. The term Q^{σ}_{s+k} makes the contribution

$$\frac{C\alpha}{(\beta')^{\frac{3}{2}}}\left(\frac{\sqrt{(s+k)n}}{\sqrt{\beta-\beta'}}+\frac{s+k}{\sqrt{\beta'}}\right) \leq C\alpha A(\beta,\beta')(s+n).$$

From assumption ii) we get

$$\|P^{(s+n)}\|_{\beta} \leq Cb^{s+n},$$

and finally we employ the identity

$$\int_0^t dt_1 \dots \int_0^{t_{n-1}} dt_n = \frac{t^n}{n!}$$

to arrive at

$$\|P^s(t)\|_{\beta'} \leq b^s \sum_{n\geq 0}\frac{t^n}{n!}(s+n)^n(C\alpha Ab)^n.$$

At this point, recall the Stirling formula for $n \geq 1$

$$n! = \sqrt{2\pi n}\left(\frac{n}{e}\right)^n e^{-\frac{\theta(n)}{12}},\ 0<\theta(n)<1,$$

and

$$(s+n)^n \leq n^n(1+\frac{s}{n})^n \leq n^n e^s,$$

which lead us to the final estimate

$$P^{(s)}(z^s,t) \leq e^{-\beta' E_s}(Cb)^s\sum_{n\geq 0}t^n(C\alpha Ab)^n. \tag{4.18}$$

Thus we can bound the series solution of the BBGKY hierarchy by a geometric series, which converges if t is small enough.

Exactly the same estimates apply to the Boltzmann hierarchy. In particular, it follows that the series solution of the Boltzmann hierarchy is bounded by the right-hand side of (4.18). This completes step 3 of our proof.

Step 4. Factorization, Propagation of Chaos, and Uniqueness. The series we used to construct a mild solution of the Boltzmann hierarchy is determined completely by the initial data, and therefore this solution is determined uniquely by the data. Now assume that $f_0^{(s)}(z^s) = \prod_{i=1}^s f_0(x_i, \xi_i)$, and that $f(x, \xi, t)$ is a (mild) solution of the Boltzmann equation to the initial value f_0, i.e.,

$$f(x, \xi, t) = S_0(t) f_0(x, \xi) + \int_0^t S_0(t - t_1)(Qf)(x, \xi, t_1) dt_1.$$

Substitute this formula for $f(\cdot, t)$ in the integral, and repeat this procedure recursively. The result is exactly the series we used to solve the first equation in the Boltzmann hierarchy. However, we just proved that this series converges! In other words, we proved that this series gives us a mild solution of the Boltzmann equation, $f(\cdot, t)$.

Now, we know (from Chapter 2) that the tensor products

$$\tilde{f}^{(s)}(z^s, t) := \prod_{i=1}^s \otimes f(x_i, \xi_i, t)$$

then give a solution of the Boltzmann hierarchy in mild form; for $\tilde{f}^{(s)}(z^s, t)$, we have

$$\frac{d}{dt}\left[S_0(-t)\tilde{f}^{(s)}(z^s, t)\right] = \left[S_0(-t)Q\tilde{f}^{(s+1)}\right](z^s, t). \tag{4.19}$$

By applying the same procedure of a formal series solution to (4.19), we see that this series is *algebraically the same* as the right-hand side of the solution series for $f^{(s)}(z^s, t)$; it follows that both series converge and that $f^{(s)}(z^s, t) = \tilde{f}(z^s, t)$, i.e., we have shown that $f^{(s)}$ factorizes and that chaos propagates.

Uniqueness of solutions of the Boltzmann hierarchy is easily proved along the lines of the estimates of step 3. We leave the details of this uniqueness proof to the reader. More on the uniqueness of solutions of the Boltzmann hierarchy can be found in Section 4.8. This completes the proof of Theorem 4.4.1. □

Remarks. The reason we end up with a geometrically convergent series in the proof is that we had to use the integral

$$\int_0^t dt_1 \ldots \int_0^{t_{n-1}} dt_n = \frac{t^n}{n!} \tag{4.20}$$

in combination with the counting factor

$$\sum_{j_1=1}^{s} \cdots \sum_{j_n=1}^{s+n} = s(s+1)\dots(s+n) = \frac{(s+n)!}{s!} = n! \binom{s+n}{s} \leq n! \cdot 2^{s+n}.$$

The number of particles increases in each step of the iteration as a consequence of the bilinearity of the Boltzmann collision operator. We conclude that our present method cannot possibly give a better than local result for general initial values. This is also true for any other fully nonlinear model equation.

In fact, we made no effort in the proof to optimize the convergence time with respect to α or $\|f_0\|_{L^\infty}$. However, it is transparent from (4.18) that every convergence time will only be a fraction of α, which in itself is proportional to the inverse of the mean free time between collisions. In other words, our validation applies only to a fraction of the average time between two collisions suffered by a particle.

Notice that otherwise our convergence time depends only on $\|f_0\|_{L^\infty}$. This means that the initial distribution need not be normalized. As a consequence, our result also applies to the physically relevant case of a gas in all of $\Re^3$, without any decay at infinity with respect to the spatial variable. In this case, one has to reformulate the problem in terms of "rescaled" correlation functions (defined in Section 4.6) expressing the mass density and the mass correlations.

A final comment. We already mentioned that the Lanford result that we formulated and proved in this section is unsatisfactory, because its validity time is unsatisfactorily short on physically relevant scales. On the other hand, the conceptual impact of the result was remarkable and persists; we have proved that a rigorous transition from reversible to irreversible dynamics is possible, and this is significant even if the time interval in question is extremely short.

It has often been questioned whether the Boltzmann equation could at all be derived from Hamiltonian dynamics. When Lanford's theorem was presented and discussed without detailed proofs almost twenty years ago, the lack of detail left many doubters questioning the result. We hope that the lengthy and detailed exposition which we have given to this subject here will dissipate any remaining doubts about the completeness and relevance of the result.

Problems

1. By representing all collision points in terms of their outgoing configuration, give the formal derivation of equation (4.4) with a minus sign in front of the collision term.
2. Prove the statement in step 2.
3. Prove, for $r_i \geq 0$, that

$$\sum_{i=1}^{s} r_i e^{-\alpha \sum_{i=1}^{s} r_i^2} \leq C\sqrt{\frac{s}{\alpha}}.$$

Hint: first prove that the left-hand side is maximized when $r_1 = r_2 = \dots$.

4. Prove (4.20).
5. Prove that any solution of the Boltzmann hierarchy satisfying a boundedness condition

$$f^{(s)}(\cdot, t) \leq b^s$$

is unique in this class.

4.5 Validity of the Boltzmann Equation for a Rare Cloud of Gas in the Vacuum

As we already mentioned, the method discussed in the previous section can only give a local result for general initial values. However, for suitable initial data one can prove global validity by replacing the smallness condition on time by a largeness condition on the mean free path; consider, for example, hard–sphere dynamics in all of $\Re^3$ and initial data that decay sufficiently fast at infinity, a situation we address in this section.

First, we note that steps 1, 2, and 4 from the proof in the previous section are completely general; therefore, a global result will be proved if we can control the series solution for the BBGKY and Boltzmann hierarchies globally.

As before, we assume a factorizing initial value

$$f_0^{(s)}(z^s) = \prod_{i=1}^{s} f_0(x_i, \xi_i)$$

and the Boltzmann–Grad limit $N \longrightarrow \infty$, $\sigma \longrightarrow 0$, $N\sigma^2 = \alpha$.

In addition, suppose that

i) the $P_\sigma^{(s)}(\cdot, 0)$ are continuous on $(\Re^3 \times \Re^3)_{\neq}^{s,\sigma}$ and

$$\lim_{N\to\infty} P_\sigma^{(s)}(\cdot, 0) = f_0^{(s)}$$

uniformly on compact subsets of $(\Re^3 \times \Re^3)_{\neq}^{s,0}$.

ii) there are constants $\beta_0 > 0$, $c > 0$ and $b > 0$ such that

$$\sup_{z^s} \left(P_\sigma^{(s)}(z^s, 0) \exp\left(\beta_0 \sum_{i=1}^{s} (\xi_i^2 + x_i^2) \right) \right) \leq cb^s.$$

Under these hypotheses, we have the following.

(4.5.1) Theorem. *Suppose that i) and ii) hold. Then, if $b \cdot \alpha$ is sufficiently small, the series solution $P_\sigma(\cdot, t)$ of the BBGKY hierarchy converges for all $t > 0$ in the Boltzmann–Grad limit almost everywhere to the series solution $f(\cdot, t)$ of the Boltzmann hierarchy. The latter factorizes as*

$$f^{(s)}(z^s, t) = \prod_{i=1}^{s} f(x_i, \xi_i, t)$$

and f is a mild global solution of the Boltzmann equation for the initial value f_0. Moreover, $P_\sigma^{(s)}$ and $f^{(s)}$ satisfy the estimates

$$0 \leq P_\sigma^{(s)}(z^s, t) \leq (c \cdot b)^s \exp(-\beta_0 \cdot I(T^{-t} z^s))$$

(where $I(z^s) = \sum_{i=1}^{s} x_i^2$, see Lemma 4.2.3),

$$0 \leq f(x, \xi, t) \leq (c \cdot b) \exp\left(-\beta_0 (x - t\xi)^2\right).$$

The constant c is independent of N and s.

Proof. As mentioned above, we only have to do step 3 of the proof of Theorem 4.4.1. Consider

$$S_\sigma(t - t_1)|Q^\sigma|S_\sigma(t_1 - t_2) \ldots S_\sigma(t_{n-1} - t_n)|Q^\sigma|S_\sigma(t_n) P_0^{n+s}(z^s) \tag{5.1}$$

where $|Q^\sigma|$ denotes the collision operator with $n \cdot (\xi_i - \xi_j)$ replaced by $|n \cdot (\xi_i - \xi_j)|$ (i.e., we give the "loss" term in the collision operator the "wrong" sign). The operator $S_\sigma(t)|Q_\sigma|S_\sigma(\tau)$ is then a monotone operator, i.e., $S_\sigma(t)|Q_\sigma|S_\sigma(\tau)P$ increases with P, and so (5.1) is an upper bound for the nth term in the series expansion.

Let $r = s + n - 1$, i.e., $r + 1 = s + n$, and focus on the last part of (5.1):

(5.2)

$$\begin{aligned} &S_\sigma(t_{n-1} - t_n)|Q^\sigma|S_\sigma(t_n) P_0^{n+s}(z^s) \\ &= \sum_{j=1}^{r} (N - r)\sigma^2 \int_{\Re^3} \int_{S^2} |n \cdot (\eta_j - \xi_{r+1})| \left[S_\sigma(t_n) P_0^{r+1}\right] (Y^r \cup z_{j,r})\, dn\, d\xi_{r+1} \end{aligned}$$

with $Y^r = (y_1 \ldots y_r, \eta_1 \ldots \eta_r) = T_\sigma^{t_n - t_{n-1}} z^r$, $z_{j,r} = (y_j + n\sigma, \xi_{r+1})$, and, accordingly,

$$Y^r \cup z_{j,r} = (y_1 \ldots y_r, y_j + n\sigma, \eta_1 \ldots \eta_r, \xi_{r+1}).$$

Recalling $I(z^s) = \sum_{i=1}^{s} x_i^2$, we use Lemma 4.3.2 to estimate

$$\begin{aligned} I(T_\sigma^{-t_n}(Y^r \cup z_{j,r})) &\geq I(T_0^{-t_n}(\ldots)) \\ &= I\left(T_0^{-t_n}(T_\sigma^{t_n - t_{n-1}} z^r)\right) + (y_j + n\sigma - t_n \xi_{r+1})^2 \\ &\geq I\left(T_0^{-t_{n-1}} z^r\right) + (y_j + n\sigma - t_n \xi_{r+1})^2. \end{aligned} \tag{5.3}$$

Furthermore, by energy conservation,

$$E(z^r) = E(T_\sigma^t z^r),$$

and using this and (5.3) we find

$$\begin{aligned}
&P_0^{r+1}\left(T_\sigma^{-t_n}\left(T_\sigma^{t_n-t_{n-1}} z^r \cup z_{j,r}\right)\right) \\
&\le c\cdot b^{r+1}\cdot \exp\left(-\beta_0 \xi_{r+1}^2\right)\cdot \exp\left(-\beta_0 E(z^r)\right) \\
&\cdot \exp\left(-\beta_0 (y_j + n\sigma - t_n \xi_{r+1})^2\right)\cdot \exp\left(-\beta_0 I\left(T_0^{-t_{n-1}} z^r\right)\right).
\end{aligned} \tag{5.4}$$

Proceeding from here as in the previous section, we have (by using energy conservation again)

$$\begin{aligned}
&\sum_{j=1}^{r} (|\eta_j| + |\xi_{r+1}|)\cdot \exp\left(-\frac{\beta_0}{n} E(z^r)\right)\cdot \exp\left(-\frac{\beta_0}{2}\xi_{r+1}^2\right) \\
&\qquad \le C(r + \sqrt{r}\sqrt{n}) \\
&\qquad \le C(s+n).
\end{aligned} \tag{5.5}$$

Here, the constant C depends on β_0. Also, of course, $(N-r)\sigma^2 \le \alpha$, and finally

$$\sup_{y_j} \int\!\!\int \exp\left(-\frac{\beta_0}{2}\xi_{r+1}^2\right) \exp(-\beta_0 (y_j + n\sigma - t\xi_{r+1})^2)\, dn\, d\xi_{r+1} \le \frac{C}{1+t^3} \tag{5.6}$$

(see Problem 1).

Now, insert (5.4–6) into (5.2). We obtain an estimate

$$\begin{aligned}
&[S_\sigma(t_{n-1} - t_n)|Q^\sigma|S_\sigma(t_n)P_0^{s+n}](z^r) \\
&\le C\cdot \alpha\cdot b^{s+n}\cdot \frac{s+n}{1+t_n^3}\cdot \exp\left(-\beta_0 \frac{n-1}{n} E(z^{s+n-1})\right) \\
&\cdot \left\{S_0(t_{n-1}) \exp\left(-\beta_0 I(z^{s+n-1})\right)\right\}.
\end{aligned} \tag{5.7}$$

If we substitute (5.7) into (5.1), the steps (5.4–6) can be repeated recursively. Ultimately, we arrive at an estimate

$$S_\sigma(t-t_1)|Q^\sigma|S_\sigma(t_1-t_2)\ldots S_\sigma(t_{n-1}-t_n)|Q^\sigma|S_\sigma(t_n)P_0^{n+s}(z^s)$$

$$\le C^n\cdot \alpha^n\cdot b^{s+n}\cdot (s+n)^n \left[\prod_{j=1}^{n}\left(\frac{1}{1+t_j^3}\right)\right]\cdot S_0(t)\left(\exp\left[-\beta_0 I(z^s)\right]\right)$$

or

$$P^{(s)}(z^s,t) \le b^s \sum_{n\ge 0} b^n\cdot C^n\cdot \alpha^n\cdot (s+n)^n$$

$$\int_0^t \cdots \int_0^{t_{n-1}} \prod_{j=1}^{n}\left(\frac{1}{1+t_j^3}\right) dt_n \ldots dt_1\cdot S_0(t)\left(\exp\left[-\beta_0 I(z^s)\right]\right).$$

Finally, we observe that

$$(5.8) \qquad \int_0^t \cdots \int_0^{t_{n-1}} \prod_{j=1}^{n} \left(\frac{1}{1+t_j^3} \right) dt_n \ldots dt_1 \leq \frac{C}{n!},$$

and by again using Stirling's formula, we get an estimate

$$(5.9) \quad P^{(s)}(z^s, t) \leq b^s \sum_{n \geq 0} C^n \cdot (b\alpha)^n \cdot \left(\frac{s+n}{n} \right)^n \cdot S_0(t)(\exp(-\beta_0 I(z^s)))$$

[the constant C here is not the same as in (5.8)]. If $(b \cdot \alpha)$ is small enough, the right hand side of (5.9) is the convergent majorant series $\sum b_n$ we set out to find. Also, we have an estimate

$$P^{(s)}(z^s, t) \leq (b \cdot c)^s (\exp(-\beta_0 I(T_0^{-t} z^s))),$$

from which we can conclude asymptotic dispersion of the gas cloud.

The same argument, with some simplifications, can be used to obtain a bound for the series solution of the Boltzmann hierarchy. This completes the proof of Theorem 4.5.1. □

Problems

1. Prove the estimate (5.6).
2. Prove the estimate (5.8).

4.6 Interpretation

We now discuss the question whether or not the Boltzmann equation actually yields any information about the time evolution of individual phase points of an N-particle gas. We will see that the answer to this is yes, and Theorem 4.4.1 from the previous section delivers the necessary result.

Suppose a gas of N hard spheres is confined to a domain Λ, with reflecting boundary conditions. Also, assume that we have a solution of the Boltzmann equation, $t \to f(\cdot, t)$, to some initial value f_0.

We ask: For a gas of N hard spheres, in what sense does $f(\cdot, t)$ describe the state at time t? The limit we have to investigate is, again, the Boltzmann-Grad limit $N \to \infty$, $\sigma \to 0$, $N\sigma^2 \to \alpha$.

For any rectangular parallelepiped $\Delta \subset \Lambda \times \Re^3$ and a phase point $z = (x_1 \ldots x_N, \xi_1 \ldots \xi_N)$, let

$$F_\Delta(z) = \frac{1}{N} \sum_{i=1}^{N} \chi_\Delta(x_i, \xi_i)$$

be the fraction of particles in Δ. As $N \to \infty$, $N\sigma^2 \to \alpha$, we expect this fraction at time t to be well represented by

$$\int_{\Delta} f(x, \xi, t)\, dx\, d\xi.$$

Here, "well represented" means "statistically close with respect to a sequence of probability measures in phase space." This sequence is, for any t, given by solving the Liouville equation. For convenience, we will refer in this section to such a sequence as "approximating sequence for f." The necessity for a statistical link between $F_\Delta(z)$ and $\int_\Delta f(x, \xi, t)\, dx\, d\xi$ is clear, because there is no a priori relationship between the two.

(4.6.1) Definition. *A sequence of symmetric probability measures $\{\mu^N\}_{N \in \mathcal{N}}$ on $\Lambda^N \times \Re^{3N}$ is called an approximating sequence for a given probability density f on $\Lambda \times \Re^3$ if for each rectangular parallelepiped $\Delta \subset \Lambda \times \Re^3$ and for each $\epsilon > 0$*

$$\lim_{N \to \infty} \mu^N \left\{ z;\, |F_\Delta(z) - \int_\Delta f(x, \xi)\, dx\, d\xi| > \epsilon \right\} = 0. \tag{6.1}$$

Let $\omega_z = \frac{1}{N} \sum_{i=1}^N \delta_{(x_i, \xi_i)}$, then

$$|F_\Delta(z) - \int_\Delta f(x, \xi)\, dx\, d\xi| = |\int_\Delta d\omega_z - \int_\Delta f(x, \xi)\, dx\, d\xi|.$$

In this formulation, we see that the statement "$\{\mu^N\}$ is an approximating sequence" means that as $N \to \infty$, the probability measures μ^N concentrate on those phase points for which the discrete probability measures ω_z are well approximated by $f\, dx\, d\xi$ in the weak-$*$-topology on the space of measures. Suppose that every μ^N is absolutely continuous, with density $P_0^{(N)} \in L^1_+(\Lambda^N \times \Re^{3N})$. The time evolution of $P_0^{(N)}$ is given by the Liouville equation

$$\frac{d}{dt}\left[P^{(N)}(T^t z, t)\right] = 0, \qquad P^{(N)}(z, 0) = P_0^{(N)}(z).$$

Again $P_0^{(s)}$ will denote the s-particle distribution function.

(4.6.2) Lemma.

a) $$\int F_\Delta(z)\, d\mu^N(z) = \int_\Delta P^{(1)}(x, \xi)\, dx\, d\xi.$$

b) $$\int (F_\Delta(z))^2\, d\mu^N(z) = \frac{1}{N} \int_\Delta P^{(1)}(x, \xi)\, dx\, d\xi + \frac{N-1}{N} \iint_{\Delta\Delta} P^{(2)}(z^2)\, dz^2.$$

Proof. a) is left to the reader. To show b) just calculate

$$\int (F_\Delta(z))^2 \, d\mu^N(z) = \frac{1}{N^2} \sum_{i=1}^{N} \sum_{j=1}^{N} \int \chi_\Delta(x_i, \xi_i) \cdot \chi_\Delta(x_j, \xi_j) \, d\mu^N(z)$$

$$= \frac{1}{N} \int \chi_\Delta(x_1, \xi_1) \, d\mu^N(z) + \frac{N(N-1)}{N^2} \int \chi_\Delta(x_1, \xi_1) \cdot \chi_\Delta(x_2, \xi_2) \, d\mu^N(z)$$

(by symmetry) and use the definition of $P^{(1)}$ and $P^{(2)}$. □

Remark. By a) $\int_\Delta P^{(1)}(x, \xi) \, dx \, d\xi$ is the expected fraction of particles in Δ, and by a) and b)

$$\frac{N-1}{N} \iint_{\Delta\Delta} P^{(2)}(z^2) \, dz^2 = E\left(F_\Delta(z)\right)^2 - \frac{1}{N} E\left(F_\Delta(z)\right)$$

($F_\Delta(z)$ is here interpreted as a random variable, with μ^N the relevant probability measure). We see that $P^{(1)}$ and $P^{(2)}$ determine the mean values and fluctuations of the occupation numbers $F_\Delta(z)$.

(4.6.3) Lemma. *$\{\mu^N\}_{N \in \mathcal{N}}$ is an approximating sequence for f if and only if*

$$a) \quad \lim_{N \to \infty} P^{(1)}(x, \xi) \, dx \, d\xi = f(x, \xi) \, dx \, d\xi$$

and

$$b) \quad \lim_{N \to \infty} P^{(2)}(z^2) \, dz^2 = f \otimes f \, dz^2$$

weak–$*$ in the sense of measures.

Proof. Suppose that $\{\mu^N\}_{N \in \mathcal{N}}$ is an approximating sequence. Let Δ be some rectangular parallelepiped; then by Lemma 4.6.2

$$\int_\Delta P^{(1)}(x, \xi) \, dx \, d\xi - \int_\Delta f(x, \xi) \, dx \, d\xi$$

$$= \int \left[F_\Delta(z) - \int_\Delta f(x, \xi) \, dx \, d\xi \right] d\mu^N(z),$$

and the last integral goes to 0 as $N \to \infty$ because $\{\mu^N\}$ is an approximating sequence.

To show that $\int_{\Delta_1 \times \Delta_2} P^{(2)}(z^2) \, dz^2 \to \int_{\Delta_1 \times \Delta_2} f \otimes f \, dz^2$ for all rectangular parallelepipeds, we can (by symmetry of $P^{(2)}$ and $f \otimes f$) assume that $\Delta_1 = \Delta_2 = \Delta$ (see Problem 1). Then, by Lemma 4.6.2,

$$\int_{\Delta\times\Delta} P^{(2)}(z^2)\,dz^2 - \int_{\Delta\times\Delta} f\otimes f\,dz^2$$

$$= \int\left[\frac{N}{N-1}F_\Delta^2(z) - \Big(\int_{\Delta\times\Delta} f\otimes f\,dz^2\Big)\right]d\mu^N(z) - \frac{1}{N-1}\int_\Delta P^{(1)}\,dx\,d\xi.$$

The right-hand side of this identity goes to zero as $N\to\infty$ because for all $\epsilon>0$ $\mu^N\left\{z;\,|F_\Delta^2(z) - \left(\int_\Delta f\,dx\,d\xi\right)^2| > \epsilon\right\}\to 0$ as $N\to\infty$, and $\frac{1}{N-1}\int_\Delta P^{(1)}\,dx\,d\xi\to 0$. Conversely, suppose that a) and b) hold. Then

$$\mu^N\left\{z;\,|F_\Delta(z) - \int_\Delta f\,dx\,d\xi|^2 > \epsilon^2\right\} \le \frac{1}{\epsilon^2}\int |F_\Delta(z) - \int_\Delta f\,dx\,d\xi|^2\,d\mu^N(z)$$

$$= \frac{1}{\epsilon^2}\left\{\int F_\Delta^2(z)\,d\mu^N(z) - 2\int F_\Delta(z)\,d\mu^N(z)\int_\Delta f\,dx\,d\xi + \int_{\Delta\times\Delta} f\otimes f\,dz^2\right\}$$

$$\longrightarrow 0$$

as $N\to\infty$, by Lemma 4.6.2. This completes the proof. □

Lemma 4.6.3 shows that factorization (in the limit $N\to\infty$) of the 2-particle distribution function $P^{(2)}$ is necessary for the μ^N to be an approximating sequence of f. In other words, the concept of "approximating sequence" implicitly contains the concept of "molecular chaos."

We give a reformulation of Theorem 4.5.1 in terms of the concept of "approximating sequence." Theorem 4.4.1 can be rephrased similarly.

(4.5.1) Theorem. (reformulated) *Assume the hypotheses of Theorem 4.5.1 (in particular, the sequence of probability measures $\mu_0^N = P_0^{(N)}\,dz$ is an approximating sequence for f_0). Then, if $b\cdot\alpha$ is sufficiently small, the Cauchy problem for the Boltzmann equation has a unique global solution $f(\cdot\,,t)$ with initial value f_0, and the sequence of measures $\mu_t^N = P_t^{(N)}\,dz$, where $P_t^{(N)}$ solves the Liouville equation with initial value $P_0^{(N)}$, is an approximating sequence for $f(\cdot\,,t)$ for all $t\ge 0$.*

Proof. This follows from the previous formulation of the theorem and Lemma 4.5.2. (The weak-$*$ convergence in the lemma follows from the a.e. convergence and the Lebesgue dominated convergence theorem.) □

Remark. In this formulation, it is clear that the validity of the Boltzmann equation is given in a statistical sense; as $N\to\infty$, $\sigma\to 0$ such that $N\sigma^2 = \alpha$, the fraction of particles in a cell Δ at time t ($F_\Delta(z)$) is, in measure with respect to $P^{(N)}(\cdot\,,t)\,dz^N$, well approximated by $\int_\Delta f(x,\,\xi,\,t)\,dx\,d\xi$.

The validity result can also be recast as a "law of large numbers": Let $\mu_0^N = P_0^{(N)}\,dz$ satisfy hypotheses i) and ii); then

$$\mu_t^N \left\{ z;\ |F_\Delta(z) - \int_\Delta f(x,\,\xi,\,t)\,dx\,d\xi| > \epsilon \right\} \longrightarrow 0.$$

But the left-hand side equals

$$\mu_0^N \left\{ z;\ |F_\Delta(T^t z) - \int_\Delta f(x,\,\xi,\,t)\,dx\,d\xi| > \epsilon \right\}$$
$$= \mu_0^N \left\{ z;\ |\frac{1}{N} \int_\Delta d\omega_{T^t z}(x,\,\xi) - \int_\Delta f(x,\,\xi,\,t)\,dx\,d\xi| > \epsilon \right\},$$

and we see that the fraction of particles in Δ at time t is, with respect to the sequence of probability measures μ_0^N, indeed better and better represented by $\int_\Delta f(\cdot\,,t)$.

The interpretation would be particularly appealing if we would choose $\mu_0^N = \prod_{i=1}^N f_0(x_i,\,\xi_i)\,dz$, but this is impossible because of the exclusion principle (particles cannot overlap) and the continuity requirement on μ_0^N along trajectories. We have to redefine and renormalize μ_0^N such as to be consistent with the physical constraint. The derived factorization of μ_0^N only emerges in the strong limit given by i) in Theorem 4.4.1.

A final remark on the normalization is in order. In many texts, correlation functions $\rho^{(s)}(z^s,\,t)$ are defined by

$$\rho^{(1)}(x_1,\,\xi_1,\,t) = N \int P^{(N)}(z,\,t)\,dz^{N-1}$$
$$\rho^{(2)}(z^2,\,t) = N(N-1) \int P^{(N)}(z,\,t)\,dz^{N-2}$$
$$\vdots$$
$$\rho^{(s)}(z^s,\,t) = N \ldots (N-s+1) \int P^{(N)}(z,\,t)\,dz^{N-s}.$$

$\int_\Delta \rho^{(1)}\,dx_1\,d\xi_1$ is then the expected value of the random variable $N_\Delta = N \cdot F_\Delta(z)$, i.e., the mean number of particles in Δ. Similarly,

$$\int_\Delta \cdots \int_\Delta \rho^{(s)}(z^s)\,dz^s$$

is the mean value of $N_\Delta(N_\Delta - 1)\ldots(N_\Delta - s + 1)$. The BBGKY hierarchy can be formulated and studied in terms of the $\rho^{(s)}(z^{(s)},\,t)$, but rescaling

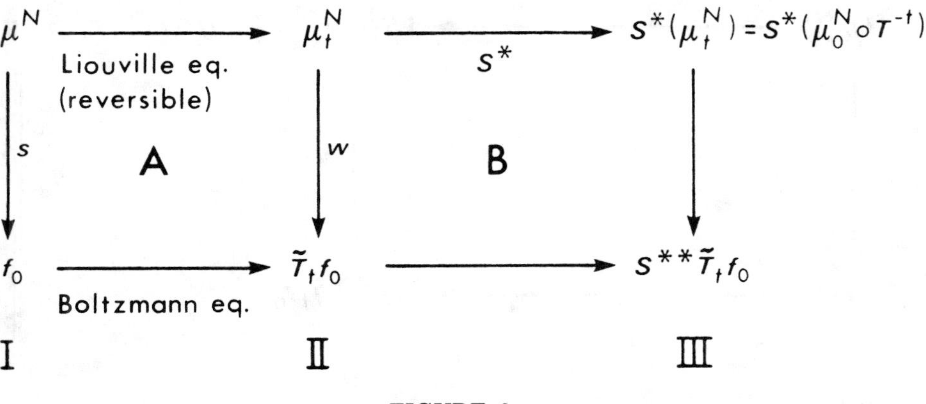

FIGURE 8.

is necessary because $\rho^{(s)} \to \infty$ as $N \to \infty$. The usual rescaling is done by setting

$$f^{(s)}(z^s, t) = N^{-s} \rho^{(s)}(z^s, t).$$

It is easily seen that in the Boltzmann-Grad limit $\lim_{N\to\infty} f^{(s)}(z^s, t)$ and $\lim_{N\to\infty} P^{(s)}(z^s, t)$ are the same. The formulation of the BBGKY hierarchy for the $f^{(s)}$ is sometimes preferred over the form for the $P^{(s)}$, because the collision operators, as acting on $f^{(s+1)}$, carry a factor $N\sigma^2$ (rather than $(N-s)\sigma^2$ for $P^{(s+1)}$).

Problem

1. Let Δ_1, Δ_2 be two disjoint subsets of $\Lambda \times \Re^3$. Show that $(\Delta_1 \times \Delta_2) \cup (\Delta_2 \times \Delta_1) = (\Delta_1 \cup \Delta_2) \times (\Delta_1 \cup \Delta_2) \backslash (\Delta_1 \times \Delta_1) \backslash (\Delta_2 \times \Delta_2)$.

4.7 The Emergence of Irreversibility

The loss of reversibility (in the mechanical sense) is now easily explained. Recall (4.2.3): $T^t S T^t z = Sz$ for all $z \in \Gamma$, $t \in \Re$. The involutive operation S of velocity inversion induces an (involutive) operator $S^* : \mu^N \to S^* \mu^N$ by $S^* \mu^N = \mu^N \circ S$. Also, if $\tilde{T}_t f_0(x, \xi) = f(x, \xi, t)$ is a solution of the Boltzmann equation, let $(S^{**} \tilde{T}_t f_0)(x, \xi) = f(x, -\xi, t)$. Consider the diagram in Fig. 8.

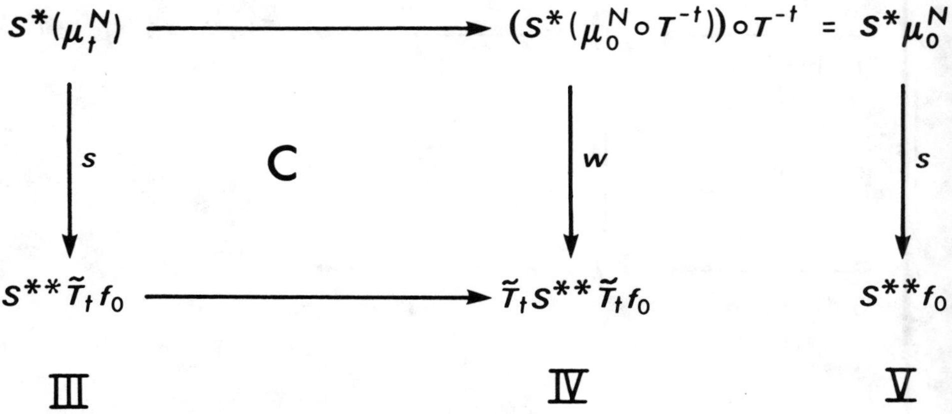

$S^*(\mu_t^N) \longrightarrow (S^*(\mu_0^N \circ T^{-t})) \circ T^{-t} = S^* \mu_0^N$

$s \downarrow$ C $\downarrow w$ $\downarrow s$

$S^{**} \tilde{T}_t f_0 \longrightarrow \tilde{T}_t S^{**} \tilde{T}_t f_0$ $S^{**} f_0$

III IV V

FIGURE 9.

Here, the horizontal arrows in part A of the diagram denote the indicated time evolutions; the vertical arrows indicate the convergences (in the Boltzmann-Grad limit) from Theorem 4.4.1: the "s" stands for the strong convergence of $P^{(s)} \to \prod_{i=1}^{s} \otimes f_0(x_i, \xi_i)$ assumed initially, and the "w" for the weaker convergence at time $t > 0$ (for t sufficiently small) given in the assertion of Theorem 4.4.1 (a.e. $P^{(s)}(\cdot, t) \to \prod_{i=1}^{s} \otimes f(x_i, \xi_i, t)$, $s \geq 2$); in other words, the left-hand side of the diagram is just a concise reformulation of the theorem.

Part B of the diagram tells us what happens if we reverse velocities: of course, the convergence "w" is preserved in this operation. Suppose that we could actually "save" the "s"-convergence in column II (and hence III) of the diagram (we already know that the series approach from Section 4 will not permit this, but let us disregard this for now). The other assumptions made for Theorem 4.4.1 can be seen to remain true anyway.

We could then use Theorem 4.4.1 once more to extend the diagram to the right as indicated in Fig. 9.

Because both s- and w-limits are unique, it would follow that

$$S^{**} f_0 = \tilde{T}_t S^{**} \tilde{T}_t f_0. \tag{7.1}$$

But the H-theorem implies that

$$\begin{aligned} H(\tilde{T}_t S^{**} \tilde{T}_t f_0) &\leq H(S^{**} \tilde{T}_t f_0) \\ &= H(\tilde{T}_t f_0) < H(f_0) \end{aligned}$$

unless f_0 is the equilibrium solution, i.e., (7.1) is in general impossible. We see that the loss in convergence quality is actually necessary to explain the irreversible behavior of the Boltzmann equation.

Conversely, this loss even explains why the decrease of the H-functional is possible.

Let

$$h(x) = \begin{cases} x \cdot \ln x & \text{for } x > 0 \\ 0 & \text{for } x = 0. \end{cases}$$

We define H-functionals $H^s(P^{(s)})$ by $H^s(P^{(s)}) := \frac{1}{s} \int h \circ P^{(s)}(z^s)\, dz^s$.

First, note that

$$H^N\Big(\prod_{i=1}^{N} \otimes f_0(x_i,\, \xi_i)\Big) = H(f_0). \tag{7.2}$$

Suppose next that we have $P_0^{(N)} \overset{s}{\longrightarrow} f_0$ (in the sense of the diagram, i.e., $P_0^{(N)}$ and f_0 satisfy the hypotheses of Theorem 4.4.1) and that *in addition*

$$H^N(P_0^{(N)}) \to H(f_0) \tag{7.3}$$

as $N \to \infty$. In view of (7.2), (7.3) can be interpreted in the sense that $P_0^{(N)}$ "almost factorizes" sufficiently fast as $N \to \infty$ (whether such "rapid" factorization is physically meaningful is an interesting question, but not of importance for our current goal).

We can set up the diagram in Fig. 10.

and conclude that

$$\liminf_{N\to\infty} H^N(P^{(N)}(\cdot\,,t)) > H(f(\cdot\,,t)), \tag{7.4}$$

with strict inequality in general. The key observation here is that the effect of collisions and the corresponding loss of convergence quality and factorization will inevitably destroy the convergence (7.3) at later times.

We summarize the relations between the $H^s(P^{(s)}(\cdot\,,t))$ and $H(f(\cdot\,,t))$ in the following theorem.

(4.7.1) Theorem. *Suppose that $P^{(N)}$ is a symmetric probability density on phase space and that $P^{(s)}$ are the s-particle density distribution functions associated with $P^{(N)}$. Then*

a) $H^1(P^{(1)}) \le H^N(P^{(N)})$ *with equality if and only if*

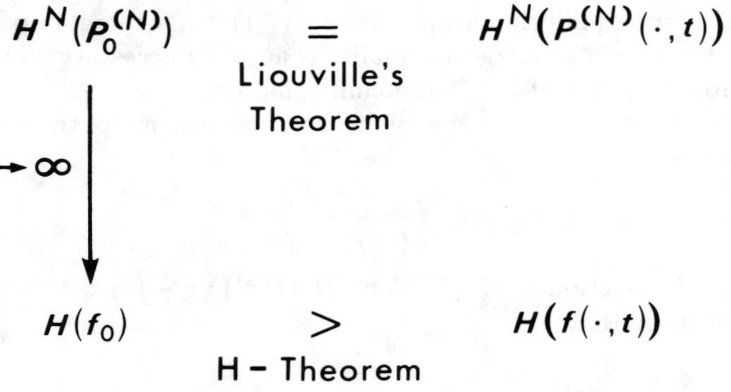

$H^N(P_0^{(N)})$ = $H^N(P^{(N)}(\cdot, t))$

Liouville's Theorem

$N \to \infty$

$H(f_0)$ > $H(f(\cdot, t))$

H - Theorem

FIGURE 10.

$$P^{(N)}(z) = \prod_{i=1}^{N} P^{(1)}(x_i, \xi_i)$$

for almost all z.

b) if $P^{(N)} \underset{w}{\longrightarrow} f$ as $N \to \infty$ in the sense that for fixed $s \in \mathcal{N}$

$$|P^{(s)}(z^s)| \leq Cb^s e^{-\beta \sum_{i=1}^{s}(x_i^2 + \xi_i^2)}$$

and

$$P^{(s)}(z^s) \to \prod_{i=1}^{s} f(x_i, \xi_i) \text{ a.e. as } N \to \infty,$$

then for every s $H^s(P^{(s)}) \to H(f)$ as $N \to \infty$, but in general

$$\liminf_{N \to \infty} H^N(P^{(N)}) > H(f).$$

Proof. The very last statement was already observed in (7.4). To prove a) note that for $x, y \geq 0$, $x - y \geq y \ln \frac{x}{y}$. (Set the right-hand side equal to zero for $y = 0$ and equal to $-\infty$ for $y > 0$, $x = 0$.)

Therefore,

$$0 = \int \left(\left(\prod_{i=1}^{N} P^{(1)}(x_i, \xi_i) \right) - P^{(N)}(z) \right) dz$$

$$\geq \int P^{(N)}(z) \cdot \ln \frac{\prod_{i=1}^{N} P^{(1)}(x_i, \xi_i)}{P^{(N)}(z)} \, dz,$$

i.e., $\int h \circ P^{(N)}(z)\, dz \geq N \cdot H^1(P^{(1)})$. Clearly, the inequality is strict exactly if $P^{(N)}$ does not factorize.

The remaining part of b) follows from the dominated convergence theorem. The almost everywhere convergence of $P^{(s)}(z^s)$ to $\prod_{i=1}^{s} f(x_i, \xi_i)$ implies the almost everywhere convergence of $h \circ P^{(s)}$ to $h \circ \prod_{i=1}^{s} f(x_i, \xi_i)$. To find a common integrable upper bound, let

$$g(x) = \begin{cases} \sqrt{x} & \text{for } 0 \leq x \leq 1 \\ 1 + x \ln x & \text{for } x > 1, \end{cases}$$

and observe that

$$|x \ln x| \leq g(x).$$

Therefore,

$$|(h \circ P^{(s)}(z^s))| \leq (g \circ P^{(s)})(z^s),$$

and from the monotonicity of g we have

$$(g \circ P^{(s)})(z^s) \leq g(Cb^s e^{-\beta \sum (x_i^2 + \xi_i^2)}),$$

and the function on the right is the common integrable upper bound. □

Remarks

1. Boltzmann[3,5] already pointed out that the entropies $H^s(t)$ associated with the s-particle distribution functions would not have to decrease as long as N was kept fixed (for $s = N$, of course, $H^N(t)$ is constant, but for $s < N$, $H^s(t)$ could possibly undergo oscillations), but would approach the strictly decreasing $H(t)$ as $N \to \infty$. The proof of the last theorem verifies this.

2. The function $h(x) = \chi_{\{x>0\}} \cdot x \cdot \ln x$ arises quite naturally in this context, because, up to a possible factor, it is the only continuous function $\varphi : \Re_+ \to \Re$ that is differentiable for $x > 0$, such that

$$\int \varphi \circ P^{(1)}(x, \xi)\, dx\, d\xi \leq \frac{1}{N} \int \dots \int \varphi \circ P^{(N)}(z)\, dz \tag{7.5}$$

for all N and for all normalized symmetric $P^{(N)} \in L^1_+$ for which the integrals on the right of (7.5) exist, with equality if $P^{(N)}(z) = \prod_{i=1}^{N} P^{(1)}(x_i,\, \xi_i)$. We leave the proof as an exercise for the reader (see Problem 1).

3. The H-functional has a suggestive physical interpretation, also discovered by Boltzmann (see also Lanford [19]). Let $\Delta_i, \dots \Delta_j$ be a finite number of nonoverlapping cells in $\Lambda \times \Re^3$, and let $\lambda^6(\Delta_i)$ be the

volume of Δ_i (assumed finite). For a given (large) N, choose integers $N_i \ldots N_j$ such that $\sum_{i=1}^{j} N_i = N$, and ask how much phase-space volume V_N in $(\Lambda \times \Re^3)^N$ is available to phase points z with $N_{\Delta_i}(z)(= N \cdot F_{\Delta_i}(z)) = N_i$. A little combinatorics and Stirling's formula ($N_i! \sim N_i^{N_i} e^{-N_i}$) shows that this volume is given by

$$\frac{N!}{\prod_{i=1}^{j} N_i!} \prod_{i=1}^{j} |\lambda^6(\Delta_i)|^{N_i} \approx \left\{ \prod_{i=1}^{j} \left(\frac{N}{N_i} |\lambda^6(\Delta_i)| \right)^{N_i/N} \right\}^N .$$

Therefore, for large N,

$$\frac{1}{N} \ln V_N = - \sum_{i=1}^{j} \frac{N_i}{N} \ln \left(\frac{N_i}{N} \frac{1}{|\lambda^6(\Delta_i)|} \right) .$$

Now, if we formally replace $\frac{N_i}{N}$ by $\int_{\Delta_i} f(x, \xi)$ (where f is the density distribution function of the gas), the last sum becomes

$$- \sum_{i=1}^{j} \left(\int_{\Delta_i} f \right) \ln \left(\frac{\int_{\Delta_i} f}{|\lambda^6(\Delta_i)|} \right) ,$$

and this turns (formally) into $-H(f)$ if we choose the Δ_i as a partition of $\Lambda \times \Re^3$ and send the mesh of this partition to zero.

The decrease of $H(f)$ with time can therefore be interpreted as motion of the system from regions of lower phase-space volume to (more likely) regions with much larger volume. [Our formal argument shows that $V_N \sim \exp(-N\,H(f))$.]

Problem

1. Prove the assertion made in Remark 2.
   Hint: Consider $P^{(1)}(x, \xi) = \left(\frac{1}{\epsilon}\right)^6 \chi_{W(\epsilon)}(x, \xi)$, where $W(\epsilon)$ is a cube of volume ϵ^6 in $\Lambda \times \Re^3$ and let $P^{(N)}(z) = \prod_{i=1}^{N} P^{(1)}(x_i, \xi_i)$.

4.8 More on the Boltzmann Hierarchy

We discuss now two related questions from Section 4.4: What happens if the P_0^N converge, at time zero, to a nonfactorizing state? Is it possible to give meaning to the Boltzmann hierarchy even for nonfactorizing solutions?

To discuss the first question, suppose that $\{f_s\}_{s=1}^{\infty}$ is a family of distribution densities, i.e., they satisfy *i)* $f_s(z) \geq 0$

ii) $\int f_s(z)\, dz = 1$

iii) $\int f_s(x_1 \dots x_s, \xi_1 \dots \xi_s)\, dx_{k+1} d\xi_{k+1} \dots dx_s d\xi_s = f_k(x_1 \dots x_k, \xi_1 \dots \xi_k)$

and assume that

$$\lim_{\sigma \to 0} P^{\sigma}_{0,s} = f_s$$

for all s, uniformly on compact sets in $\left(\Re^3 \times \Re^3\right)^{s,0}_{\neq}$ (see the hypotheses for Theorem 4.4.1). Suppose also that condition ii) there is verified. Then the same arguments used in Sections 4.4 and 4.5 are applicable to prove (under suitable smallness conditions) the convergence of $P^{\sigma}_s(t)$ to $f_s(t)$ a.e., where $f_s(t)$, $s = 1, \dots, \infty$ is a series solution of the Boltzmann hierarchy.

We want to investigate the meaning of a solution to the Boltzmann hierarchy for general nonfactorizing states $\{f_s(t)\}_{s=1}^{\infty}$.

Consider first the following simple example. Let

$$M_{\beta_i}(\xi) = \frac{1}{(2\pi\beta_i)^{\frac{3}{2}}} \exp(-\beta_i \xi^2), \tag{8.1}$$

$i = 1, 2$ be two Maxwellians at inverse temperatures β_1 and β_2, $\beta_1 \neq \beta_2$. Then

$$g_s(\xi_1 \dots \xi_s) = \lambda \prod_{i=1}^{s} M_{\beta_1}(\xi_i) + (1 - \lambda) \prod_{i=1}^{s} M_{\beta_2}(\xi_i) \tag{8.2}$$

for $\lambda \in (0, 1)$ is a nonfactorizing state satisfying the stationary Boltzmann hierarchy

$$Q_s g_s = 0; \ \sum_{i=1}^{s} \xi_i \partial_{x_i} g_s = 0 \tag{8.3}$$

(this is obvious, because the hierarchy is linear).

The state expressed by equation (8.2) describes a physical situation in which a gas is in thermal equilibrium at inverse temperature β_1 with probability λ and in thermal equilibrium at inverse temperature β_2 with probability $1 - \lambda$. The relevance and effective feasibility of states of this type in rarefied gas dynamics is not clear. However, from a mathematical point of view, such mixtures of states make pefect sense, and they arise quite naturally in the discussion of the validation of the Boltzmann equation. Therefore, it seems worthwhile to continue a deeper analysis of the Boltzmann hierarchy.

The previous example suggests that solutions of the Boltzmann hierarchy descibe mixtures of solutions of the Boltzmann equation, or, in other words, "statistical solutions" of the Boltzmann equation. This is indeed true, as pointed out by H. Spohn [22]. We explain the argument, beginning with the concept of statistical solutions in the case of an ordinary differential system of equations.

Let $T_t : \Omega \to \Omega$ be a one-parameter group of transformations on a differential manifold Ω, generated by a vector field F, and μ a Borel probability measure on Ω that represents the indeterminacy of the initial value. A statistical solution of the system

$$\dot{x} = F(x) \tag{8.4}$$

is a map $t \to \mu_t$ such that

$$\mu_t(\Phi) = \mu(\Phi \circ T_{-t}) \tag{8.5}$$

for all smooth test functions Φ on Ω, where

$$\mu(\Phi) = \int_\Omega \Phi(x)\, d\mu(x). \tag{8.6}$$

The measure function μ_t then satisfies

$$\frac{d}{dt}\mu_t(\Phi) = \mu_t(\nabla\Phi \cdot F). \tag{8.7}$$

To generalize these considerations to the Boltzmann equation, set $\Omega^0 = L^1_+$ and let μ be a probability measure on Ω^0. μ is assumed to be a Borel measure with respect to the topology induced by the weak convergence of measures. We denote by Ω the closure of Ω^0 with respect to this topology. The Boltzmann vector field is given by

$$F(f) = -\xi \cdot \partial_x f + Q(f, f), \tag{8.8}$$

$f \in \Omega^0$. Statistical solutions associated with the Boltzmann equation take the form

$$\frac{d}{dt}\mu_t(\Phi) = \mu_t\Big(\frac{\delta\Phi}{\delta f}[-\xi \cdot \partial_x f + Q(f, f)]\Big), \tag{8.9}$$

where $\frac{\delta\Phi}{\delta f}$ is the functional derivative of the "smooth" functional Φ on Ω^0, seen as a linear operator acting on $[-\xi \cdot \partial_x f + Q(f, f)]$. The information content of equation (8.9) depends on which class of "good" functionals Φ, for which (8.9) has to be true, is chosen. A natural choice is the algebraic closure of the functionals of the form

$$\Phi^z(f) = \prod_{(x,\xi)\in z} f(x, \xi), \tag{8.10}$$

i.e., the factorized functionals. As

$$\frac{\delta\Phi^z(f)}{\delta f} = \sum_{(x,\xi)\in z} \prod_{(y,\eta)\neq(x,\xi)} f(y, \eta) \tag{8.11}$$

equation (8.9) reduces to

$$\begin{aligned}\frac{d}{dt}\mu_t\Big(\prod_{(x,\xi)\in z} f(x,\xi)\Big) &= -\mu_t\Big(\sum_{(x,\xi)\in z}\ \prod_{(y,\eta)\neq(x,\xi)} f(y,\eta)\xi\cdot\partial_x f(x,\xi)\Big)\\ &\quad+\mu_t\Big(\sum_{(x,\xi)\in z}\ \prod_{(y,\eta)\neq(x,\xi)} f(y,\eta)Q(f,f)(x,\xi)\Big).\end{aligned}\tag{8.12}$$

Observe now that defining $f_s(t)$ by

$$f_s(t) = \int \mu_t(df)\, f^{\otimes s},\tag{8.13}$$

we get a family of distribution densities satisfying i), ii), and iii). Moreover, (8.12) reduces to the Boltzmann hierarchy. Thus we see that any statistical solution of the Boltzmann equation (in the sense of (8.9), with a suitable family of test functions) induces via formula (8.13) a solution of the Boltzmann hierarchy.

The converse is also true. Given a family of distribution densities (satisfying i), ii), and iii)) there exists a Borel measure μ on L^1_+ for which (8.13) holds. This is the content of the Hewitt-Savage theorem (see, e.g., Dunford and Schwarz [7]), which explains how a generic state can be decomposed in terms of pure or factorizing states. Starting now from a solution $f_s(t)$ of the Boltzmann hierarchy, by the Hewitt-Savage theorem we know the existence of a measure function μ_t, which is then easily seen to be a statistical solution of the Boltzmann equation.

The interpretation of the Boltzmann hierarchy is therefore that it describes the Boltzmann flow whenever there is some indeterminacy with respect to the initial datum f_0. If f_0 is not known, but only given by a probability distribution on Ω^0, say μ_0, the indeterminacy is still present at time t and is described by a measure function μ_t satisfying equation (8.9), or equivalently, the distribution densities given by (8.13) satisfy the hierarchy.

We return to the validation problem. If the $P^{\sigma}_{0,s}$ converge to a non-factorizing state, this means that the sequences of measure-valued random variables

$$\frac{1}{N}\sum_{i=1}^{N}\delta_{(x_i,\xi_i)}(x,\xi)$$

do not converge (for μ^N almost all z) to a single distribution f but to a random variable $f\in L^1_+$, distributed according to μ_0. Because we have so little control about the Boltzmann flow, our discussion on the meaning of the hierarchy must remain largely formal. However, some natural questions concerning the hierarchy are easily answered.

Consider

$$f_s = \int \mu_0(df)\, f^{\otimes s}$$

as the initial datum for the Boltzmann hierarchy. Suppose that μ_0 is concentrated on a subset of Ω^0, say $\tilde{\Omega}$, for which the Boltzmann flow T_t is well defined. We already know that $\tilde{\Omega}$ is not empty, including, for example, small perturbations of the vacuum (and as we shall see in Chapters 6 and 7, homogeneous data or states sufficiently close to the global Maxwellian equilibrium).

By defining

$$f_s(t) = \int \mu_0(df)(T_t f)^{\otimes s}$$

we realize that $f_s(t)$ is a solution of the Boltzmann hierarchy. Thus the existence of solutions of the Boltzmann hierarchy follows trivially from support properties of μ_0 and by a good control of the Boltzmann flow.

Next we address the question of whether this solution is unique. This turns out to be a very delicate problem. Consider the following example. Let

$$\begin{cases} \dot{a}_j(t) = j \cdot a_{j+1}(t), \; j = 1, 2, \dots \\ a_j(0) = b_j \end{cases} \tag{8.14}$$

be an infinite system of coupled ordinary differential equations. Assume that

$$b_j \leq C^j.$$

Then arguments similar (actually easier) to those used in Section 4.4 show the existence of a "unique" solution up to a small time t_0 (inversely proportional to C). This solution is unique in the class of all solutions $a_j(t)$ that satisfy bounds

$$\sup_{0 \leq t \leq t_0} |a_j(t)| \leq C_1^j$$

for some $C_1 < \infty$. However, the solution is not expected to be unique in a larger class of solutions. In fact, choose a function $t \to \varphi(t)$ such that $\frac{d^k}{(dt)^k}\varphi(t)|_{t=0} = 0$ and define recursively

$$\begin{cases} d_{j+1} = \dfrac{1}{j} \cdot \dot{d}_j \\ d_1 = \varphi. \end{cases}$$

Then $a_j(t) = d_j(t)$ and $a_j = 0$ are two solutions of the initial value problem (8.14) with initial data $b_j = 0$. Of course, $d_j(t)$ must grow faster than exponentially in j.

Obviously, our example is not particularly to the point. All of the probabilistic structure contained in the Boltzmann hierachy is lost. However, the example shows that it is certainly difficult to prove uniqueness of the Boltzmann hierarchy in a class of solutions larger than those satisfying bounds like $f_j \leq C^j$. For such solutions one can try to apply the bootstrap argument from Section 4.4 to get uniqueness.

We conclude this section by mentioning two uniqueness results on solutions of the hierarchy for situations close to the equilibrium (see R. Esposito and M. Pulvirenti [8]) and for the spatially homogeneous hierarchy (L. Arkeryd, S. Caprino, and N. Ianiro [2]).

References

1. R. K. Alexander, "The infinite hard sphere system," Ph.D. thesis, Department of Mathematics, University of California at Berkeley (1975).
2. L. Arkeryd, S. Caprino and N. Ianiro, "The homogeneous Boltzmann hierarchy and statistical solutions to the homogeneous Boltzmann equation," *J. Stat. Phys.* **63**, 345–361 (1991).
3. L. Boltzmann, "Weitere Studien über das Wärmegleichgewicht unter Gasmolekülen," *Sitzungsberichte der Akademie der Wissenschaften, Wien* **66**, 275–370 (1872).
4. Broadwell, "Shock structure in a simple discrete velocity gas," *Phys. Fluids* **7**, 1243–1247 (1964).
5. S. G. Brush (Ed.), *Kinetic theory,* Pergamon (1966).
6. S. Caprino, A. DeMasi, E. Presutti, and M. Pulvirenti, "A derivation of the Broadwell equation," *Commun. Math. Phys.* **135**, 443–465 (1991).
7. N. Dunford and J. T. Schwarz, *Linear Operators, I.* Interscience (1963).
8. R. Esposito and M. Pulvirenti, "Statistical solutions of the Boltzmann equation near the equilibrium," *Transport Theory Stat. Phys.* **18 (1)**, 51–71 (1989).
9. R. Gatignol, "Théorie cinétique des gas à repartition discrète des vitesses," *Lecture Notes in Physics* **36**, Springer-Verlag (1975).
10. V. I. Gerasimenko, "On the solution of a Bogoliubov hierarchy for one–dimensional hard sphere particle system," *Teor. Math. Fiz.* **91**, 120–128 (1992).
11. V. I. Gerasimenko and D. Ya. Petrina, "Existence of Boltzmann–Grad limit for an infinite hard sphere system," *Teor. Math. Fiz.* **83**, 92-114 (1990).
12. R. Illner, "Finiteness of the number of collisions in a hard sphere particle system in all space, II: Arbitrary diameters and masses," *Transport Theory Stat. Phys.* **19 (6)**, 573–579 (1990).
13. R. Illner, "On the nunmber of collisions in a hard sphere particle system in all space," *Transport Theory Stat. Phys.* **18 (1)**, 71–86 (1989).
14. R. Illner, "Derivation and validity of the Boltzmann equation: Some remarks on reversibility concepts, the H-functional and coarse-graining," in *Material Instabilities in Continuum Mechanics and Related Mathematical Problems,* J. M. Ball (Ed.), Clarendon Press, Oxford 1988.
15. R. Illner and H. Neunzert, "The concept of irreversibility in the kinetic theory of gases," *Transport Theory Stat. Phys.* **16(1)**, 89–112 (1987).
16. R. Illner and T. Platkowski, "Discrete velocity models of the Boltzmann equation: A survey on the mathematical aspects of the theory," *SIAM Review* **30 (2)**, 213–255 (1988).
17. R. Illner and M. Pulvirenti, "Global validity of the Boltzmann equation for a two- dimensional rare gas in vacuum," *Commun. Math. Phys.* **105**, 189–203 (1986).
18. R. Illner and M. Pulvirenti, "Global validity of the Boltzmann equation for two- and three-dimensional rare gases in vacuum: Erratum and improved result," *Commun. Math. Phys.* **121**, 143–146 (1989).
19. O. Lanford III, "Time evolution of large classical systems," *Lecture Notes in Physics* **38**, E. J. Moser (ed.), 1–111. Springer–Verlag (1975).
20. C. Marchioro, A. Pellegrinotti, E. Presutti, and M. Pulvirenti, "On the dynamics of particles in a bounded region: A measure theoretical approach," *J. Math. Phys.* **17**, 647–652 (1976).
21. D. Ya. Petrina, V. I. Gerasimenko, and P. V. Malyshev, *Mathematical Foundations of Classical Statistical Mechanics,* Gordon and Breach Sci. Publ. (1989).

22. H. Spohn, "Fluctuation theory for the Boltzmann equation," in *Nonequilibrium Phenomena I: The Boltzmann Equation.* J. L. Lebowitz and E. W. Montroll (Eds.), 225–251, North–Holland (1983).
23. H. Spohn, *Large Scale Dynamics of Interacting Particles,* Springer-Verlag (1991).
24. K. Uchiyama, "Derivation of the Boltzmann equation from particle dynamics," *Hiroshima Math. J.* **18(2)**, 245–297 (1988).
25. K. Uchiyama, "On the Boltzmann–Grad limit for the Broadwell model of the Boltzmann equation," *J. Stat. Phys.* **52 1&2**, 331–355 (1988).
26. L. Vaserstein, "On systems of particles with finite-range and/or repulsive interactions," *Commun. Math. Phys.* **69**, 31–56 (1979).

Appendix 4.A
More About Hard-Sphere Dynamics

In this appendix we sketch a proof of Theorem 4.2.1 and discuss properties of N- particle dynamics that are not directly relevant for the validation problem but are of some intrinsic interest.

In both the derivation of the BBGKY hierarchy and the problem of multiple collisions of hard-sphere dynamics, we take advantage of coordinates known as "special flow representation." We first explain these coordinates in a general context.

Consider a smooth divergence–free vector field $\mathcal{F} : \Re^n \longrightarrow \Re^n$, $\operatorname{div}\mathcal{F} = 0$, and the flow generated by this field, i.e.,

$$\frac{d}{dt}\varphi_t(x) = \mathcal{F}(\varphi_t(x))$$
$$\varphi_0(x) = x.$$

By the Liouville theorem, this flow preserves Lebesgue measure.

We assume the existence of a relatively compact invariant set Λ and a smooth manifold Σ of codimension one in $\Re^n$, such that under the action of the flow φ_t all the points in Λ have crossed Σ in the past and will do so again in the future; see Fig. 11.

If dy denotes the surface element on Σ and $n(y)$ is a normal to Σ at y, we define a measure $d\sigma(y)$ on Σ by

$$d\sigma(y) = |\mathcal{F}(y) \cdot n(y)|\, dy$$

[the sign of $n(y)$ is of no importance]. Finally, we define a function α : $\Sigma \longrightarrow \Re_+$ by

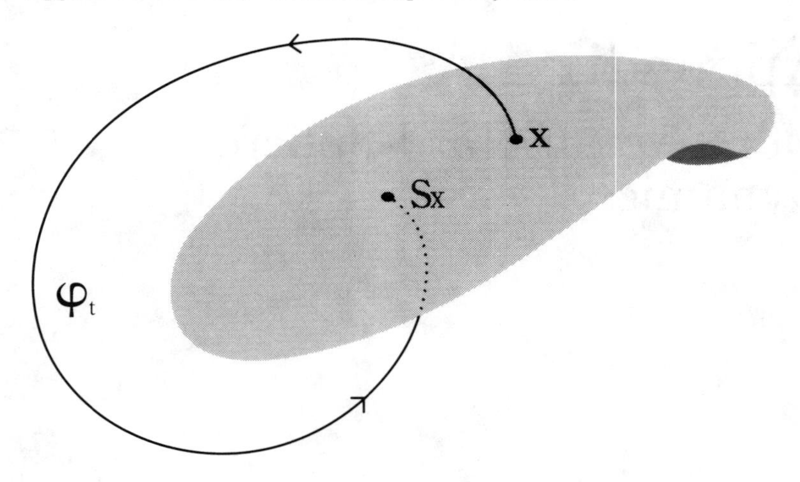

FIGURE 11.

$$\alpha(y) = \min\{t > 0;\ \varphi_t(y) \in \Sigma\}.$$

We refer to α as a "ceiling function" for reasons that will become clear momentarily.

The one–to–one mapping

$$(A.1) \qquad \Psi : \Lambda \longrightarrow \tilde{\Lambda} = \{(y,t);\ y \in \Sigma,\ 0 \leq t < \alpha(y)\},$$

defined by

$$x = \varphi_t(y),$$

induces an automorphism

$$\Psi : (\Lambda, dx) \longrightarrow (\tilde{\Lambda}, d\sigma dt)$$

in the sense of the measure spaces. This is a consequence of the Liouville theorem (see Problem 1). Fig. 12 visualizes the mapping Ψ and explains why we refer to α as a "ceiling function."

A major advantage of the special flow representation is that the representation of the flow φ_t becomes trivial. If $\Psi(x) = (y,t)$ and we define $T : \Sigma \longrightarrow \Sigma$ by $Ty = \varphi_{\alpha(y)}y$, then

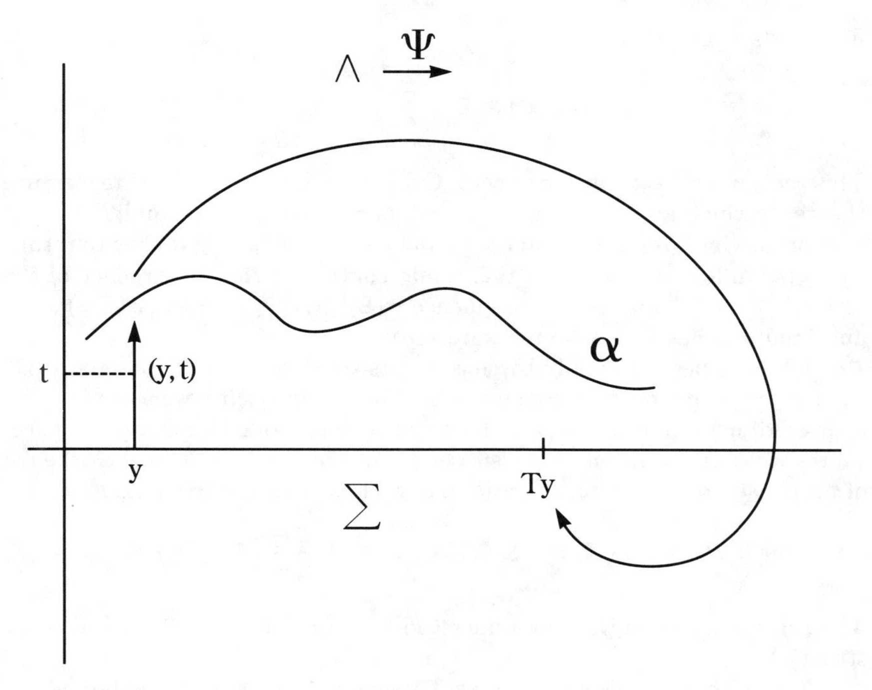

FIGURE 12.

$$\Psi(\varphi_s(x)) = \begin{cases} (y, t+s) \text{ if } t+s < \alpha(y) \\ (Ty, t+s-\alpha(y) \text{ if } t+s-\alpha(y) < \alpha(Ty) \\ \dots \end{cases} .$$

The mapping $T : \Sigma \longrightarrow \Sigma$ has the property to be σ-preserving (see Problem 2).

We can now prove that the set B of all points crossing Σ infinitely many times in a finite time interval has zero Lebesgue measure. To this end, let

$$\Sigma_B = \{y \in \Sigma; \sum_{k=1}^{\infty} \alpha(T^k y) < \infty\}.$$

Then

$$\Psi(B) = \{(y, t); \ y \in \Sigma_B, \ 0 \le t < \alpha(y)\},$$

and

$$\int_B dx = \int_{\Sigma_B} d\sigma(y) \int_0^{\alpha(y)} dt = \int_{\Sigma_B} d\sigma(y)\alpha(y).$$

For $a > 0$, consider the set

$$\Sigma_B(a) = \{y \in \Sigma_B; \sum_{k=1}^{\infty} \alpha(T^k y) > a\}.$$

This set must have σ–measure zero: Otherwise, by the Poincaré recurrence theorem (which applies to our situation because $\sigma(\Sigma) < \infty$ and T is measure preserving), for almost all the points $y \in \Sigma_B$, $T^n y$ would have to return to $\Sigma_B(a)$ infinitely often, and this would contradict the convergence of the series $\sum_{k=1}^{\infty} \alpha(T^k y)$. As a consequence $\sigma(\Sigma_B(a)) = 0$, hence $\sigma(\Sigma_B) = 0$, and finally B has Lebesgue measure zero.

We next begin to apply these concepts to hard-sphere dynamics. Consider N hard spheres in a domain Λ, which, for simplicity, we assume to be a three–dimensional torus. In order to be able to work in a compact space, we restrict our attention to the subset Γ_E of phase space defined as the set of all phase points having (kinetic) energy less than the fixed value E:

$$\Gamma_E = \{(x_1 \dots x_N, \xi_1 \dots \xi_N); x_i \in \Lambda, \frac{1}{2} \sum_i \xi_i^2 \leq E\}.$$

The set Γ_E will be invariant under the flow (which we still have to construct).

Notice that in this situation we have no collisions with the boundary. Also, we can safely disregard triple collisions: The set of all phase points for which at least three particles are in contact is a manifold of codimension two, and therefore the set of all phase points leading to multiple collisions in the past or in the future has codimension one and is of measure zero. Therefore, to construct an almost everywhere defined flow, all we have to prove is that all phase points leading to infinitely many collisions in a finite time interval have measure zero.

We prove this by using, as above, the special flow representation and the Poincaré recurrence theorem. Let

$$F_{ij}^{+(-)} = \{z \in \Gamma_E; |x_i - x_j| = \sigma, \ (x_i - x_j) \cdot (\xi_i - \xi_j) > 0 \, (< 0)\}$$

and

$$\Sigma^{+(-)} = \cup_i \cup_{j \neq i} F_{ij}^{+(-)}.$$

Σ^+ and Σ^- are those phase points in Γ_E for which there are two particles in contact in ingoing or outgoing configuration respectively.

There is a natural mapping

$$R : \Sigma^- \longrightarrow \Sigma^+$$

that is defined by transforming the ingoing velocities ξ_i ,ξ_j into the outgoing velocities ξ_i', ξ_j'. All the functions we consider are continuous along trajectories, such that $y \in \Sigma^-$ and $Ry \in \Sigma^+$ are essentially the same point in phase space. If $y \in \Sigma^+$, let $\alpha(y)$ be the first time at which there is a new collision. Then $\varphi_{\alpha(y)} \in \Sigma^-$ (here, φ_t denotes the free flow). We define the mapping $T : \Sigma^+ \longrightarrow \Sigma^-$ by

$$Ty = R^{-1}\varphi_{\alpha(y)}(y).$$

T is not defined on those points leading to no collisions in the future ($\alpha(y) = \infty$), but for such points there is no problem in defining the flow globally for all future.

According to the previous definition, we introduce a measure $d\sigma$ on Σ, whose restriction onto F^+_{ij} is

$$d\sigma = dx_1 \dots dx_{j-1}dx_{j+1} \dots dx_N d\xi_1 \dots d\xi_N dy_{ij} n_{ij} \cdot (\xi_i - \xi_j),$$

where $n_{ij} = \frac{(x_i - x_j)}{\|x_i - x_j\|}$ and dy_{ij} is the surface element over the sphere of radius σ centered in x_i.

As before, we represent the points of $x \in \Gamma_E$ that experienced at least one collision in the past as a pair (y, t), where $x = \varphi_t(y)$, $y \in \Sigma^+$, $t < \alpha(y)$. In the special flow representation, the Lebesgue measure becomes $d\sigma dt$.

The Poincaré recurrence theorem is applicable to our situation because σ is invariant for the flow. The theorem implies that

$$\sigma\{y \in \Sigma^+; \sum_{k=1}^{\infty} \alpha(T^k y) < \infty\} = 0,$$

and from this we conclude that phase points leading to infinitely many collisions in a finite time interval must have Lebesgue measure zero.

The cases of reflecting boundary conditions and $\Lambda = \Re^3$ can be treated similarly. After this construction, the statement of Proposition 4.2.5 is immediate.

The existence of the dynamics of a finite hard-sphere particle system was first established by Alexander[1]. Another proof that the set of phase points leading to infinitely many collisions in finite time has measure zero was given by Uchiyama [25]. The ideas of the proof presented here are due to Marchioro et al. [20]

With the results and methods now at our disposal, we can show yet another interesting property of a hard-sphere particle system in all space, namely, the existence of a last collision. Following a proposal by Sinai, L. Vaserstein [26] proved this result in 1978. The proof we present here is taken from Refs. 12 and 13.

The collision transformation has yet another useful property. Let

$$e(t) = \frac{x(t)}{\|x(t)\|}, \; e^0(t) = \frac{x^0(t)}{\|x^0(t)\|},$$

where $x(t) = T^t x$, $x^0(t) = T_0^t x$ are here the spatial components of the interacting and free flow phase points at time t respectively.

(A.1) Lemma.

$$\lim_{t\to\infty} e^0(t) = \frac{\xi(0)}{\|\xi(0)\|}$$

if $\xi(0) \neq 0$. *The angle between* $x^0(t)$ *and* $\xi^0(t) = \xi(0)$ *is decreasing.*

Proof. Exercise. Draw a figure!

We now show that $e(t)$ also has a limit as $t \to \infty$. To this end, we investigate the sum of angles through which $e(t)$ turns in between collisions. If we abbreviate $\mathrm{var}_{t=0}^T e(t)$ for this sum, then clearly $\mathrm{var}_{t=0}^T e(t) = \sum_{k=0}^j \gamma_k$, where γ_k is the angle through which $e(t)$ turns between the kth and the $k+1$st collision instants in $[0,T]$. However, $\gamma_k = \beta_k - \alpha_{k+1}$, where β_k is the outgoing angle between $x(t)$, $\xi(t)$ at the kth collision instant and α_{k+1} the ingoing angle between $x(t)$ and $\xi(t)$ at the $k+1$st collision instant (see Fig. 13).

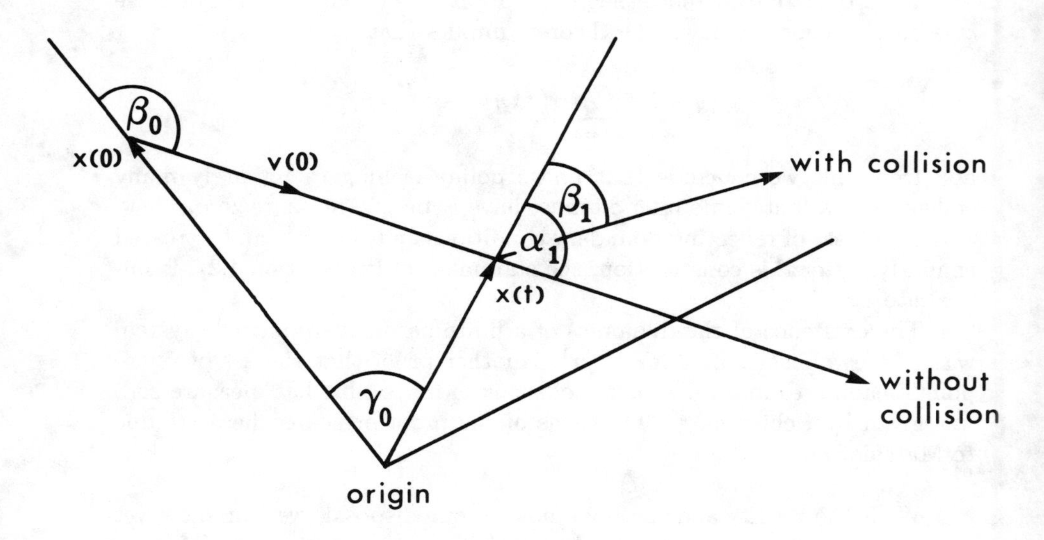

FIGURE 13.

Let t_k be the kth collision instant. One or several collisions occur at t_k, and we denote by $\xi^-(t_k)$, $\xi^+(t_k)$ the vectors of pre- and post-collisional velocities, respectively. By direct inspection, we see that the condition $n \cdot (\xi_i - \xi'_i) > 0$ (for an ingoing collision) is equivalent to

$$x(t_k) \cdot \xi^-(t_k) \leq x(t_k) \cdot \xi^+(t_k),$$

and as $\|\xi^-(t_k)\| = \|\xi^+(t_k)\|$ (by energy conservation), this implies that $\alpha_k > \beta_k$. Therefore,

$$\text{var}_{t=0}^T e(t) = \sum_{k=0}^{j} \gamma_k = \sum_{k=0}^{j} (\beta_k - \alpha_{k+1})$$

$$\leq \sum_{k=0}^{j} (\beta_k - \beta_{k+1}) = \beta_0 - \beta_{j+1} \leq \beta_0 \leq \pi$$

(β_0 is the angle between x and ξ at time 0, and $\alpha_{j+1} = \beta_{j+1}$ is the angle between x and ξ at time T).

We have proved the following.

(A.2) Lemma. $\text{var}_{t=0}^T e(t) \leq \text{var}_{t=0}^\infty e^0(t) \leq \pi$. *In particular, there is an* $e \in S^{3N-1}$ *such that* $\lim_{t\to\infty} e(t) = e$.

(A.3) Theorem. *If* z *is such that* $T^t z$ *never leads to a multiple collision for* $t > 0$, *then there is a time* $t^*(z) < \infty$ *such that there are no collisions at all after* $t^*(z)$.

Proof. First, we remark that if $T^t z$ leads to infinitely many collisions in a finite time interval, then there must be a multiple collision at a cluster point of the collision instant. Hence this situation is ruled out by assumption. By Lemma 4.2.2 we can choose coordinates such that $\sum \xi_i(t) = \sum x_i(t) = 0$. The case $\|\xi(0)\| = 0$ is trivial, so let us assume that $\|\xi(0)\| \neq 0$. Lemma 4.2.3 then implies that $\|x(t)\| \to \infty$, and $\sum x_i(t) = 0$ is equivalent to $x(t) \perp H$, where $H = \{(h_1, \ldots, h_N) \in \Re^{3N}, h_i = h_j, i, j = 1, \ldots, N\}$.

We proceed by induction. The assertion is clearly true for $N = 2$. Assuming that it is true for all particle numbers strictly less than N, all we have to show is that after finite time the system will break up into at least two noninteracting clusters.

Suppose this is wrong; label the particles $1, \ldots, N$ and connect the vertices i and j by an edge (i, j) exactly if the particles labeled i and j have infinitely many collisions with each other. This defines a graph G that is connected exactly if the system does not break up into noninteracting clusters.

Consider an edge (i, j) in G. By definition, there is a sequence $t_k \to \infty$ such that

$$|x_i(t_k) - x_j(t_k)| = \sigma$$

($|\cdot|$ denotes the Euclidean norm in $\Re^3$) for all k, i.e.,

$$\lim_{k\to\infty} \left| \frac{x_i(t_k)}{\|x(t_k)\|} - \frac{x_j(t_k)}{\|x(t_k)\|} \right| = |e_i - e_j| = 0.$$

It follows that $e_i = e_j$ if $(i, j) \in G$. If G is connected, we conclude that $e_i = e_j$ for all i, j, i.e., $e \in H$. However, this contradicts $x(t) \perp H$ and $\|x(t)\| \to \infty$. The proof is complete. □

Problems

1. Prove the identity

$$\int_{\Lambda} = \int_{\Sigma} \int_{0}^{\alpha(y)} dt f(\Psi^{-1}(y,t)),$$

where Ψ is defined by (A.1).

Hint: Use the Liouville theorem to prove that the Lebesgue measure of the "tube" $\Psi^{-1}(A \times [0, t_0))$, where $A \subset \Sigma$ is a measurable set and $t_0 < \inf_{y \in A} \alpha(y)$, is $\sigma(A)t_0$.

2. Prove that the map T is σ–preserving.
Hint: Use the Liouville theorem.

Appendix 4.B
A Rigorous Derivation of the BBGKY Hierarchy

Here, we want to prove Theorem 4.3.1. As in Section 4.3, we write

$$P^{(s)}(z^s,t) = \int P(z^s, z^{N-s}, t) dz^{N-s}$$

where $P(\cdot, t)$ is defined by (3.5). As for P_0, we make assumptions 1 and 2 from Section 4.3.

Also, we introduce a class of test functions $C^1_{T_s}$ which is the family of functions u_s of $6s$ variables such that $u_s \in L^\infty$ and such that $t \to u_s(T^t z^s)$ is differentiable for almost all z_s, with a distributional derivative in L^∞.

We will make use of the special flow representation introduced in Appendix 4.A.

Proof of Theorem 4.3.1

Step 1: The special flow representation. The following special coordinates are well suited to the problem. Fix $s < N$, and let

$$F^{+(-)}_{ij} = \{z \in \Gamma; |x_i - x_j| = \sigma, \quad (x_i - x_j) \cdot (\xi_i - \xi_j) > 0 (< 0)\}$$

(for some $i \in \{i, \dots, s\}$, $j \in \{s+1, \dots, N\}$), be the set of all phase points that display an outgoing (ingoing) collision between the particles with labels i, j. Let

$$F^{+(-)} = \bigcup_{i=1}^{s} \bigcup_{j=s+1}^{N} F^{+(-)}_{ij}, \qquad F = F^+ \cup F^-.$$

Elements of $F^+ (F^-)$ will be referred to as outgoing (ingoing) "s-interacting" states.

We split Γ into two disjoint sets: $\Gamma = \Gamma^{\infty} \cup \tilde{\Gamma}$, where $\Gamma^{\infty} = \{(z^s, z^{N-s}); T^t z = (T^t z^s, T^t z^{N-s})$ for all $t \in \Re\}$ is the set of all phase points that never pass through an s-interacting state, and $\tilde{\Gamma} = \Gamma - \Gamma^{\infty}$ are the phase points that do. The special flow representation is a representation of $\tilde{\Gamma}$. We split F^+ into four disjoint subsets as follows.

$$
\begin{aligned}
F_0^+ &: = \{z \in F^+;\ T^{-t} z \notin F^+,\ T^t z \notin F^- \text{ for all } t > 0\},\\
F_1^+ &: = \{z \in F^+;\ T^{-t} z \notin F^+ \text{ for all } t > 0,\\
&\qquad \text{but there is an } r > 0 \text{ such that } T^r z \in F^-\},\\
F_2^+ &: = \{z \in F^+;\ \text{there is a } \ t > 0 \text{ such that } T^{-t} z \in F^+,\\
&\qquad \text{and there is an } r > 0 \text{ such that } T^r z \in F^-\}.
\end{aligned}
$$

The smallest possible r in the definitions of F_1^+, F_2^+ will be denoted by $\alpha_i(z)$, $i = 1, 2$. We set $\alpha_0(z) = \alpha_3(z) = \infty$.

$$
\begin{aligned}
F_3^+ &:= \{z \in F_;^+ \text{ there is a } t > 0 \text{ such that } T^{-t} z \in F^+,\\
&\qquad \text{but } T^r z \notin F^- \text{ for all } r > 0\}.
\end{aligned}
$$

This decomposition of F^+ induces a natural partition of $\tilde{\Gamma}$ by

$$
\begin{aligned}
\tilde{\Gamma}_0 &: = \{z \in \tilde{\Gamma};\ z = T^t y \text{ for some } y \in F_0^+ \text{ and some } t \in (-\infty, \infty)\},\\
\tilde{\Gamma}_1 &: = \{z \in \tilde{\Gamma};\ z = T^t y \text{ for some } y \in F_1^+, \text{ some } t \in (-\infty, \alpha_1(y))\},\\
\tilde{\Gamma}_2 &: = \{z \in \tilde{\Gamma};\ z = T^t y \text{ for some } y \in F_2^+, \text{ some } t \in [0, \alpha_2(y))\},
\end{aligned}
$$

and

$$
\tilde{\Gamma}_3 : = \{z \in \tilde{\Gamma};\ z = T^t y \text{ for some } y \in F_3^+, \text{ some } t \in [0, \infty)\}.
$$

The mapping $\psi : (y, r) \to T^r y$, defined on

$$
\begin{aligned}
&(F_0^+ \times \Re) \cup \{(y, r);\ y \in F_1^+, r \in (-\infty, \alpha_1(y))\}\\
&\qquad \cup \{(y, r);\ y \in F_2^+, r \in [0, \alpha_2(y))\} \cup (F_3^+ \times [0, \infty)),
\end{aligned}
$$

with values in $\tilde{\Gamma}$, is then one to one, measurable, and has a measurable inverse. We extend ψ to $F^+ \times \Re$ by $\psi(y, r) = T^r(y)$.

Let $\gamma_0 = \gamma_1 = -\infty$, $\gamma_2 = \gamma_3 = 0$. Then, for $f \in L^1(\tilde{\Gamma})$,

$$
\begin{aligned}
\int_{\tilde{\Gamma}} f(z)\, dz &= \sum_{i=0}^{3} \int_{\tilde{\Gamma}_i} f(z)\, dz\\
&= \sum_{i=0}^{3} \int_{F_i^+} \int_{\gamma_i}^{\alpha_i(y)} f(\psi(y, r))\, dr\, d\sigma^+(y).
\end{aligned}
$$

The measure $d\sigma^+$ is defined on F^+ and involves the Jacobian determinant of the transformation ψ. On F^+_{ij}, $d\sigma^+$ is

(B.1)
$$d\sigma^+ = dz^s \, dx_{s+1} \dots dx_{j-1} dx_{j+1} \dots dx_N \, d\xi_{s+1} \dots d\xi_N \, dy_{ij} n_{ij} \cdot (\xi_i - \xi_j)$$

with $n_{ij} := \frac{x_i - x_j}{|x_i - x_j|} \cdot dy_{ij}$ denotes the Lebesgue measure on the sphere with radius σ and center x_i, and dz^s denotes Lebesgue measure on Γ_s.

(B.1) is a special case of the measure $d\sigma(y, \cdot)$, which we introduced and described in Appendix 4.A; here we verify its form in detail. It is enough to study a two-particle phase point $z = (x_1, \xi_1, x_2, \xi_2)$, because only pair collisions are being considered. Suppose that $z = \psi(y, r)$, where $y = (y_1, y_1 + n\sigma, \xi_1, \xi_2)$ is in outgoing collision configuration ($n \cdot (\xi_2 - \xi_1) > 0$). We have $x_1 = y_1 + r\xi_1$, $x_2 = y_1 + n\sigma + r\xi_2$. The Jacobian determinant $\left|\frac{\partial z}{\partial(y,r)}\right|$ is a 12×12 determinant (six variables are needed to describe a particle), but the only nontrivial part comes from the spatial components. We can write

$$\left|\frac{\partial z}{\partial(y,r)}\right| = \begin{vmatrix} \frac{\partial x_1}{\partial y_1} & \frac{\partial x_1}{\partial n} & \frac{\partial x_1}{\partial r} \\ \frac{\partial x_2}{\partial y_1} & \frac{\partial x_2}{\partial n} & \frac{\partial x_2}{\partial r} \end{vmatrix}.$$

Let $n = (\sin\phi\cos\theta, \sin\phi\sin\theta, \cos\phi)$ and let D be the matrix

$$\left(\frac{\partial x_2}{\partial \phi}, \frac{\partial x_2}{\partial \theta}\right);$$

then

$$\left|\frac{dz}{\partial(y,r)}\right| = \sigma^2 \begin{vmatrix} 1 & 0 & 0 & 0 & 0 & \xi_{11} \\ 0 & 1 & 0 & 0 & 0 & \xi_{12} \\ 0 & 0 & 1 & 0 & 0 & \xi_{13} \\ 1 & 0 & 0 & \cos\phi\cos\theta & -\sin\phi\sin\theta & \xi_{21} \\ 0 & 1 & 0 & \cos\phi\sin\theta & \sin\phi\cos\theta & \xi_{22} \\ 0 & 0 & 1 & -\sin\phi & 0 & \xi_{23} \end{vmatrix}$$

$$= \sigma^2 \begin{vmatrix} 1 & 0 & 0 & 0 & 0 & \xi_{11} \\ 0 & 1 & 0 & 0 & 0 & \xi_{12} \\ 0 & 0 & 1 & 0 & 0 & \xi_{13} \\ & & & & & \xi_{21} - \xi_{11} \\ & \mathrm{O} & & & \mathrm{D} & \xi_{22} - \xi_{12} \\ & & & & & \xi_{23} - \xi_{13} \end{vmatrix}$$
$$= \sigma^2 \cdot \sin\phi \, n_{12} \cdot (\xi_1 - \xi_2)$$

[compare with (B.1)].

The right-hand side of (B.1) also defines a negative measure on F^-, which we denote by $d\sigma^-$.

Step 2: The BBGKY hierarchy. Now, multiply the solution $P(z, t)$ of the Liouville equation by a test function $u_s \in C^1_{T_s}$ (u_s depends only on z^s) and calculate

$$\frac{d}{dt} \int_{\tilde{\Gamma}} u_s(z) P(z, t)\, dz$$

$$= \frac{d}{dt} \sum_{i=0}^{3} \int_{F_i^+} \int_{\gamma_i}^{\alpha_i(y)} u_s(\psi(y, r) P_0\big(T^{-t}(\psi(y, r))\big)\, dr\, d\sigma^+(y).$$

But $P_0\big(T^{-t}(\psi(y, r))\big) = P_0\big(\psi(y, r - t)\big)$; after substituting $\tau = r - t$, we find

$$\frac{d}{dt} \sum_{i=0}^{3} \int_{F_i^+} \int_{\gamma_i - t}^{\alpha_i(y) - t} u_s\big(\psi(y, t + \tau)\big) P_0\big(\psi(y, \tau)\big)\, d\tau\, d\sigma^+(y).$$

The function $t \to u_s(T^t z)$ is in general discontinuous at an s-interaction, because u_s does not depend on the last $N - s$ particles. Therefore $u_s(\psi(y, t + \tau))$, $\tau \in [\gamma_i - t, \alpha_i(y) - t]$, can "jump" at $\tau = -t$. To take account of these jumps, we split the inner integrals for F_0^+ and F_1^+ at $-t$ and differentiate.

$$\frac{d}{dt} \int_{F_0^+} \left(\int_{-\infty}^{-t} + \int_{-t}^{\infty} \right) u_s\big(\psi(y, t + \tau)\big) P_0\big(\psi(y, \tau)\big)\, d\tau\, d\sigma^+(y)$$

$$= - \int_{F_0^+} u_s\big(\psi(y, 0-)\big) P\big(\psi(y, 0), t\big)\, d\sigma^+(y)$$

$$+ \int_{F_0^+} u_s\big(\psi(y, 0+)\big) P\big(\psi(y, 0), t\big)\, d\sigma^+(y)$$

$$+ \int_{F_0^+} \left(\int_{-\infty}^{-t} + \int_{-t}^{\infty} \right) \frac{d}{dt} u_s\big(\psi(y, \tau + t)\big) P_0\big(\psi(y, \tau)\big)\, d\tau\, d\sigma^+(y). \tag{B.2}$$

$$\frac{d}{dt} \int_{F_1^+} \left(\int_{-\infty}^{-t} + \int_{-t}^{\alpha_1(y) - t} \right) u_s\big(\psi(y, t + \tau)\big) P_0\big(\psi(y, \tau)\big)\, d\tau\, d\sigma^+(y)$$

$$= - \int_{F_1^+} u_s \cdot P\big(\psi(y, \alpha_1, (y)-), t\big) d\tau\, d\sigma^+(y)$$

$$+\int_{F_1^+}\left[u_s(\psi(y,\,0+)) - u_s(\psi(y,\,0-))\right]P(\psi(y,\,0),\,t)d\tau\;d\sigma^+(y)$$

$$\text{(B.3)}\qquad +\int_{F_1^+}\int_{-\infty}^{\alpha_1(y)-t}\frac{d}{dt}u_s(\psi(y,\,t+\tau))P_0(\psi(y,\,\tau))d\tau\;d\sigma^+(y).$$

No splitting is necessary for F_2^+ and F_3^+. By direct differentiation, we get

$$\frac{d}{dt}\int_{F_2^+}\int_{-t}^{\alpha_2(y)-t} u_s(\psi(y,\,t+\tau))P_0(\psi(y,\,\tau))\;d\tau\;d\sigma^+(y)$$

$$= -\int_{F_2^+}\left[u_s(\psi(y,\,\alpha(y)-))P(\psi(y,\,\alpha_2(y)),\,t)\right.$$

$$\left.-u_s(\psi(y,\,0+))P(\psi(y,0),t)\right]\;d\sigma^+(y)$$

$$\text{(B.4)}\qquad +\int_{F_2^+}\int_{-t}^{\alpha_2(y)-t}\frac{d}{dt}u_s(\psi(y,\,t+\tau))P_0(\psi(y,\,\tau))\;d\tau\;d\sigma^+(y)$$

and

$$\frac{d}{dt}\int_{F_3^+}\int_{-t}^{\infty} u_s(\psi(y,\,t+\tau))P_0(\psi(y,\,\tau))\;d\tau\;d\sigma^+(y)$$

$$= \int_{F_3^+} u_s(\psi(y,\,0+))P(\psi(y,0),t)\;d\sigma^+(y)$$

$$\text{(B.5)}\qquad +\int_{F_3^+}\int_{-t}^{\infty}\frac{d}{dt}u_s(\psi(y,\,t+\tau))P_0(\psi(y,\,\tau))\;d\tau\;d\sigma^+(y)$$

We collect all the boundary terms in these equations that correspond to limits from below:

$$-\int_{F_0^+}(u_s\cdot P)(\psi(y,\,0-),\,t)\;d\sigma^+(y)$$

$$-\int_{F_1^+}(u_s\cdot P)(\psi(y,\,\alpha_1(y)-),\,t)\;d\sigma^+(y)$$

$$-\int_{F_1^+} (u_s \cdot P)\big(\psi(y, 0-)), t\big)\, d\sigma^+(y)$$

$$-\int_{F_2^+} (u_s \cdot P)\big(\psi(y, \alpha_2(y)-), t\big)\, d\sigma^+(y). \tag{B.6}$$

F^- can be split in analogy to the partition of F^+; in fact, this partition is already given by the integration variables in the various terms in (B.6). For example, $\{(y, 0-);\ y \in F_0^+\}$ corresponds bijectively to the set of all ingoing s-interactions such that $\{T^t y;\ t \in \Re\}$ contains exactly one such interaction, $\{(y, \alpha_1(y)-);\ y \in F_1^+\}$ corresponds to those ingoing s-interactions that have seen exactly one s-interaction before, $\{(y, 0-);\ y \in F_1^+\}$ corresponds to those that are first s-interactions but not last ones, and $\{(y, \alpha_2(y)-);\ y \in F_2^+\}$ corresponds to those that have seen at least two s-interactions in the past.

Therefore, (B.6) is equal to

$$\int_{F^-} (u_s \cdot P)\big(\psi(y, 0-), t\big)\, d\sigma^-(y). \tag{B.7}$$

The other boundary terms in (B.2–B.5) add up to

$$\int_{F^+} (u_s \cdot P)\big(\psi(y, 0+)\big)\, d\sigma^+(y). \tag{B.8}$$

If we add (B.7) and (B.8) and use (B.1), we find

$$\sum_{i=1}^{s} \sum_{j=s+1}^{N} \int_{F_{ij}} u_s(z^s) P(x_1 \ldots x_i \ldots x_{j-1} y_j x_{j+1} \ldots x_N, \xi_1 \ldots \xi_N)$$
$$dz^s\, dx_{s+1} \ldots dx_{j-1}\, dx_{j+1} \ldots dx_N d\xi_{s+1} \ldots d\xi_N\, dy_{ij} \cdot n_{ij} \cdot (\xi_j - \xi_i). \tag{B.9}$$

By assumption, we know that $P(\cdot\,, t)$ is $d\sigma^+(d\sigma^-)$ almost everywhere defined on $F^+(F^-)$. Hence $P^{s+1}(x_1, \ldots, x_s, x_i - n\sigma, \xi_i, \ldots, \xi_{s+1})$ is almost everywhere defined with respect to $dz^s\, dy_{i,s+1}\, d\xi_{s+1}$, and because of the symmetry assumption, we rewrite (B.9) as $\int u_s(z^s)(Q_{s+1}^{\sigma} P^{s+1})(z^s, t)\, dz^s$. The remaining terms in the identities (B.2–B.5) add up to

$$\sum_{i=0}^{3} \int_{F_i^+} \int_{\gamma_i}^{\alpha_i(y)} D_\tau[u_s(\psi(y, \tau))] P(\psi(y, \tau), t)\, d\tau\, d\sigma^+(y),$$

where $D_\tau[u_s(\psi(y, \tau))]$ is set equal to zero at discontinuity points.

If $z = \psi(y, \tau)$, let $\mathcal{L}_s u_s(z) = D_\tau[u_s(\psi(y, \tau))]$. We can then write

$$\frac{d}{dt}\int u_s(z^s)P^s(z^s,\, t)\, dz^s$$

$$= \frac{d}{dt}\int_{\Gamma^\infty} u_s(z)P(z,\, t)\, dz + \frac{d}{dt}\int_{\tilde{\Gamma}} u_s(z)P(z,\, t)\, dz$$

$$= \frac{d}{dt}\int_{\Gamma^\infty} u_s(T_s^t z^s)P_0(z)\, dz + \frac{d}{dt}\int_{\tilde{\Gamma}} u_s P(z,\, t)\, dz$$

$$= \int_{\Gamma^\infty} (\mathcal{L}_s u_s)(z)\cdot P(z,\, t)\, dz + \int_{\tilde{\Gamma}} (\mathcal{L}_s u_s)(z)P(z,\, t)\, dz$$

$$+ \int u_s(z^s)(Q_{s+1}^\sigma P^{s+1})(z^s,\, t)\, dz^s$$

$$= \int (\mathcal{L}_s u_s)(z^s)P^s(z^s,\, t)\, dz^s + \int u_s(z^s)(Q_{s+1}^\sigma P^{s+1})(z^s,\, t)\, dz^s.$$

Step 3: Uniqueness and the BBGKY hierarchy in mild form. We have so far established that the s-particle distribution functions $P^s(\cdot\,, t)$ satisfy a weak version of the BBGKY hierarchy equations, namely,

$$\frac{d}{dt}\int u_s(z^s)P^s(z^s,\, t)\, dz^s = \int (\mathcal{L}_s u_s)(z^s)P^s(z^s,\, t)\, dz^s + \int u_s(z^s)(Q_{s+1}^\sigma P^{s+1})(z^s,\, t)\, dz^s. \tag{B.10}$$

Now we prove that the $P^s(\cdot\,, t)$ are the only solutions of (B.10) for the given initial data such that $t \to P^s(T_s^t z^s,\, t)$ is continuous for almost all z^s. Clearly this is true for $P^N(\cdot\,, t) = P(\cdot\,, t)$, which is just the solution of the Liouville equation. Suppose then that P^s and P_1^s are two solutions of (B.10) such that $P^N \equiv P_1^N$; we consider $s = N-1$, and set $h(z^{N-1},\, t) = (P^{N-1} - P_1^{N-1})(z^{N-1},\, t)$. Then h satisfies

$$\frac{d}{dt}\int u_{N-1}(z^{N-1})h(z^{N-1},\, t)\, dz^{N-1}$$

$$= \int (\mathcal{L}_{N-1}u_{N-1})(z^{N-1})h(z^{N-1},\, t)\, dz^{N-1}$$

for all u_{N-1} such that $t \to u_{N-1}(T_{N-1}^t z^{n-1})$ is differentiable. Choose a family of test functions $u_{N-1,\tau}(z^{N-1})$, defined by $u_{N-1,\tau}(z^{N-1}) = u_{N-1}(T_{N-1}^{-\tau} z^{N-1})$, then

$$\text{(B.11)}\quad \frac{d}{dt}\int u_{N-1,t}(z^{N-1})h(z^{N-1},\,t)\,dz^{N-1}$$
$$= \int (\mathcal{L}_{N-1}u_{N-1,t}(z^{N-1}))h(z^{N-1},t)\,dz^{N-1}$$
$$+ \int \frac{d}{dt}\left[u_{N-1,t}(z^{N-1})\right]h(z^{N-1},t)\,dz^{N-1} = 0,$$

because

$$\frac{d}{dt}\left[u_{N-1,t}(z^{N-1})\right] = -\mathcal{L}_{N-1}u_{N-1,t}(z^{N-1}).$$

(B.11) implies that $h(z^{N-1},\,t) = 0$ almost everywhere.

The continuity of $t \to P^{N-1}(T^t_{N-1}z^{N-1},\,t)$, $P_1^{N-1}(\ldots,t)$ almost everywhere implies that $(Q^\sigma_{N-1}h)(z^{N-2},\,t) = 0$ for almost all z^{N-2}; by repeating the argument, we see that $P^{N-2}(z^{N-2},t) = P_1^{N-2}(z^{N-2},t)$ almost everywhere, etc.

The proof that the s-particle density functions $P^s(z^s,\,t)$ actually satisfy the continuity $t \to P^s(T^t_s z^s,\,t)$ is a simple consequence of assumptions 3 and 4 on P_0. The details are left to the reader (see Problem 2).

Finally, we note that the family of functions $P^s(\cdot\,,t)(s = 1,\ldots,N)$ defined by (4.7) is a solution of the BBGKY hierarchy in the weak sense; because the sum in (4.7) is actually finite, it follows that $t \to P^s(T^t_s z^s,\,t)$ is absolutely continuous for almost all z^s; by the uniqueness just proved, the $P^{(s)}$ given by (4.7) must coincide with the s-particle distribution functions, and by the absolute continuity of $t \to P^s(T^t_s z^s,\,t)$, it follows that the s-particle distribution functions satisfy the hierarchy equations in the mild sense. The proof is complete.

Problems

1. What is the special flow representation in the case $N = 2$, $s = 1$?
2. Show that hypotheses 3 and 4 on P_0 imply that $t \to P^s(T^t_s z^s,t)$ is continuous for almost all z^s.

Appendix 4.C
Uchiyama's Example

We now present an example of a discrete velocity Boltzmann equation for which the rigorous validation from Section 4.4 fails. The example was discovered by Uchiyama [25] in the mid-1980s and was, at the time, a shattering blow to attempts of rigorously deriving discrete velocity models from hierarchy equations.

Discrete velocity models of the Boltzmann equation were introduced to get analytically more tractable equations [9] and to be able to get explicit solutions for physically interesting situations [4]. For a detailed discussion of the subject, we refer the reader to Refs. 9 or 16.

The particular example we focus on is known as the four-velocity plane Broadwell model. Consider a gas of two-dimensional "hard diamonds" of diagonal length $\sigma > 0$ (see Fig. 14), each of which can move with one of the four admissible velocities

$$\xi_1 = (1, 0),\ \xi_2 = -\xi_1,\ \xi_3 = (0, 1),\ \xi_4 = -\xi_3.$$

The one-particle distribution density function $P^1(x, \xi, t)$ takes the form

$$P^1(x, \xi, t) = \sum_{i=1}^{4} P_i^1(x, t)\, \delta_{\xi_i}(\xi)\,,$$

such that $P_i^1(x, t)$ is the density function describing only the particles that move with velocity ξ_i.

It is useful to visualize the collision laws between the particles by means of the diamond shapes. Two types of collisions are possible, as depicted in Figs. 15A and 15B.

$\longleftarrow \sigma \longrightarrow$

FIGURE 14.

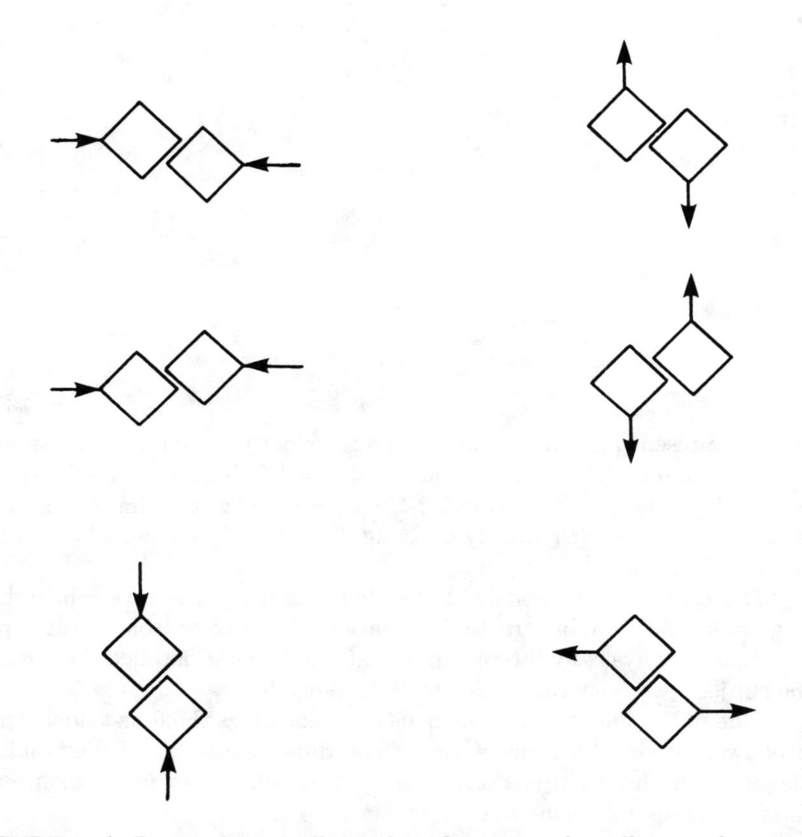

FIGURE 15A. Pre- and post-collisional configurations for collisions of type 1.

Collisions of type 2 amount to velocity exchange between the particles. In the limit $\sigma \to 0$, these collisions become indistinguishable from non-collisions and are expected to make no contributions to the collision term

FIGURE 15B. Pre– and post-collisional configuration for a collision of type 2.

[as is clearly visible in (C.1)]. This holds actually rigorously. Collisions of type 2 are not the reason for the pathology we want to explain.

Formally, it is expected that in the Boltzmann-Grad limit $N \to \infty$, $\sigma \to 0$, $N\sigma \to \alpha$ (we have a two-dimensional gas) the BBGKY hierarchy for this system of particles turns into the Boltzmann hierarchy associated with the discrete velocity equations

$$(C.1)\qquad \begin{aligned}(\partial_t + \partial_x) f_1 &= c(f_3 f_4 - f_1 f_2)\\ (\partial_t - \partial_x) f_2 &= c(f_3 f_4 - f_1 f_2)\\ (\partial_t + \partial_y) f_3 &= c(f_1 f_2 - f_3 f_4)\\ (\partial_t - \partial_y) f_4 &= c(f_1 f_2 - f_3 f_4).\end{aligned}$$

The correlation functions

$$P^{(s)}(z^s, t) = \sum_{i_1,\dots,i_s=1}^{4} P^{(s)}_{i_1,\dots,i_s}(x_1,\dots,x_s,t)\,\delta_{\xi_{i_1}}(\xi_1)\delta_{\xi_{i_2}}(\xi_2)\dots\delta_{\xi_{i_s}}(\xi_s)$$

satisfy a hierarchy of equations much like the BBGKY hierarchy for hard spheres. We need some notation to write it down.

Let Λ be the diamond with vertices $(\pm 1, 0)$ and $(0, \pm 1)$. Particles at x and y are, by definition, in a collision configuration if

$$l := \frac{1}{\sigma}(y - x) \in \partial\Lambda \backslash \{(\pm 1, 0), (0, \pm 1)\}.$$

(We disregard what happens in measure zero situations like $l = (1, 0)$; see Fig. 16.)

FIGURE 16.

Let ξ be the velocity of the particle at x, η the velocity of the particle at y. A collision will change the velocities ξ, η only if $l \cdot \xi > 0$ *and* $l \cdot \eta < 0$ (see the examples in Fig. 17).

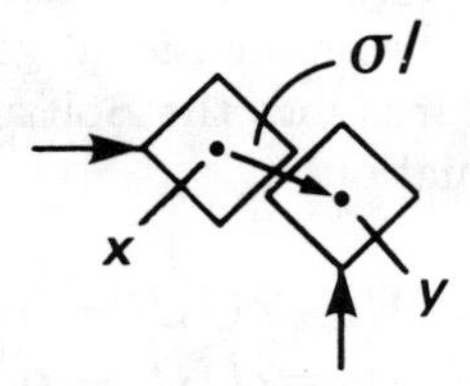

FIGURE 17A. Collision with change of velocities.

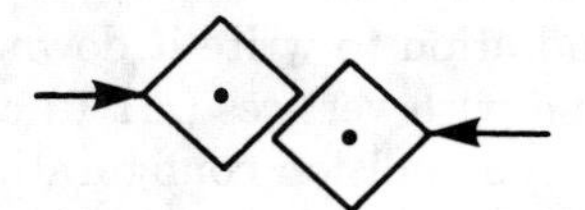

FIGURE 17B. Collision with change of velocities.

Let i denote the counterclockwise rotation by $\frac{\pi}{2}$ in $\Re^2$ [e.g., $i(1,0) = (0,1)$], then the post-collisional velocities ξ^*, η^* are given as follows:

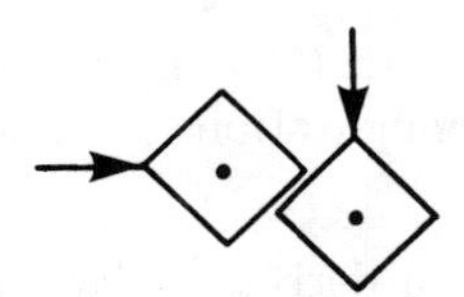

FIGURE 17C. Collision without change of velocities.

$$
(C.2)\qquad
\begin{aligned}
&\xi^* = i\xi\,,\ \eta^* = i\eta \text{ if } \xi\cdot\eta = -1 \text{ and } l\cdot(i\xi) < 0\\
&\xi^* = -i\xi\,,\ \eta^* = -i\eta \text{ if } \xi\cdot\eta = -1 \text{ and } l\cdot(i\xi) > 0\\
&\qquad \text{(see Fig. 17B)}\\
&\xi^* = \eta\,,\ \eta^* = \xi \text{ if } \xi\cdot\eta = 0\\
&\qquad \text{(velocity exchange, see Fig. 17C)}
\end{aligned}
$$

We will not use these explicit formulas in the sequel and gave them only for the sake of completeness. (C.2) will, of course, be referred to as the collision transformation for the present context. In a slight abuse of notation, we use T_s^t (as in Sections 4.3 and 4.4) to denote now the evolution group for a system of s "flat hard diamonds." The BBGKY hierarchy becomes, in mild form,

$(C.3)$

$$
P^{(s)}(z^s,t) = P_0^{(s)}(T_s^{-t}z^s) + (N-s)\sigma\int_0^t \left(C_{s+1}^{\sigma}P^{(s+1)}\right)\left(T_s^{-(t-t_1)}z^s,t_1\right)dt_1
$$

with

$$
(C.4)\qquad
\begin{aligned}
&\left(C_{s+1}^{\sigma}P^{(s+1)}\right)(z^s,t_1) = \sqrt{2}\sum_{k=1}^{s}\int_\eta\int_{l\in\partial\Lambda(\eta,\xi_k)}\\
&\left\{P^{(s+1)}(x_1,\dots,x_s,x_k-\sigma l,\xi_1,\dots,\xi_k^*,\dots,\xi_s,\eta^*;t_1)\right.\\
&\left.-\,P^{(s+1)}(x_1,\dots,x_s,x_k+\sigma l,\xi_1,\dots,\xi_k,\dots,\xi,\eta;t)\right\}dl\,d\eta
\end{aligned}
$$

with $\partial\Lambda(\eta,\xi) = \{l\in\partial\Lambda; l\cdot\eta > 0 \text{ and } l\cdot\xi < 0\}$.

Notice that we have in the collision integral already distinguished between in- and outgoing configurations. Assuming as usual continuity of $P_0^{(N)}$ along N- trajectories, one can equivalently write

$(C.5)$

$$
\int_\eta\int_{l,-l\in\partial\Lambda(\eta,\xi_k)} P^{(s+1)}(x_1,\dots,x_s,x_k+\sigma l,\xi_1,\dots,\xi_k,\dots,\xi_s,\eta;t)\,dl\,d\eta.
$$

The expression (C.5) reduces to (C.4) by rewriting the part where $-l \in \partial\Lambda(\eta, \xi_k)$ —the outgoing configurations— in terms of ingoing configurations.

We can now show why a derivation like the one done in Section 4.4 must fail for Eqs. (C.1). We use the notation from Section 4.4, but it is understood that T_s^t and C_{s+1}^σ represent the flow and collision operator introduced in this appendix.

Let $s = 1$ and consider the series solution to the hierarchy given in (C.2–4). We demonstrate that step 1 in the proof of Theorem 4.4.1 fails for $n = 3$ (i.e., three particles are adjoined in the process). Spelling out (4.14) for this situation, we find

$$(N-1)(N-2)(N-3)\sigma^3 \sum_{k_1=1}^{1} \sum_{k_2=1}^{2} \sum_{k_3=1}^{3}$$

$$\int d\xi_2 d\xi_3 d\xi_4 \int dl_1 dl_2 dl_3 \, P_{\sigma,0}^4(Y^4),$$

where

$$\begin{aligned} Y^4 &= T_\sigma^{-t_3}(Y^3 \cup z_{1,3}^{k_3}) \\ Y^3 &= T_\sigma^{t_3-t_2}(Y^2 \cup z_{1,2}^{k_2}) \\ Y^2 &= T_\sigma^{t_2-t_1}(Y^1 \cup z_{1,1}^{k_1}) \\ Y^1 &= T_\sigma^{t_1-t} z, \end{aligned}$$

or, in one formula,

$$Y^4 = T_\sigma^{-t_3}(T_\sigma^{t_3-t_2}(T_\sigma^{t_2-t_1}(T_\sigma^{t_1-t} z \cup z_{1,1}^{k_1}) \cup z_{1,2}^{k_2}) \cup z_{1,3}^{k_3}).$$

Recall the meaning of all the indices: If $k_2 = 1$, then the notation $z_{1,2}^1$ means that $z_{1,2}^1$ denotes position and velocity of the second particle, which gets adjoined in the process, and it is adjoined to the particle whose coordinates are listed in first position.

For the model under consideration, it is true that there are choices $z_{1,1}^{k_1}$, $z_{1,2}^{k_2}$ and $z_{1,3}^{k_3}$ *of positive measure*, such that on *a subset of positive measure* of the three-dimensional simplex $0 \le t_3 \le t_2 \le t_1 \le t$

$$Y^4 \not\longrightarrow Y_0^4 = T_0^{-t_3}(T_0^{t_3-t_2}(T_0^{t_2-t_1}(T_0^{t_1-t} z \cup z_{1,1}^{k_1}) \cup z_{1,2}^{k_2}) \cup z_{1,3}^{k_3})$$

as $\sigma \to 0$.

To see this, consider a particle with label 1 in position x, moving with velocity (0,1) at time t; so $z = (x, \xi_3)$ (see Fig. 18).

FIGURE 18. 1) The particle represented by z.

FIGURE 19. 2) $Y^1 \cup z^1_{1,1}$.

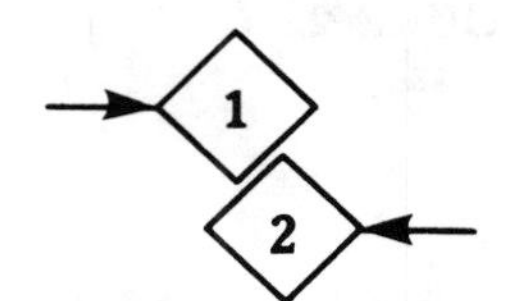

FIGURE 20.

At time $t - t_1$ earlier, the particle has coordinates $T^{t_1-t}_\sigma z$, and particle 2, with coordinates $z^1_{1,1} = (x - (t - t_1)\xi_3 + \sigma l, \xi_4)$ is adjoined ($k_1 = 1$) (see Fig. 19).

In precollisional configuration, the configuration from Fig. 19 is as in Fig. 20,

and at time $t_1 - t_2$ earlier this was the configuration Y^2, to which particle 3 with coordinates $z^1_{1,2}$ ($k_2 = 1$) is adjoined (see Fig. 21).

FIGURE 21. 4) $Y^2 \cup z_{1,2}^1$.

At time $t_2 - t_3$ earlier, particle 4 gets adjoined, with coordinates $z_{1,3}^2$ ($k_3 = 2$) (see Fig. 22).

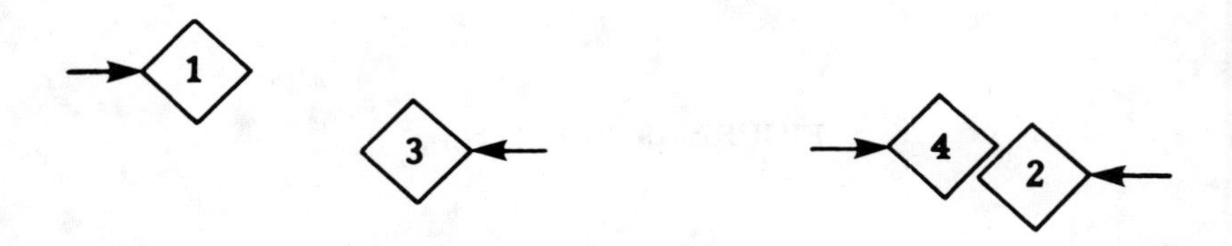

FIGURE 22. 5) $Y^3 \cup z_{1,3}^2$.

Up to this level, the convergence

$$\begin{aligned} &T_\sigma^{t_3-t_2}\left(T_\sigma^{t_2-t_1}\left(T_\sigma^{t_1-t} z \cup z_{1,1}^1\right) \cup z_{1,2}^1\right) \cup z_{1,3}^2 \\ \longrightarrow &T_0^{t_3-t_2}\left(T_0^{t_2-t_1}\left(T_0^{t_1-t} z \cup z_{1,1}^1\right) \cup z_{1,2}^1\right) \cup z_{1,3}^2 \end{aligned}$$

is true as $\sigma \to 0$ because there are no collisions other than the ones described via the adjoinment of particles. However, if we apply the operator $T_\sigma^{-t_3}$ to the situation in Fig. 22, we have a positive probability of an encounter between particles 3 and 4, i.e., at time t_3 earlier we have the situation in Fig. 23.

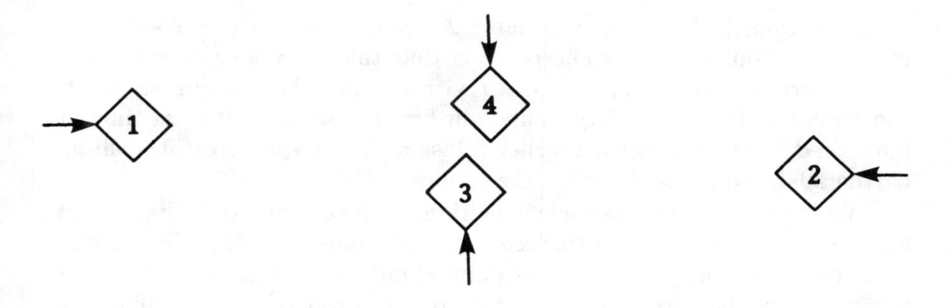

FIGURE 23. 6) $T_\sigma^{-t_3}\left(Y^3 \cup z_{1,3}^2\right)$.

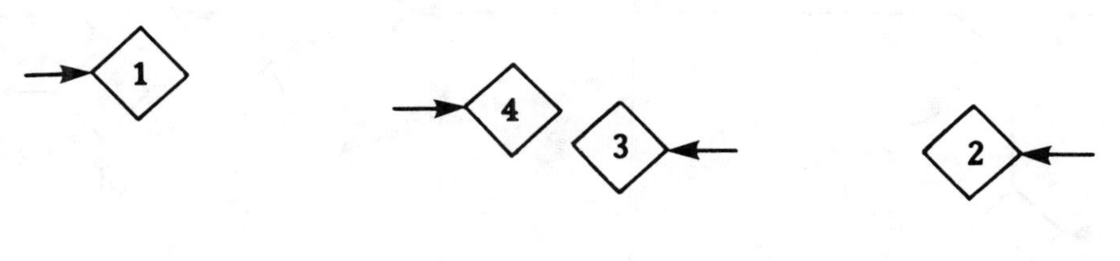

FIGURE 24.

In contrast, $T_0^{-t_3}\left(Y^3 \cup z_{1,3}^2\right)$ would be the configuration in Fig. 24, (particles 3 and 4 ignore each other), and it is clear that step 1 from the proof of Theorem 4.4.1 is not true. Notice that this happens no matter how small σ is.

The pathology clearly arises from the discrete velocity structure in the considered model. If the adjoined particles can have any velocity and the velocity distribution is absolutely continuous, then the probability of an earlier collision between particles 3 and 4 becomes negligible in the limit.

We emphasize that the difficulty arises on the level of the BBGKY hierarchy for flat diamonds. Notice that the adjoinment of particles in the construction was, for the particles labeled 3 and 4, done in ingoing configurations. Therefore, the picture does not depict a real evolution; for a real evolution, the particle with label 2 would emerge with velocity (1,0) from the encounter in 5). The point is that this encounter only happens with a certain probability, and ditto with all the other encounters in the construction. The probability to find our first particle in state z at time t is influenced by the compulsive earlier collision between the possible collision partners labeled 3 and 4; this is the problem.

We know of no *real* particle evolution, with continuous or discrete velocity distribution, in which this recollision phenomenon occurs with a non-negligible probability. However, as pointed out by Uchiyama, particles engaged in a collision of type 2 have a positive probability to recollide, as demonstrated in Fig. 25.

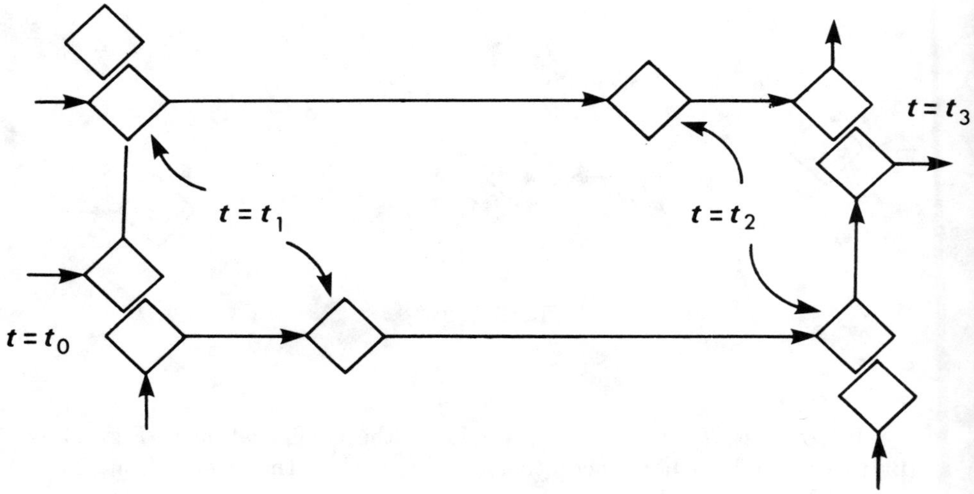

FIGURE 25.

However, as pointed out before, these collisions become in the Boltzmann-Grad limit indistinguishable from noncollisions.

We note that the problem of recollisions could be avoided by "adding noise" to the system. For example, a particle system with four velocities in a lattice, subject to a suitable stochastic process, yields, in the right scaling limit, the Broadwell equations (C.1). See Ref. 6.

As a final remark, we note that the pathology occurring in the discrete velocity models must be related to the following simple mechanical fact. The outgoing configurations in a diamond collision are completely determined by the incoming velocities, whereas in the hard-sphere case one has to specify the impact parameter n, which is closely related to the relative position of the two particles at the instant of collision.

5
Existence and Uniqueness Results

5.1 Preliminary Remarks

Existence and uniqueness theorems play a very central part in the theory of partial differential equations, particularly in the context of mathematical physics. The well-posedness of a Cauchy or boundary value problem is of tantamount importance for the physical interpretation and/or practical application of the equation under consideration. For instance, numerical calculations become a touchy business in the absence of uniqueness or continuous dependence on the data, and the function spaces in which existence theorems can be proved usually contain intrinsic useful information about the solutions. Moreover, having in mind that the Boltzmann equation is a schematization of the reality (described at a more detailed level by the Newton laws) expected to be valid only in the asymptotic regime when a gas is extremely rarefied, a good existence theorem for the solutions of such an equation is, at least, the first check of the validity of the mathematical model under investigation.

By the well-posedness of the initial value problem (IVP) for the Boltzmann equation we mean the construction of a unique nonnegative solution preserving the energy and satisfying the H-theorem, from a positive initial datum with finite energy and entropy. However, for general initial data, it is difficult, and until now not known, whether such a well-behaved solution can be constructed globally in time. The difficulty in doing this is obviously related to the nonlinearity of the collision operator and the apparent lack of conservation laws or a priori estimates preventing the solution from becoming singular in finite time.

A complete validity discussion for the Boltzmann equation will automatically contain existence and uniqueness results. Consequently, by the discussion in Chapter 4, we already have some existence and uniqueness theorems (we will revisit these in Section 5.2). Unfortunately, a validity proof involves several hard additional steps beyond existence and uniqueness, like estimates for the BBGKY hierarchy. Therefore, the Boltzmann equation has been validated rigorously only in the few simple situations we discussed in Chapter 4 (locally in time and globally for a rare gas cloud in all space).

Existence (and in some situations uniqueness) of solutions to the Cauchy problem is known for a much larger variety of cases, and it is our purpose in this chapter to present these results.

The state of the art of the global existence theory for the Boltzmann equation can be summarized as follows:

1. *The homogeneous equation*

When the distribution function of a gas is not depending on the space variable, the equation is considerably simplified. The collision operator is basically Lipschitz continuous in L^1_+ and the equation becomes globally solvable in time. Moreover, uniqueness, asymptotic behavior, and a theory of classical solutions have been established. The theory for the spatially homogeneous Boltzmann equation began in the early 1930s (see Section 5 in Chapter 6 for references) and can be considered rather complete. However, the validation problem is hard and open. We present the homogeneous theory in Chapter 6.

2. *Perturbations from Maxwellian equilibria*

If the solution is initially sufficiently close to a Maxwellian, it is possible to prove that a solution can be constructed globally in time, and we have uniqueness and asymptotic behavior. The approach is based on the analysis of the linearized Boltzmann operator, which leads us to a differential inequality of the type

$$\frac{d}{dt}y \leq -ky + y^2$$

where $y = y(t)$ is some norm of the deviation of the solution from the Maxwellian and k is a positive number. Therefore, if $y(0)$ is sufficiently small, we can control the solution for all times. As we said, the basic ingredient is good control of the linearized Boltzmann operator. This theory will be discussed in Chapter 7.

3. *Perturbations from vacuum*

This case has already been discussed in Chapter 4 and will be revisited in the next section, together with the local existence theory.

4. *Solutions with small gradients*

If the initial value is close to a homogeneous distribution, a solution starting from it can be constructed globally in time. Uniqueness and asymptotic behavior can also be proved. The main ingredient in proving such a result is good control of the solutions of type 1 and 2. The main idea will be explained in Chapter 7, Section 7.

Except the first one, all these results are of a perturbative character. The knowledge of particular solutions helps to construct other solutions close to the original ones. The general IVP is poorly understood, although a significant and somewhat unexpected step was done in the late 1980s.

5. *Solutions for general L^1_+ data with finite energy and entropy*

We will illustrate this result. Consider a regularized version of the Boltzmann equation for which we have conservation of mass and energy and the H-theorem. Denote the solutions by $f^\epsilon(t)$. Here, ϵ is the regularization parameter for which the solutions formally converge to a solution of the Boltzmann equation in the limit $\epsilon \longrightarrow 0$. The conservation laws yield the existence of a weak cluster point denoted by $f(t)$. However, since the collision operator is essentially a product, it cannot be weakly continuous. Thus it does not follow by general arguments that $f(t)$ solves the Boltzmann equation. Nevertheless, some smoothness gained by the stream operator (whenever the velocity space is continuous!) gives enough compactness to prove that $f(t)$ actually solves the Boltzmann equation in the mild sense.

The method gives neither uniqueness nor energy conservation, but the entropy is seen to decrease along the solution trajectories. This approach will be illustrated in Section 5.3.

5.2 Existence from Validity, and Overview

The validity theorems from Section 4 contain the following existence and uniqueness results for the Boltzmann equation.

(5.2.1) Theorem. (Local existence and uniqueness.) *Suppose that $f_0 \in L^1_+(\Lambda \times \Re^3)$ and a.e. $0 \le f_0(x, \xi) \le Ce^{-\beta_0 \xi^2}$ for some $C, \beta_0 > 0$, and impose the specular reflection boundary condition for $x \in \partial\Lambda$: If n is the inner normal to $\partial\Lambda$ at x, set $f(x, \xi, t) = f(x, \xi - 2n(n \cdot \xi), t)$ whenever $n \cdot \xi < 0$. Then there is a $t_0 > 0$ (depending on C, β_0) such that the Cauchy problem for the Boltzmann equation with initial value f_0 has an a.e. non-negative mild solution $f(x, \xi, t)$, defined for $t \in [0, t_0)$. In particular, $t \to f(T^t(x, \xi), t)$ is absolutely continuous for almost all (x, ξ).*

Remarks We outlined the proof of this in Remark 2 at the end of Section 4.4. It is clear that the proof generalizes to many other boundary conditions (the only essential condition is that particles do not gain kinetic energy upon encounters with the wall).

Non-negativity of f is a consequence of the fact that the one-particle distribution function $P^{(1)}(x, \xi, t)$ is non-negative by construction, and that $P^{(1)} \to f$ in the Boltzmann-Grad limit.

Local existence and uniqueness (but not validity) have been shown for many other boundary conditions (including the important stochastic boundary conditions, see Ref. 2) and for much more general collision kernels. For details, we refer the reader to Kaniel-Shinbrot [14] and Babovsky [2]. The technicalities arising from these generalizations lead to complications; here, we avoid most of these because they are of marginal interest for the objective of this book. The only exception we make concerns the iteration scheme due to Kaniel-Shinbrot [14], which nicely exploits the monotonicity properties of the collision operators. We sketch the idea.

In mild form, the Boltzmann equation can be written as

$$\frac{d}{dt}f^{\#}(x,\,\xi,\,t) + f^{\#}\cdot R^{\#}(f)(x,\,\xi,\,t) = Q_{+}^{\#}(f,f)(x,\,\xi,\,t), \tag{2.1}$$

where

$$\begin{aligned} f^{\#}(x,\,\xi,\,t) &= f(T^{t}(x,\,\xi),t),\\ Q_{+}(f,f)(x,\,\xi,\,t) &= \iint\limits_{S^2} |n\cdot(\xi-\xi_*)|f(x,\,\xi',\,t)f(x,\,\xi'_*,\,t)\;dn\;d\xi_*,\\ R(f)(x,\,\xi,\,t) &= \iint\limits_{S^2} |n\cdot(\xi-\xi_*)|f(\xi_*)\;dn\;d\xi_*. \end{aligned}$$

Clearly, if $g \geq f$, $Q_+(g,g) \geq Q_+(f,f)$, and $R(g) \geq R(f)$. The Kaniel-Shinbrot iteration scheme solves (2.1) by cleverly exploiting these monotonicities.

Suppose we already have a finite sequence of lower bounds satisfying a.e. $l_0 \leq l_1 \leq \ldots \leq l_{n-1}$ $(l_i = l_i(x,\,\xi,\,t))$ and a finite sequence of upper bounds $u_{n-1} \leq u_{n-2} \leq \ldots \leq u_0$ $(u_i = u_i(x,\,\xi,\,t))$ to a solution of (2.1). Then we define l_n and u_n by

$$\begin{aligned} &\frac{d}{dt}l_n^{\#} + l_n^{\#}R^{\#}(u_{n-1}) = Q_{+}^{\#}(l_{n-1},l_{n-1})\\ &\frac{d}{dt}u_n^{\#} + u_n^{\#}R^{\#}(l_{n-1}) = Q_{+}^{\#}(u_{n-1},u_{n-1}) \end{aligned} \tag{2.2}$$

with initial conditions $l_n(\cdot\,,0) = u_n(\cdot\,,0) = f_0$. Eqs. (2.2) are a pair of (readily solvable) linear differential equations for l_n and u_n; the solution formulas, the monotonicity properties of Q_+ and R, and the (inductive) assumption $l_{n-2} \leq l_{n-1} \leq u_{n-1} \leq u_{n-2}$ together imply that

$$l_{n-1} \leq l_n \leq u_n \leq u_{n-1} \qquad \text{a.e.}$$

Two steps remain. First, and crucial, we have to find a start to the iteration, i.e., we have to choose l_0 and u_0 such that

$$0 \leq l_0 \leq l_1 \leq u_1 \leq u_0. \tag{2.3}$$

If we choose $l_0 = 0$ and u_0 a suitable upper bound for f_0 (say, $u_0 = 2Ce^{-\beta_0\xi^2}$), the equations for l_1 and u_1 become

$$\frac{d}{dt}l_1^{\#} + l_1^{\#}R(u_0)^{\#} = 0$$
$$\frac{d}{dt}u_1^{\#} = Q_+^{\#}(u_0, u_0).$$

It is immediate that

$$0 = l_0 \leq l_1 \leq u_1 \qquad \text{a.e.},$$

but it is also clear that the inequality

$$u_1^{\#}(x, \xi, t) \leq u_0(x, \xi) \quad (= u_0^{\#}(x, \xi)) \qquad \text{a.e.}(x, \xi) \tag{2.4}$$

will in general hold only locally, i.e., for a short time. This is the reason why this method leads for large data only to a local theorem. There have been many attempts to choose l_0 and u_0 in a smarter way to prove global existence, but success has only come in a few exceptional cases, like the rare gas cloud in all space, which we already considered in Chapter 4. See also Ref. 13.

The second remaining step is to prove that $\lim_{n\to\infty} l_n = l$ and $\lim_{n\to\infty} u_n = u$ exist and are identical; if they are, they are clearly the solution of our Cauchy problem. This step follows essentially from Gronwall's inequality.

Details of this approach can be found in Ref. 14. One nice feature of the procedure is that all the l_n, u_n (and hence their limits) are automatically non-negative.

(5.2.2) Theorem. (Global existence and uniqueness for a rare gas cloud in all space.) *Suppose that $f_0 \in L^1_+(\Re^3_x \times \Re^3_\xi)$ and that a.e.*

$$0 \leq f_0(x, \xi) \leq b\, e^{-\beta_0(x^2+\xi^2)}$$

for some $b > 0$, $\beta_0 > 0$. Then, if $b \cdot \alpha$ is sufficiently small, the Cauchy problem for the Boltzmann equation has a unique global mild solution, which satisfies

$$0 \leq f(x, \xi, t) \leq (C \cdot b)e^{-\beta_0(x-t\xi)^2} \qquad \text{a.e.} \tag{2.5}$$

Proof. This result is part of Theorem 4.5.1. Non-negativity follows either from the validity or from a different proof, for example, by using the Kaniel-Shinbrot iteration scheme (see Ref. 13). □

Remarks

1. **On the asymptotic behavior.** Estimate (2.5) entails in particular that $\lim_{t\to\infty} f(x, \xi, t) = 0$ if $\xi \neq 0$. This is physically meaningful because the limit state for a rare gas cloud dispersing in vacuum is just vacuum itself. The method of proof from Theorem 4.4.1 also shows that for almost all (x, ξ), $f^{\#}(x, \xi, t)$ has a limit as $t \to \infty$ (this is because $\sum_{n=1}^{\infty} \int_0^t \cdots \int_0^{t_{n-1}} S_0(t-t_1)QS_0(t_1-t_2)\dots QS_0(t_n)f_0\, dt_n \dots dt_1$ has a limit as $t \to \infty$ if $b \cdot \alpha$ is small enough).

2. **On "traveling Maxwellians."** The functions

$$T_n(x, \xi, t) = Ce^{-\beta_0(x-\xi t)^2} e^{-\alpha_0 \xi^2},$$

where C, β_0, $\alpha_0 > 0$, are all explicit solutions of the Boltzmann equation (see Problem 1). We refer to them as "traveling Maxwellians." Toscani [17], by using the Kaniel-Shinbrot iteration scheme, has shown that the Boltzmann equation has global solutions for initial data that are sufficiently close to a traveling Maxwellian initially.

3. **On a major drawback.** There is a severe limitation to all the methods and results we have presented so far. It is that we haven't *really* taken advantage of the cancellation properties of the collision term $Q(f, f)$ that follow from the minus sign in front of $f \cdot R(f)$. In fact, the upper bound on the series solution from Chapter 4 was obtained by changing this minus sign to a plus sign. In other words, the methods discussed so far can be equally used to prove local existence and uniqueness (and global existence for suitably small data) for equations like

$$\frac{\partial}{\partial t} f + \xi \cdot \nabla_x f = Q_+(f, f) \tag{2.6}$$

or

$$\frac{\partial}{\partial t} f + \xi \cdot \nabla_x f = Q_+(f, f) + fR(f). \tag{2.7}$$

This feature of the methodology used so far is also present in the Kaniel-Shinbrot iteration scheme. For the (reasonable) choice $l_0 = 0$, u_1 is obtained by solving $\frac{d}{dt} u_1^{\#} = Q_+^{\#}(u_0, u_0)$, and the loss term is not present in this equation. Loosely speaking, we can say that we know how to solve the Boltzmann equation if we can solve the Boltzmann equation without the loss term.
It is known [12] that the Boltzmann equation without the loss term does not in general have a global solution. Consequently, it appears hopeless to extend these methods to significantly more general data.

4. **On related results.** There are, nonetheless, many related existence results. We mention in particular the work by Bellomo and Toscani [3], in which a global existence result is proved. Again the result is for a rare gas cloud in vacuum, but the data do not have to decay like

$e^{-\beta x^2}$ with respect to the space variable; instead, polynomial decay (like $(1+x^2)^{-p/2}$, $p>1$) is sufficient. The additional property of the collision transformation that enters crucially into the proof in Ref. 3 is that $\xi - \xi'$ and $\xi - \xi'_*$ are orthogonal (Problem 2). Also, the result is for more general interaction kernels. Related results were also proved by Hamdache [11] and Polewczak [16]. Polewczak stated correctly in what spaces the global solutions exist (there are some mistakes in the original papers [3,12] — the spaces used there are slightly too small) and he proved, in addition, some regularity results.

A global existence result of a slightly different flavor concerns the existence and uniqueness of homoenergetic affine flows described by the Boltzmann equation, proved in Ref. 4.

Problems

1. Verify that $Ce^{-\beta_0(x-\xi t)^2}e^{-\alpha_0\xi^2}$ is an explicit solution of the Boltzmann equation.
2. Check that $\xi - \xi' \perp \xi - \xi'_*$.
3. Show that the Cauchy problem

$$\frac{d}{dt}f(\xi,t) = \iint\limits_{\Re^3 S^2} f(\xi',t)f(\xi'_*,t)\,dn\,d\eta$$

$$f(\xi,0) = \varphi_0(\xi), \qquad \varphi_0 \in L^1_+(\Re^3)$$

has no global solution unless $\varphi_0 = 0$.

5.3 A General Global Existence Result

The DiPerna-Lions result

For decades, there was no global existence theorem for the Boltzmann equation for general initial data. Of course, it is easy to solve (explicitly) the transport equation

$$(\partial_t + \xi\cdot\nabla_x)f = 0,$$

and as we saw in Chapter 6, the spatially homogeneous Boltzmann equation

$$\partial_t f = Q(f,f)$$

is quite well understood and has been so for quite a while. The difficulty arises when $\xi\cdot\nabla_x f$ and the collision term $Q(f,f)$ occur together. It becomes hard to work in the (natural) space L^1, because the space-dependent collision operator is undefined for general L^1-functions (because of the pointwise product with respect to the x-variable).

On the other hand, there is some evidence that the flow term $\xi \cdot \nabla_x$ should actually have a smoothing effect. For example, we saw in the previous sections that the equation

$$\partial_t f + \xi \cdot \nabla_x f = Q_+(f, f) \tag{3.1}$$

(the "gain-term-only" Boltzmann equation) has a global solution if the data are small in a suitable sense. The spatially homogeneous counterpart to this equation, on the other hand,

$$\partial_t f = Q_+(f, f), \tag{3.2}$$

displays blowup of the solution even for arbitrarily small (isotropic) data different from zero (see Ref. 12), and we observe that the flow term in (3.1) actually dampens the local solution enough to let it exist for all time.

Smoothing effects of the flow term for large data were anticipated (and speculated on) for a long time. In 1987, Golse et al. [10] were able to prove quantitative results on such effects; these results have become known as "velocity averaging lemmas," and we present some of them in this chapter. Also in 1987, DiPerna and Lions used these lemmas and other careful estimates to prove the first general global existence theorem for the Boltzmann equation [7]. We present their result, with proof, in this section, following the excellent review article by Gérard [9]. We split the discussion into fourteen steps.

Step 1 We begin by clarifying some notation. Let $d \in \mathcal{N}$. If $\Omega \subset \Re^d$ is open, $L^p_{\text{loc}}(\Omega) = \{f : \Omega \to \Re, \ \ f\,|_U \in L^p(U)$ for all $U \subset \Omega$ that are open and relatively compact$\}$. If $\Omega \subset \Re^d$, $\Omega' \subset \Re^l$, the space of all measurable functions on $\Omega' \times \Omega$ whose restrictions to $\Omega' \times U$ is in $L^p(\Omega' \times U)$ for each open and relatively compact $U \subset \Omega$ will be denoted by $L^p(\Omega' \times \Omega_{\text{loc}})$. $S(\Re^d)$ denotes the Schwartz space of rapidly decreasing C^∞-functions on $\Re^d$. For each $s \in \Re_+$, $H^s(\Re^d)$ is the usual Sobolev space, i.e. the completion of $S(\Re^d)$ with respect to the norm

$$\|f\|_{H^s} := \left(\int (1 + |z|^{s/2})^2 |\hat{f}(z)|^2 \, dz \right)^{1/2}$$

Sometimes, we will use T as an abbreviation for the transport operator $\partial_t + \xi \cdot \nabla_x$.

Step 2 As in previous sections, our main interest lies in the analysis of gases of hard spheres; however, the method of proof used in the sequel applies to much more general collision kernels, and there is a technical reason (the approximation by solutions to a truncated equation; see Lemma 5.3.3) why such kernels must be studied. In fact, we first prove some results about a generalized Boltzmann equation

(3.3)
$$(\partial_t + \xi \cdot \nabla_x) f = \int_{\Re^d} \int_{S^{d-1}} q(x,\, \xi - \xi_*,\, n)[f' f'_* - f f_*]\, dn\, d\xi_*$$

with the usual convention $f_* = f(\cdot,\, \xi_*,\, \cdot)$, $f' = f(\cdot,\, \xi',\, \cdot)$, etc. Of course, for hard spheres $q(\ldots) = |n \cdot (\xi - \xi_*)|$, and $d = 3$ for the physically relevant case.

(5.3.1) Lemma. *Suppose that q is a non-negative measurable function in $L^\infty_{\text{loc}}(\Re^d \times \Re^d \times S^{d-1})$ that depends only on x, $|\xi - \xi_*|$, and $|(\xi - \xi_*) \cdot n|$ and grows at most polynomially with respect to x and $\xi - \xi_*$. Then, if $f = f(x,\, \xi,\, t) \in C^1\left(\Re_+,\, S(\Re^d \times \Re^d)\right)$ is a positive solution of (3.3) such that $|\ln f|$ grows at most polynomially in (x, ξ), uniformly on compact time intervals in $\Re_+$, we have*

(3.4)
$$\iint f(\cdot\,, t)\, dx\, d\xi = \iint f(\cdot\,, 0)\, dx\, d\xi,$$

(3.5)
$$\iint f(\cdot\,, t) \xi^2\, dx\, d\xi = \iint f(\cdot\,, 0) \xi^2\, dx\, d\xi,$$

(3.6)
$$\iint f(\cdot\,, t) |x - t\xi|^2\, dx\, d\xi = \iint f(\cdot\,, 0) x^2\, dx\, d\xi,$$

(3.7)
$$\iint f \ln f(\cdot\,, t)\, dx\, d\xi + \int_0^t \iint e(f)(\cdot,\, s)\, dx\, d\xi\, ds = \iint f \ln f(\cdot\,, 0)\, dx\, d\xi,$$

where
$$e(f)(x,\, \xi,\, t)$$
$$= \tfrac{1}{4} \iint (f' f'_* - f f_*) \ln\left(\frac{f' f'_*}{f f_*}\right)(x, \xi, \xi_*, n, t) \cdot q(x,\, \xi - \xi_*,\, n)\, d\xi_*\, dn.$$

These identities imply the estimates

(3.8)
$$\iint f(\cdot\,, t)(1 + |x|^2 + |\xi|^2)\, dx\, d\xi \le \iint f(\cdot\,, 0)(1 + 2|x|^2 + (2t^2 + 1)|\xi|^2)\, dx\, d\xi$$

and

(3.9)
$$\iint f(\cdot\,, t) |\ln f(\cdot\,, t)|\, dx\, d\xi + \int_0^\infty \iint e(f)(\cdot,\, s)\, ds\, dx\, d\xi$$
$$\le \iint f(\cdot\,, 0)(|\ln f(\cdot\,, 0)| + 2|x|^2 + 2|\xi|^2)\, dx\, d\xi + C_d$$

where the constant C_d depends only on the dimension d.

Proof. First, note that if q has only the dependencies required in the lemma, if $\varphi(\xi)$ is of polynomial growth, and if $g(\xi)$ decreases rapidly, then

(3.10) $$\int Q(g,g)\varphi\, d\xi = -\tfrac{1}{4}\iiint (g'g'_* - gg_*)(\varphi' + \varphi'_* - \varphi - \varphi_*)q\, d\xi\, d\eta\, dn.$$

This follows by the usual arguments, using the collision transformation and the changes of variables $(\xi, \xi_*) \to (\xi_*, \xi)$, $(\xi, \xi_*) \to (\xi'_*, \xi')$; all these transformations leave $|\xi - \xi_*|$ and $|(\xi - \xi_*)\cdot n|$ invariant. The right-hand side of (3.10) is zero if φ is a collision invariant.

Now suppose that $\psi = \psi(t, x, \xi) \in C^1(\Re_+ \times \Re^d \times \Re^d)$ is such that ψ and $\frac{\partial\psi}{\partial t}$ grow at most polynomially in (x, ξ), uniformly for t in compact sets. If in addition $T\psi = 0$, it follows that

$$\begin{aligned}\frac{d}{dt}\iint f\psi\, dx\, d\xi &= \iint (Tf)\psi\, dx\, d\xi \\ &= \iint Q(f,f)\psi\, dx\, d\xi,\end{aligned}$$

and if $\psi(t, x, \cdot)$ is a collision invariant for each (t, x),

$$\frac{d}{dt}\iint f\psi\, dx\, d\xi = 0.$$

Equations (3.4–3.6) follow by choosing successively $\psi \equiv 1$, $|\xi|^2$, $|x - t\xi|^2$. Similarly, calculate

$$\frac{d}{dt}\iint f\ln f\, dx\, d\xi = \iint (1 + \ln f)Q(f,f)\, dx\, d\xi$$

and let $\varphi = \ln f$ in (3.10) to arrive at (3.7). The estimate (3.8) is immediate from (3.4–3.6), whereas (3.9) follows from (3.7) and the observation that

(3.11) $$\iint_{f\le 1} f|\ln f| \le \iint_{\Omega} f|\ln f| + C'_d$$

with $\Omega = \{(x,\xi); e^{-|x-t\xi|^2-|\xi|^2} \le f(x, \xi, t) \le 1\}$. The last integral in (3.11) is bounded by $\iint (|x - t\xi|^2 + |\xi|^2)f\, dx\, d\xi$, and the assertion finally follows from (3.5) and (3.6). □

Remarks

a) We remind the reader that $e(f)$ is always non-negative (H-theorem).
b) Other possible weight functions ψ, which will lead to conservation equations, are ξ_i, $1 \le i \le d$ (momentum), $x_i\xi_j - x_j\xi_i$, $1 \le i \le j \le d$ (angular momentum), $x_i - t\xi_i$, $1 \le i \le d$, (center of mass), and $(x - t\xi)^2$ (moment of inertia).

Step 3 We now specify the assumptions on the collision kernel $q(\xi - \xi_*, n)$ for which the general existence result will be proved. For convenience, let $V =$

$\xi - \xi_*$. Notice that we assume no dependence of q on x; such a dependence only enters when we construct approximate solutions a little later.

Suppose that $q \in L^1_{\text{loc}}(\Re^d \times S^{d-1})$, $q \geq 0$, and that q depends only on $|V|$ and $|V \cdot n|$. Let

$$A(V) = \int_{S^{d-1}} q(V, n)\, dn.$$

Suppose, furthermore, that for every $R > 0$

$$\text{(3.12)} \qquad \frac{1}{1+|\xi|^2} \int_{|\xi_*| \leq R} A(\xi - \xi_*)\, d\xi_* \to 0 \qquad \text{as} \quad |\xi| \to \infty$$

and that

$$\text{(3.13)} \qquad A \in L^\infty_{\text{loc}}(\Re^d)$$

(this last assumption was not made by DiPerna and Lions, but, as noticed by Gérard, it simplifies certain technicalities of the proof).

It is easy to check that the collision kernel for hard spheres,

$$q = |(\xi - \xi_*) \cdot n|,$$

which is our main case of interest in this book, satifies both (3.12) and (3.13).

As in earlier chapters, we shall again split $Q(f, f) = Q_+(f, f) - fR(f)$ and write $Q_-(f, f) = fR(f)$. Note that $R(f) = A * f$.

Step 4 Solution concepts. The existence of a global classical solution of the Boltzmann equation is unknown. One of the crucial steps in our present endeavor is to introduce weaker solution concepts that lighten the burden of proof but are still strong enough to guarantee that the collision terms are defined. As before, we write

$$g^\#(x, \xi, t) = g(x + \xi t, \xi, t)$$

for each measurable g on $[0, \infty) \times \Re^d \times \Re^d$. First we reformulate the mild solution concept, with minimal integrability constraints on the collision terms.

(5.3.2) Definition. *A measurable function $f = f(x, \xi, t)$ on $[0, \infty) \times \Re^d \times \Re^d$ is a mild solution of the Boltzmann equation to the (measurable) initial value $f_0(x, \xi)$ if for almost all (x, ξ)*

$$Q_\pm(f, f)^\#(x, \xi, \cdot)$$

are in $L^1_{\text{loc}}[0, \infty)$, and if for each $t \geq 0$

$$f^{\#}(x,\,\xi,\,t) = f_0(x,\xi) + \int_0^t Q(f,f)^{\#}(x,\,\xi,\,s)\,ds.$$

One of the key ideas of DiPerna and Lions was to relax the solution concept even further, such that the bounds (3.8) and (3.9) could be put to best use, and then to regain mild solutions via a limit procedure. They called the relaxed solution concept "renormalized solution" (this concept of renormalization is different from the usual use of the word "renormalization" in modern physics).

(5.3.3) Definition. *A function $f = f(x,\,\xi,\,t) \in L^1_+(\Re^+_{\text{loc}} \times \Re^d \times \Re^d)$ is called a renormalized solution of the Boltzmann equation if*

$$\frac{Q_\pm(f,f)}{1+f} \in L^1_{\text{loc}}(\Re_+ \times \Re^d \times \Re^d) \tag{3.14}$$

and if for every Lipschitz continuous function $\beta : \Re_+ \to \Re$ that satisfies $|\beta'(t)| \le \frac{C}{1+t}$ for all $t \ge 0$ one has

$$T\beta(f) = \beta'(f)Q(f,f) \tag{3.15}$$

in the sense of distributions.

Remark. The division by $1+f$ is natural inasmuch as it leads to a "quasi-linearization" of $Q(f,f)$. There were earlier attempts to use (3.15) to obtain global solutions of the Boltzmann equation, but DiPerna and Lions were the first to notice that renormalization would actually give mild solutions.

(5.3.4) Lemma. *Let $f \in (L^1_{\text{loc}} \times \Re^d \times \Re^d)$.*

i) If f satisfies (3.14) and (3.15) with $\beta(t) = \ln(1+t)$, then f is a mild solution of the Boltzmann equation.

ii) If f is a mild solution of the Boltzmann equation and if $\frac{Q_\pm(f,f)}{1+f} \in L^1_{\text{loc}}(\Re_+ \times \Re^d \times \Re^d)$, then f is a renormalized solution.

The proof of Lemma 5.3.4 is presented in Appendix 5.A.

Step 5 The result.

(5.3.5) Theorem. (DiPerna and Lions, 1988) *Suppose that $f_0 \in L^1_+(\Re_+ \times \Re^d \times \Re^d)$ is such that*

$$\iint f_0(1+|x|^2+|\xi|^2)\,dx\,d\xi < \infty$$

and

$$\iint f_0|\ln f_0|\,dx\,dv < \infty.$$

Then there is a renormalized solution of the Boltzmann equation such that $f \in C(\Re_+, L^1(\Re^d \times \Re^d))$, $f\mid_{t=0}= f^0$, *and (3.8) and (3.9) hold.*

Step 6 Solving a truncated equation. The renormalized solution f will be found as a limit of functions solving truncated equations. For some $\delta > 0$ and some modified non-negative collision kernel $\bar{q} \in C_0^\infty(\Re^d \times S^{d-1})$ such that $\bar{q}$ vanishes for $(\xi - \xi_*) \cdot n < \delta$, let

$$\bar{Q}(g,g) = \iint \bar{q}(g'g'_* - gg_*)\,dn\,d\xi_*$$

and

$$\tilde{Q}(g,g) = (1 + \delta \int |g|\,d\xi)^{-1}\bar{Q}(g,g). \tag{3.16}$$

(5.3.6) Lemma. *Let* $f_0 \in S(\Re^d \times \Re^d)$ *be non-negative such that* $|\ln f_0|$ *grows at most polynomially. Then the Cauchy problem*

$$Tf = \tilde{Q}(f,f), \qquad f\mid_{t=0}= f_0 \tag{3.17}$$

has a unique global solution f *that satisfies the hypotheses of Lemma 5.3.1. It also satisfies the estimates (3.8) and (3.9).*

Proof. Lemma 5.3.6 is in itself of interest, but the assertion is an example for results for a modified Boltzmann-type equation, for which the contraction mapping principle can be invoked. In fact, because $\tilde{Q}$ grows only linearly in $|g|$ we have estimates

$$\int_{\Re^d} |\tilde{Q}(g,g)|\,d\xi \le C_\delta \int_{\Re^d} |g|\,d\xi, \tag{3.18}$$

$$\|\tilde{Q}(g,g)\|_{L^\infty(\Re^d_\xi)} \le C_\delta \|g\|_{L^\infty(\Re^d_\xi)}, \tag{3.19}$$

and

$$\int_{\Re^d} |\tilde{Q}(g,g) - \tilde{Q}(f,f)|\,d\xi \le C_\delta \int_{\Re^d} |f - g|\,d\xi. \tag{3.20}$$

C_δ stands for various constants independent of g and f. The estimates (3.18) and (3.20) are easily checked. For (3.19), it is clearly enough to prove that

$$|\int_{\Re^d}\int_{S^2} \bar{q}(\xi,\xi_*,n)g(\xi')dnd\xi_*| \le C_\delta \|g\|_{L^1_\xi}.$$

A simple shift shows that it is sufficient to consider the case $\xi = 0$. Let $v = |n \cdot \xi_*|$. We interpret v as the radial part of a representation of ξ' in polar coordinates, with n representing the angular part. The integral over $dnd\xi_*$ is then replaced by the integral $\int_{S^2}\int_0^\infty\int_W \ldots dwdvdn$, where W denotes the plane through $n(n\cdot\xi_*)$ and orthogonal to n. Note that we have to insert a factor v^2 for the volume integral in spherical coordinates. The integral in question becomes

$$\int_{S^2}\int_0^\infty\int_W \frac{\bar{q}(\ldots)}{v^2} dwg(v,n)v^2 dvdn.$$

From the truncation for $\bar{q}$ and the assumption that q vanishes for v near zero we get that

$$\int_W \frac{\bar{q}(\ldots)}{v^2} dw \le C_\delta,$$

and (3.19) follows.

The estimates (3.18–3.20) can be used to show that the iteration

$$Tf^{n+1} = \tilde{Q}(f^n, f^n),\ f^n\mid_{t=0} = f_0,\ Tf^0(x,\,\xi,\,t) = 0$$

will converge in $C([0,T]; L^1(\Re^d\times\Re^d))$. We also get an $L^1\cap L^\infty(\Re^d\times\Re^d)$ estimate uniform in $t\in[0,T]$. In fact, the convergence is even in $C(\Re^+; S(\Re^d\times\Re^d))$, because higher moments and derivatives of f^{n+1} can be readily estimated in terms of higher moments and derivatives of f^n. This is easy, but lengthy and tedious, so we omit it (some details are given by DiPerna and Lions). Non-negativity of the solution of (3.17) follows as a side result of the proof that $|\ln f|$ grows at most polynomially. In fact, our hypothesis on f_0 entails that $f_0(x,\xi) \ge Ke^{-C_1(|x|^k+|\xi|^k)}$ for some constants $K>0$, $C_1>0$, $k\in\mathcal{N}$. Also, if f^n is non-negative (which is certainly true for at least a short time), we have

$$Tf^{n+1} \ge -\frac{\overline{A}*f^n}{1+\delta\int f^n\,d\xi}\cdot f^n \ge -C_2 f^n,$$

where $C_2\ge 0$. Because of the bounds on the f^n, we can send $n\to\infty$ to find

$$Tf \ge -C_2 f$$

(at least locally). This implies that

$$f(x,\,\xi,\,t) \ge Ke^{-C_1(|x-\xi t|^k+|\xi|^k)}\cdot e^{-C_2 t}.$$

The proof of Lemma 5.3.6 is complete. □

Step 7 Preparations. Let $q_n\in C^\infty_{0,+}(\Re^d\times S^{d-1})$ satisfy (3.12) and (3.13) (uniformly for all n) and suppose that $q_n\to q$ a.e. Furthermore, we approximate f_0 in $L^1_+(\Re^d\times\Re^d)$ by a sequence $\{f_0^n\}_n\subset S(\Re^d\times\Re^d)$ such that

$$\forall n\quad f_0^n \ge \mu_n e^{-|x|^2-|\xi|^2}\qquad (\mu_n>0),$$

$$\iint f_0^n(1+|x|^2+|\xi|^2\, dx\, d\xi \longrightarrow \iint f_0(1+|x|^2+|\xi|^2)\, dx\, dv,$$

$$\iint f_0^n|\ln f_0^n|\, dx\, d\xi \longrightarrow \iint f_0|\ln f_0|\, dx\, d\xi.$$

Let $\delta_n \searrow 0$, and let Q^n be $\tilde{Q}$ (from Step 6) with $\delta = \delta_n$, $\bar{q} = q_n$. Then Lemma 5.3.6 assures us that there is a sequence $\{f^n\}$ such that

$$Tf^n = Q^n(f^n, f^n), \qquad f^n|_{t=0} = f_0^n,$$

and (by (3.8) and (3.9))

$$\forall T>0 \sup_{t\in[0,T]} \sup_n \iint f^n(1+|x|^2+|\xi|^2)\, dx\, d\xi < \infty, \tag{3.21}$$

$$\forall T>0 \sup_{t\in[0,T]} \sup_n \iint f^n|\ln f^n|\, dx\, d\xi < \infty, \tag{3.22}$$

$$\sup_n \int_0^\infty \iint e_n(f^n)\, dx\, d\xi\, dt < \infty, \tag{3.23}$$

where
(3.24)

$$e_n(f^n) = \tfrac{1}{4}(1+\delta_n \int f^n\, d\xi)^{-1} \cdot \iint (f^{n'} f_*^{n'} - f^n f_*^n) \ln\left(\frac{f^{n'} f_*^{n'}}{f^n f_*^n}\right) q_n\, d\xi_*\, dn.$$

Step 8 Weak compactness. We recall the Dunford-Pettis criterion for weak compactness in L^1 (see Ref. 8). Let $\{f_n\}_{n\in\mathcal{N}} \subset L^1(\Re^d)$. Then the following i) and ii) are equivalent.

i) $\{f_n\}$ is contained in a weakly sequentially compact set of $L^1(\Re^d)$.
iia) $\{f_n\}$ is bounded in $L^1(\Re^d)$.
iib) $\forall \epsilon>0 \quad \exists \delta>0$ such that $\forall E \subset \Re^d$ (E measurable) with $\lambda(E)<\delta$,

$$\sup_n \int_E |f_n|\, dx \le \epsilon.$$

iic) $\forall \epsilon > 0 \exists K$ compact, $K \subset \Re^d$, such that $\sup_n \int_{\Re^d - K} |f_n|\, dx \le \epsilon$.

We will apply the criterion to the following situation.
If $h \in C(\Re_+, \Re_+)$ and $w \in L^\infty_{\rm loc}(\Re^d, \Re_+)$ are such that $h(t)/t \to \infty$ $(t\to\infty)$ and $w(x) \to \infty$ $(|x| \to \infty)$, then the inequality

$$\sup_n \int_{\Re^d} [h(|f_n|) + |f_n|(1+w)]\, dx < \infty \tag{3.25}$$

implies that $\{f_n\}_{n\in\mathcal{N}}$ satisfies ii).

A major problem with weak convergence is that nonlinear functions are in general not weakly continuous. The standard example is the sequence $\sin(nx)\cdot\chi_{[0,1]}(x)$, which converges weakly to 0 (Riemann-Lebesgue lemma). However, $\sin^2(nx)\cdot\chi_{[0,1]}(x) \to \frac{1}{2}\chi_{[0,1]}(x)$, such that in this case with $F(x) = x^2$, $F \circ f_n \underset{w}{\not\longrightarrow} F \circ \lim_w f_n$. A useful property for our objective is the fact that convex functions are at least lower semicontinuous. If $F : \Re \to \Re$ is convex and if $f_n \underset{w}{\longrightarrow} f$ in L^1, then

$$\int F \circ f \, dx \leq \liminf_{n\to\infty} \int F \circ f_n \, dx$$

(the above example confirms this assertion). The proof of this well-known fact can be found, e.g., in Ref. 5. Also, if one of the factors in a product converges a.e. and the other factor converges weakly, then the product is compact in the weak topology. Specifically, let $f_n \underset{w}{\longrightarrow} f$ in L^1, let $\{g_n\} \subset L^\infty$ be bounded and let $g_n \to g$ a.e., then

$$f_n \cdot g_n \underset{w}{\longrightarrow} fg \text{ in } L^1.$$

This follows because for every $\epsilon > 0$ there is a compact set K such that $\sup_n \int_{\Re^d\setminus K} (|f_n g_n| + |fg|) \, dx \leq \epsilon$, and by Egorov's theorem, there is a set $E \subset K$ such that $\sup_n \int_E |f_n| \, dx \leq \epsilon$ and such that $g_n \to g$ uniformly on $K\setminus E$. The details are left as an exercise.

Step 9 Weak compactness of the collision terms. We now work with the "approximating sequence of solutions to modified equations" given in Step 7. Q_+^n, Q_-^n, A_n all refer to this situation. The collision kernel in $Q_{+,-}^n$ is really x- (and t-) dependent, and given by

$$q_n(x, V, n) = \frac{1}{1 + \delta_n \int f_n \, d\xi} q_n(V, n).$$

(5.3.7) Lemma. *For all* $T > 0$, $R > 0$, *the sequences*

$$\frac{Q_+^n(f_n, f_n)}{1 + f_n} \quad \text{and} \quad \frac{Q_-^n(f_n, f_n)}{1 + f_n}$$

are contained in weakly compact subsets of $L^1\left((0,T) \times \Re^d \times B_R\right)$, *where* $B_R = \{\xi \in \Re^d; \|\xi\| \leq R\}$.

Proof. For Q_-^n, we have $0 \leq \frac{Q_-^n(f_n,f_n)}{1+f_n} \leq A_n * f_n$. We verify the Dunford-Pettis criterion (iia,b, and c) for $A_n * f_n$, using (3.21), (3.22), and (3.12) (which holds uniformly in n for all A_n):

For iia) note that by (3.12)

$$\int_{\Re^d}\int_{B_R} A_n * f^n \, d\xi \, dx = \int_{R^d}\int_{R^d}\int_{B_R} A_n(\xi - z) \, dz \, f^n \, d\xi \, dx$$

$$\leq \int_{\Re^d}\int_{\Re^d} C_1(1+|\xi|^2) f^n \, d\xi \, dx \leq C,$$

For the proof of iib) and iic) we first focus on the simpler case where A is integrable, i.e., $\|A\|_{L^1(\Re^d)} < \infty$. Then, without restricting the generality, $a_n := \|A_n\|_{L^1(\Re^d)}$ can be assumed to be bounded uniformly in n.

Let

$$\phi(t) = \begin{cases} t\ln t & \text{for } t \geq 1 \\ 0 & \text{for } 0 \leq t < 1. \end{cases}$$

ϕ is convex and satisfies the inequality

$$\phi(L) \leq a\phi\left(\frac{L}{a}\right) + L|\ln a|$$

for all $a \geq 0$. By this and Jensen's inequality,

$$\begin{aligned}
&\int_{\Re^d}\int_{B_R} \phi(A_n * f^n) \, d\xi \, dx \\
&\leq a_n \int_{\Re^d}\int_{B_R} \phi\left(\frac{A_n * f^n}{a_n}\right) \, d\xi \, dx + |\ln a_n| a_n \|f^n\|_{L^1(\Re^d\times\Re^d)} \\
&\leq \int_{\Re^d}\int_{B_R} A_n * \phi(f^n) \, d\xi \, dx + |\ln a_n| a_n \|f^n\|_{L^1(\Re^d\times\Re^d)} \\
&\leq a_n \int_{\Re^d}\int_{\Re^d} \phi(f^n) \, d\xi \, dx + |\ln a_n| a_n \|f^n\|_{L^1(\Re^d\times\Re^d)},
\end{aligned}$$

iib) then follows from the bounds on $\int_{R^d}\int_{B_R} \phi(A_n * f^n) \, d\xi \, dx$. (Actually, we could have set $R = \infty$ for this argument; we shall have to use $R < \infty$ when we generalize to a nonintegrable A.)

To show iic) we consider $\int_{\Re^d}\int_{|\xi|\geq K} A_n * f^n \, d\xi \, dx$ and estimate

$$\begin{aligned}
\int_{\Re^d}\int_{|\xi|\geq K} A_n * f^n \, d\xi \, dx &\leq \int_{\Re^d}\int_{\Re^d}\int_{\Re^d} A_n(\xi-\xi_*) f^n(x,\xi_*,t) \chi_{\{|\xi_*|\geq K/2\}} \, d\xi_* \, d\xi \, dx \\
&+ \int_{\Re^d}\int_{\Re^d}\int_{\Re^d} \chi_{\{|\xi|\geq K/2\}} \cdot \chi_{\{|\xi_*|\leq K/2\}} A_n(\xi-\xi_*) f^n(x,\xi_*,t) \, d\xi_* \, d\xi \, dx \\
&\leq \frac{4a_n}{K^2} \int_{\Re^d}\int_{\Re^d} |\eta|^2 f^n(x,\xi_*,t) \, d\xi_* \, dx + \left(\int_{|z|\geq K/2} A_n(z) \, dz \right) \cdot \|f^n\|_{L^1(\Re^d\times\Re^d)}.
\end{aligned}$$

The right-hand side becomes (uniformly in n and t) small as $K \to \infty$, and iic) follows.

Next we remove the uniform integrability condition on A_n [the only assumption we retain is (3.12)]. Notice that the reasoning given earlier remains valid if all A_n were supported in some compact set, say

$$A_n(z) = A_n(z) \cdot \chi_{\{|z| \le K\}}. \tag{3.26}$$

Otherwise, let $A_{n,K}$ be defined by the right hand side of (3.26). Weak compactness of $A_n * f^n$ will follow if we can show that

$$\sup_n \|A_{n,K} * f^n - A_n * f^n\|_{L^\infty(0,T;\, L^1(\Re^d \times B_R))} \to 0 \quad \text{as} \quad K \to \infty.$$

But

$$\|A_{n,K} * f^n - A_n * f^n\|_{L^1(\Re^d \times B_R)}$$
$$= \int_{\Re^d} \int_{B_R} \int_{\Re^d} A_n(\xi - \xi_*) \cdot \chi_{\{|\xi - \xi_*| \ge K\}} f^n(x, \xi_*, t)\, d\xi_*\, d\xi\, dx.$$

By (3.12)

$$\int_{B_R} A_n(\xi - \xi_*)\, d\xi \le \epsilon(1 + |\xi_*|^2) + C_\epsilon,$$

and $\{\xi_*; |\xi - \xi_*| \ge K\} \subset \{\xi_*; |\xi_*| \ge K - R\}$ if $|\xi| \le R$ and if $K > R$.

$(|\xi_*| = |\xi_* - \xi + \xi| \ge |\xi - \xi_*| - |\xi| \ge K - R)$.

So,

$$\int_{\Re^d} \int_{B_R} \int_{\Re^d} A_n(\xi - \xi_*) \cdot \chi_{\{|\xi - \xi_*| \ge K\}} f^n\, d\xi_*\, d\xi\, dx$$
$$\le \epsilon \iint (1 + |\xi_*|^2) f^n + \frac{C_\epsilon}{(K - R)^2} \iint |\xi_*|^2) f^n,$$

and the assertion follows by first sending K to ∞, then ϵ to 0_+.

Boundedness and weak compactness of $\left\{\frac{Q_+^n(f^n, f^n)}{1+f^n}\right\}$ in $L^1\left((0,T) \times \Re^d \times B_R\right)$ is now deduced from the H-theorem. Recall that if

$$e_n(f^n) = \iint q_n(\xi - \xi_*, n)(f^{n'} f_*^{n'} - f^n f_*^n) \ln \frac{f^{n'} f_*^{n'}}{f^n f_*^n},$$

then, by (3.7), $e_n(f^n)$ is bounded in $L^1\left((0,T) \times \Re^d \times \Re^d\right)$, and we have for all $K > 1$

$$Q_+^n(f^n, f^n) \le K Q_-^n(f^n, f^n) + \frac{1}{\ln K} e_n. \tag{3.27}$$

We leave the proof of (3.27) until later. Because we already checked the weak compactness of $\frac{Q_-^n(f^n, f^n)}{1+f^n}$, (3.27) readily implies that

(3.28)

$$\sup_n \int_0^T \int_{\Re^d}\int_{\Re^d} \left(\chi_A(x,\xi) + \chi_{\{|x|+|\xi|\geq R\}}(x,\xi)\right) \frac{Q_+^n(f^n,f^n)}{1+f^n}\, d\xi\, dx\, dt \to 0$$

as $\lambda(A) \to 0$ and $R \to \infty$.

To prove (3.27), let $A_K^n = \left\{(\xi,\,\xi_*,\,n)\colon f^{n'} f_*^{n'} \geq K f^n f_*^n\right\}$ and $B_K^n = (A_K^n)^c$, then

$$Q_+^n = \iint q_n f^{n'} f_*^{n'}$$

$$\leq K \iint q_n f^n f_*^n \chi_{B_K^n} + \iint q_n \chi_{A_K^n} (f^{n'} f_*^{n'} - f^n f_*^n) + \iint q_n \chi_{A_K^n} f^n f_*^n.$$

On A_K^n, $\ln \frac{f^{n'} f_*^{n'}}{f^n f_*^n} \geq \ln K$, and (3.27) follows easily.

This completes the proof of Lemma 5.3.7. □

Remark. In the same way, one proves that

(3.29)
$$Q_-^n \leq K Q_+^n + \frac{4}{\ln K} e_n.$$

Step 10 Extracting a weakly convergent subsequence, and first properties of the limit. Because $\{f^n\}$ has uniformly bounded entropy and second moments, (3.25) implies that we can extract a subsequence (again denoted by $\{f^n\}$), which converges weakly in $L^1((0,T)\times\Re^d\times\Re^d)$,

$$f^n \underset{w}{\longrightarrow} f.$$

Let, as in the Appendix, $g_\delta^n \colon= \frac{1}{\delta}\ln(1+\delta f^n)$. The uniform bounds on entropy and second moments for f^n easily imply that

(3.30)
$$\sup_{t\in[0,T]} \sup_n \|f^n(\cdot\,,t) - g_\delta^n(\cdot\,,t)\|_{L^1(\Re^d\times\Re^d)} \longrightarrow 0 \qquad \text{as} \quad \delta\to 0.$$

Also, because

$$T g_\delta^n = \frac{1}{1+\delta f^n} Q^n(f^n,f^n),$$

$$g_\delta^{n\#}(t+h) - g_\delta^{n\#}(t) = \int_t^{t+h} \frac{Q^n(f^n,f^n)^\#(s)}{1+\delta f_n^\#(s)}\, ds.$$

By the compactness spelled out in Lemma 5.3.7, $\forall\delta>0\ \forall T>0\ \forall R>0$

$$\sup_{t\in[0,T]} \sup_n \|g_\delta^{n\#}(t+h) - g_\delta^{n\#}(t)\|_{L^1(\Re^d\times B_R)} \longrightarrow 0$$

as $h\to 0$. We next estimate, by (3.30) and (3.8),

$$\sup_t \|f^{n\#}(t+h) - f^{n\#}(t)\|_{L^1(\Re^d\times\Re^d)}$$

$$\leq 0(\delta) + \sup_t \|g_\delta^{n\#}(t+h) - g_\delta^{n\#}(t)\|_{L^1(\Re^d\times B_R)}.$$

This easily entails

$$\sup_{t\in[0,T]} \sup_n \|f^{n\#}(t+h) - f^{n\#}(t)\|_{L^1(\Re^d\times\Re^d)} \underset{h\to 0}{\longrightarrow} 0,$$

and a standard equicontinuity argument shows that the (weak) limit f must then satisfy

$$f^\# \in C(\Re_+;\, L^1(\Re^d \times \Re^d)) \tag{3.31}$$

and, for all $T > 0$

$$\sup_{t\in[0,T]} \|f^\#(t+h) - f^\#(t)\|_{L^1} \underset{h\to 0}{\longrightarrow} 0.$$

Actually, by using an elementary argument from integration theory,

$$f \in C(\Re_+;\, L^1(\Re^d \times \Re^d)).$$

Also, by using the convexity of the function $x \cdot \max(\ln x, 0)$,

$$\forall t \quad \iint f|\ln f|\, d\xi\, dx + \limsup_n \int_0^t\!\!\iint e_n(f^n)\, d\xi\, dx \leq \iint f_0(|\ln f_0| + 2|x|^2 + 2|\xi|^2)\, d\xi\, dx + C_d, \tag{3.32}$$

and

$$\forall t \quad \iint f(1 + |x|^2 + |v|^2)\, d\xi\, dx \leq \iint f_0(1 + 2|x|^2 + (2t^2+1)|\xi|^2)\, d\xi\, dx.$$

Step 11 Velocity averaging. By now, we have a weakly convergent sequence $f_n \to f$, and the limit f is in

$$C([0,T];\, L^1).$$

Subsequences of $\frac{Q^n_{+,-}(f^n,f^n)}{1+f^n}$ will also converge weakly (by Lemma 5.3.7), but we cannot say a priori whether the limits will by $\frac{Q_{+,-}(f,f)}{1+f}$, because nonlinear functionals are in general not weakly continuous. This problem was first overcome by DiPerna and Lions by skillful application of results known as "velocity averaging lemmas" (see Ref. 10). We present these now (actually, we confine our discussion to a simplified situation, which is all we need).

(5.3.8) Lemma. *Let $u \in L^2(\Re \times \Re^d \times \Re^d)$ have compact support, and suppose that $Tu \in L^2(\Re \times \Re^d \times \Re^d)$. Then*

$$\int u\, d\xi \in H^{1/2}(\Re \times \Re^d),$$

and the $H^{1/2}$-norm of $\int u\, d\xi$ is bounded in terms of $\|u\|_{L^2}$, $\|Tu\|_{L^2}$, and the support of u.

Remark. The function $\int u\, d\xi$ is a "velocity average" of u.

Proof. Let $\hat{u} = \hat{u}(z, \xi, \tau)$ be the Fourier transform of u with respect to t and x. By assumption, $\hat{u}$ and $(\tau + \xi \cdot z)\hat{u}$ are in $L^2(\Re \times \Re^d \times \Re^d)$ and have compact support in ξ. We now split the integration domain in

$$\int \hat{u}(z, \xi, \tau)\, d\xi$$

in a suitable way: Let $\rho = (\tau^2 + |z|^2)^{1/2}$, $\tau_0 := \frac{\tau}{\rho}$, $z_0 := \frac{z}{\rho}$,

$$\begin{aligned} \int \hat{u}\, d\xi &= \int_{\{|\tau_0 + \xi z_0| \le \frac{1}{\rho}\}} \hat{u}\, d\xi + \int_{\{|\tau_0 + \xi z_0| \ge \frac{1}{\rho}\}} \hat{u}\, d\xi \\ &= \quad \mathrm{I} \quad + \quad \mathrm{II}. \end{aligned}$$

To estimate I, observe that for every compact $K \subset \Re^d$ there are $\epsilon_0 > 0$ and $C > 0$ such that for all $\epsilon \in (0, \epsilon_0)$

$$\sup_{(\tau_0, z_0) \in S^d} \lambda\{\xi \in K;\ |\tau_0 + \xi \cdot z_0| \le \epsilon\} \le C\epsilon,$$

where λ denotes the Lebesgue measure in $\Re^d$. Then, by the Cauchy-Schwarz inequality, with $\epsilon = \frac{1}{\rho}$,

$$|\mathrm{I}|^2 \le \frac{C}{\rho} \int |\hat{u}|^2\, d\xi$$

and

$$\begin{aligned} |\mathrm{II}|^2 &\le \frac{1}{\rho^2} \left(\int_{\{\xi \in K; |\tau_0 + \xi z_0| \ge \frac{1}{\rho}\}} |\tau_0 + \xi z_0|^{-2}\, d\xi \right) \cdot \left(\int |\tau + \xi z|^2 |\hat{u}|^2\, d\xi \right) \\ &\le \frac{C'}{\rho} \int |\tau + \xi z|^2 |\hat{u}|^2\, d\xi. \end{aligned}$$

The last inequality follows because by an elementary integration

$$\int_{\{\xi \in K; |\tau_0 + \xi z_0| \ge \frac{1}{\rho}\}} |\tau_0 + \xi z_0|^{-2}\, d\xi \le C' \cdot \rho. \tag{3.33}$$

The constant C' depends on the support of $\hat{u}$ with respect to ξ. To check (3.33), consider first $\tau_0 = 0$, $z_0 = (1,0,0)$. Together, the estimates for I and II imply that

$$\int (\tau^2 + |z|^2)^{1/2} \left| \int \hat{u}(z,\, \xi,\, \tau)\, d\xi \right|^2 dz\, d\tau < \infty,$$

$$\text{i.e.,} \quad \int u\, d\xi \in H^{1/2}.$$

□

We will use Lemma 5.3.8 to pass from weak to strong convergence in L^1-settings. The next lemma is the crucial one.

(5.3.9) Lemma. *Suppose that* $\{g_n\} \subset L^1\big((0,T) \times \Re^d \times \Re^d\big)$ *is weakly relatively compact, and that* $\{Tg_n\}$ *is weakly relatively compact in* $L^1_{\text{loc}}\big((0,T) \times \Re^d \times \Re^d\big)$. *Then, if* $\{\psi_n\}$ *is a bounded sequence in* $L^\infty\big((0,T) \times \Re^d \times \Re^d\big)$ *that converges a.e., then* $(\int g_n \psi_n \, d\xi)$ *is compact in the norm topology in* $L^1((0,T) \times \Re^d)$.

Before proving this, we note an immediate corollary.

Corollary. *Under the hypotheses of Lemma 5.3.9, if* $g_n \underset{w}{\longrightarrow} g$ *in* $L^1((0,T) \times \Re^d \times \Re^d)$ *and* $\psi_n \to \psi$ *a.e., then*

$$\left\| \int g_n \psi_n \, d\xi - \int g\psi \, d\xi \right\|_{L^1((0,T)\times\Re^d)} \to 0.$$

Proof. (Of Lemma 5.3.9.) Because for each $\epsilon > 0$ there is a compact set $K \subset (0,T) \times \Re^d \times \Re^d$ such that for all n

$$\iiint_{K^c} (|g_n \psi_n| + |g\psi|) < \epsilon,$$

we may assume that all the g_n (and g) are supported in a fixed compact set. Also, because by Egorov's theorem $\psi_n \to \psi$ uniformly except on a set of arbitrary small measure, we may as well take $\psi_n = \psi$ for all n. Moreover, it is enough to take $\psi = 1$: If ψ is smooth enough, $\{g_n \psi\}$ satisfies the same hypotheses as $\{g_n\}$, and if ψ is in L^∞, we approximate it by C^∞ functions ψ_k such that $\|\psi_k - \psi\|_{L^1} \to 0$ and $\sup_k \|\psi_k\|_{L^\infty} < \infty$. Then

$$\iint \left| \int \psi_k g_n \, d\xi - \int \psi g_n \, d\xi \right| dx\, dt \le \iiint |\psi_k - \psi|\, |g_n| \, dt\, dx\, d\xi \longrightarrow 0$$

(by the Dunford-Pettis criterion).

After this reduction, define u_n and h_n by

$$Tu_n = Tg_n \cdot \chi_{\{(x,\,\xi,\,t);\,|Tg_n|\le M\}}$$
$$Th_n = Tg_n \cdot \chi_{\{(x,\,\xi,\,t);\,|Tg_n|\ge M\}}$$

where $u_n|_{t=0} = h_n|_{t=0} = 0$. Clearly, then, $g_n = u_n + h_n$, because $T(u_n + h_n) = Tg_n$ and g_n is the unique solution of $Tf = Tg_n$, $f|_{t=0} = 0$. Because $\{Tg_n\}$ is weakly compact and because

$$h_n(x + t\xi,\, \xi,\, t) = \int_0^t Tg_n(x + \tau\xi,\, \xi,\, \tau) \cdot \chi_{\{|Tg_n|\ge M\}}(x + \tau\xi,\, \xi\tau)\, d\tau,$$

it follows that uniformly with respect to n

$$\int_0^T \int_{\Re^d}\int_{\Re^d} |h_n(x,\, \xi,\, t)|\, dx\, d\xi\, dt \underset{M\to\infty}{\longrightarrow} 0.$$

On the other hand $\{u_n\}$ and $\{Tu_n\}$ are bounded sequences in L^2, so that $\{\int u_n\, d\xi\}$ is bounded in $H^{1/2}$ (by Lemma 5.3.8), i.e., compact in L^2, and, because it is compactly supported, compact in L^1. The proof is complete. □

(5.3.10) Lemma. *Let $\{f_n\}$ be a relatively compact sequence in $L^1((0,T) \times \Re^d \times \Re^d)$, and suppose that there is a family of real-valued uniformly Lipschitz continuous functions $\{\beta_\delta\}_{\delta>0}$, $\beta_\delta(0) = 0$ for all δ, such that*
i) $\beta_\delta(s) \to s$ *as* $\delta \to 0$, *uniformly on compact subsets of* $\Re_+$,
ii) *the sequence $\{T(\beta_\delta(f^n))\}$ is, for every δ, weakly relatively compact in $L^1_{\mathrm{loc}}((0,T) \times \Re^d \times \Re^d)$. Then, if $f^n \overset{w}{\longrightarrow} f$ in L^1_+, $\{\psi_n\}_n$ is bounded in $L^\infty((0,T) \times \Re^d \times \Re^d)$ and $\psi_n \to \psi$ a.e.,*

$$\lim_{n\to\infty} \left\| \int f^n \psi_n\, d\xi - \int f\psi\, d\xi \right\|_{L^1} = 0.$$

Proof. The weak compactness of $\{f^n\}$ implies that

$$\sup_n \|f^n - \beta_\delta(f^n)\|_{L^1} \to 0 \tag{3.34}$$

as $\delta \to 0$. $\{\beta_\delta(f^n)\}$ is also a weakly relatively compact sequence, so that Lemma 5.3.9 applies to $g_\delta^n := \beta_\delta(f^n)$ for each $\delta > 0$, and Lemma 5.3.10 follows easily from (3.34). Note that we can extract a subsequence of $\{\beta_\delta(f^n)\}$, say $\{\beta_\delta(f^{n_i})\}$, which converges to some g_δ. It is in general *false* that $g_\delta = \beta_\delta(f)$, but (3.34) guarantees that $\|f - g_\delta\|_{L^1} \to 0$ as $\delta \to 0$. □

Step 12 Passing to the limit. We return to the setting of Step 10.

(5.3.11) Lemma. *Let $\{f^n\}$ be the sequence of solutions to approximating problems as in Step 10. There is a subsequence such that for each $T > 0$*
i) $\int f^n \, d\xi \to \int f \, d\xi$ *a.e. and in* $L^1((0,T) \times \Re^d)$,
ii) $A_n * f^n \to A * f$ *in* $L^1((0,T) \times \Re^d \times B_R)$ *for all* $R > 0$, *and a.e.,*
iii) for each compactly supported function $\varphi \in L^\infty((0,T) \times \Re^d \times \Re^d)$,

$$\left(\frac{\int Q^n_\pm(f^n, f^n)\varphi \, d\xi}{1 + \int f^n \, d\xi}\right) \longrightarrow \left(\frac{\int Q_\pm(f, f)\varphi \, d\xi}{1 + \int f \, d\xi}\right)$$

in $L^1((0,T) \times \Re^d)$.

Proof. Recall Lemma 5.3.7 and apply Lemma 5.3.10 with $\beta_\delta(s) = \frac{1}{\delta}\ln(1 + \delta s)$. i) is immediate. ii) requires a vector-valued variant of Lemma 5.3.7. Let $\psi_n(\eta) = A_n(\xi - \eta) \cdot \chi_{\{|\xi| \le \Re\}} \in L^\infty_{\text{loc}}(\Re^d, L^1(B_R))$ and use the hypothesis on A and the estimate $\sup_n \int f^n(1 + |\xi|^2) \, d\xi < \infty$ to reduce the problem to bounded domains with respect to η.

For iii) and Q_-, take $\psi_n = \frac{A_n * f_n}{1 + \int f_n \, d\xi} \cdot \varphi$ and use i) and ii) and Lemma 5.3.10. For Q_+, the same reasoning applies, because by using the collision transformation we can write

$$\frac{\int Q^n_+(f^n, f^n)\varphi \, d\xi}{1 + \int f^n \, d\xi} = \frac{\iiint q_n f^n f^n_* \varphi'}{1 + \int f^n \, d\xi}$$

and choose

$$\psi_n(x, \xi, t)$$

$$= \int_{\Re^d}\int_{S^{d-1}} q_n(\ldots) f^n(x, \xi_*, t)\varphi(x, \eta', t) \, dn \, d\xi_* \Big/ \left(1 + \int f^n \, d\xi\right).$$

ψ_n is bounded in $L^\infty((0,T) \times \Re^d \times \Re^d)$ and can be assumed to converge in L^1 (by the vector-valued variant of Lemma 5.3.9), and the assertion follows again from Lemma 5.3.9. □

Remark. Unfortunately, part iii) of Lemma 5.3.10 *cannot* be changed to

$$\frac{Q^n_\pm(f^n, f^n)}{1 + f^n},$$

because the renormalizing factor $1/(1 + f^n)$ leads to a nonlinearity that cannot be controlled in the weak topology. If it weren't for this difficulty, the remainder of the proof would be short.

Step 13. Consider now T^{-1}, defined by $u = T_g^{-1}$, i.e.,
$Tu = g$ with $u|_{t=0} = 0$:

$$T^{-1}g(x,\xi,t)=\int_0^t g(x-(t-s)\xi,\xi,s)\,ds.$$

T^{-1} is, as one checks immediately, continuous and weakly continuous from $L^1((0,T)\times\Re^d\times\Re^d_{\text{loc}})$ into $C([0,T]);\ L^1(\Re^d\times\Re^d_{\text{loc}})$, and if $g\geq 0$, also $T^{-1}g\geq 0$. We use T^{-1} to rewrite the Boltzmann equation in yet another form.

Suppose that $F\in C\big([0,T];\ L^1(\Re^d\times\Re^d_{\text{loc}})\big)$, $TF\geq 0$. The operator $T_F^{-1}:=e^{-F}T^{-1}e^F$ is then continuous (and weakly continuous) from $L^1((0,T)\times\Re^d\times\Re^d_{\text{loc}})$ into $C([0,T];\ L^1(\Re^d\times\Re^d_{\text{loc}}))$.

If $\{F_n\}$ is a bounded sequence in $C\big([0,T];L^1(\Re^d\times\Re^d_{\text{loc}})\big)$ such that $TF_n\geq 0$, $F_n(x,\xi,t)\to F(x,\xi,t)$ for all t and almost all (x,ξ), and if $g_n\underset{w}{\longrightarrow}g$ in $L^1\big((0,T)\times\Re^d\times\Re^d_{\text{loc}}\big)$, then, for all $t\in[0,T]$,

$$T_{F_n}^{-1}g_n(t)\overset{w}{\longrightarrow}T_F^{-1}g(t) \tag{3.35}$$

in $L^1(\Re^d\times\Re^d_{\text{loc}})$. (This is easily proved by using the explicit solution formula for T^{-1}.)

To use (3.35), let $F_n=T^{-1}(A_n*f^n)$, where f^n, A_n are from the modified Boltzmann equation (see Step 7). This equation can be written as

$$Tf^n+(A_n*f^n)f^n=Q_+^n(f^n,f^n)$$

or (after multiplication with e^{F_n} and observing that

$$T(f^ne^{F_n})=(Tf^n)e^{F_n}+f^n(TF_n)e^{F_n}=e^{F_n}Q_+^n(f^n,f^n),$$

$$f^n=f_0^ne^{-F_n}+T_{F_n}^{-1}Q_+^n(f^n,f^n). \tag{3.36}$$

By Lemma 5.3.11 ii) and the preceding remarks, $\{F_n\}$ is a bounded sequence in

$$C\big((0,T);L^1(\Re^d)\times\Re^d_{\text{loc}}\big),$$

and for all $t\in\Re_+$

$$F_n\to F=T^{-1}(A*f)\qquad\text{a.e.}$$

(5.3.12) Lemma. *For all $t\in\Re_+$, we have $T_F^{-1}Q_+(f,f)\in L^1(\Re^d\times\Re^d_{\text{loc}})$ and*

$$f=f_0e^{-F}+T_F^{-1}Q_+(f,f). \tag{3.37}$$

Proof. Now let $\beta_m(t)=\min(t,m)$, where $t\geq 0$ and $m\in\mathcal{N}$. Without restricting the generality, we can assume (by considering subsequences if necessary) that

$$g_m^n:=\beta_m\circ f^n\underset{(n\to\infty)}{\overset{}{\underset{w}{\longrightarrow}}}g_m$$

in $L^1\big((0,T)\times\Re^d\times\Re^d\big)$ for all $T>0$, and, by (3.34)

$$g_m \underset{(m\to\infty)}{\longrightarrow} f$$

strongly and monotone increasing in $L^1((0,T)\times\Re^d\times\Re^d)$. For every fixed m, $\{g_m^n\}$ satisfies the hypotheses of Lemma 5.3.9 ($Tg_m^n = Q_n(f^n, f^n)$ if $g_m^n < m$, $= 0$ otherwise) and $|g_m^n| \le m$ for all n. Therefore, and by arguments similar to those used in Step 12, $Q_+^n(g_m^n, g_m^n) \underset{w}{\longrightarrow} Q_+(g_m, g_m)$ as $n \to \infty$ in $L^1((0,T)\times\Re^d\times B_R)$, for each $R > 0$.

Now observe that from (3.36)

$$f^n \ge f_0^n e^{-F_n} + T_{F_n}^{-1} Q_+^n(g_m^n, g_m^n). \tag{3.38}$$

By taking here the weak limit on both sides [observe (3.35)], we see that

$$f \ge f_0 e^{-F} + T_F^{-1} Q_+(g_m, g_m). \tag{3.39}$$

Finally, we use the monotone convergence theorem, the fact that $g_m \nearrow f$ in L^1 and (3.39) to conclude that $T_F^{-1}Q_+(f,f) \in L^1((0,T)\times\Re^d\times\Re^d)$ and

$$f \ge f_0 e^{-F} + T_F^{-1} Q_+(f, f). \tag{3.40}$$

For the reverse inequality, we consider now the functions

$$h_m^n := m \ln(1 + f^n/m).$$

The h_m^n satisfy, as one readily checks,

$$\begin{aligned} h_m^n &= m\ln(1+f^n/m)\, e^{-F_n} \\ &+ T_{F_n}^{-1}\left(\frac{Q_+^n(f^n, f^n)}{1+f^n/m}\right) + T_{F_n}^{-1}\left(A_n * f^n \left(h_m^n - \frac{f^n}{1+f^n/m}\right)\right). \end{aligned} \tag{3.41}$$

As before, by extracting subsequences if necessary, we can assume that

$$h_m^n \underset{\substack{n\to\infty \\ w}}{\longrightarrow} h_m, \qquad h_m \underset{m\to\infty}{\longrightarrow} f \text{ in } L^1$$

$$\frac{f^n}{1+f^n/m} \underset{\substack{n\to\infty \\ w}}{\longrightarrow} l_m, \qquad l_m \underset{m\to\infty}{\longrightarrow} f \text{ in } L^1$$

and

$$\frac{Q_+^n(f^n, f^n)}{1+f^n/m} \underset{\substack{n\to\infty \\ w}}{\longrightarrow} Q_{+,m}. \tag{3.42}$$

Now recall Lemma 5.3.11 part iii)

$$\frac{\int Q_+^n(f^n, f^n)\varphi\, d\xi}{1+\int f^n\, d\xi} \longrightarrow \frac{\int Q_+(f,f)\varphi\, d\xi}{1+\int f\, d\xi} \tag{3.43}$$

for each compactly supported φ in L^∞. Multiply the left-hand side of (3.42) by $\frac{\varphi}{1+\int f^n\, d\xi}$ ($\varphi \ge 0$) and integrate. By (3.42), the limit as $n \to \infty$ is

$\frac{\int Q_{+,m}\varphi\,d\xi}{1+\int f\,d\xi}$. By (3.43), this limit is less than $\frac{\int Q_{+}(f,f)\varphi\,d\xi}{1+\int f\,d\xi}$, and it follows that

$$Q_{+,m} \leq Q_{+}(f,f) \qquad \text{a.e.}$$

Taking now the weak limit as $n \to \infty$ in (3.41), we find

$$h_m \leq m \ln\left(1+\frac{f_0}{m}\right) e^{-F} + T_F^{-1} Q_{+}(f,f) + T_F^{-1}(A * f(h_m - l_m)).$$

As $m \to \infty$,

$$f \leq f_0 e^{-F} + T_F^{-1} Q_{+}(f,f),$$

and this and (3.40) complete the assertion of the lemma:

$$f = f_0 e^{-F} + T_F^{-1} Q_{+}(f,f). \tag{3.44}$$

□

Step 14 ***f* is a mild solution.** Eq. (3.44) is already saying that f satisfies the Boltzmann equation in some sense. We will now simply check that it satisfies the criteria for a renormalized solution (as given in step 4).

First, it is easy to show that for every $T < \infty$

$$\frac{Q_{-}(f,f)}{1+f} \in L^1([0,T] \times \Re^d \times \Re^d_{\text{loc}}) \tag{3.45}$$

(just use the condition on A and that

$$\sup_{t\in[0,T]} \sup_n \iint f^n(1+|x|^2+|\xi|^2)\,dx\,d\xi < \infty).$$

As for $\frac{Q_{+}(f,f)}{1+f}$, recall that [see (3.27)]

$$\begin{aligned} Q^n_{\pm}(f^n,f^n)\left(1+\delta\int f^n\,d\xi\right)^{-1} \\ \leq 2Q^n_{\mp}(f^n,f^n)\left(1+\delta\int f^n\,d\xi\right)^{-1} + \frac{4e_n}{\ln 2}\left(1+\delta\int f^n\,d\xi\right)^{-1} \end{aligned} \tag{3.46}$$

and

$$\sup_n \int_0^\infty \iint e_n(f^n)\,dx\,d\xi\,ds < \infty. \tag{3.47}$$

Because of the non-negativity of $e_n(f^n)$ and (3.47), we can assume that $e_n(f^n)$ converges weakly (in $\mathcal{D}'$, or in the vague topology on the bounded measures) to a bounded non-negative measure μ by Lemma 5.3.11; we also know that the other two terms in (3.46) converge weakly in L^1, and so

$$\frac{Q_\pm(f,f)}{1+\delta\int f\,d\xi} \le \frac{2Q_\mp(f,f)}{1+\delta\int f\,d\xi} + \frac{4}{\ln 2(1+\delta\int f\,d\xi)}\mu. \tag{3.48}$$

(3.48) remains true if we replace μ by its absolutely continuous part $e \in L^1\big((0,T)\times\Re^d\times\Re^d\big)$, and by taking $\delta\to 0$, it follows that

$$Q_\pm(f,f) \le 2Q_\mp(f,f) + E \tag{3.49}$$

with $E \in L^1\big((0,T)\times\Re^d\times\Re^d\big)$. (3.45) and (3.49) now entail that

$$\frac{Q_+(f,f)}{1+f} \in L^1\big([0,T]\times\Re^d\times\Re^d_{\text{loc}}\big).$$

To show that $Q_+(f,f)(x,\,\xi,\,\cdot) \in L^1(0,T)$ for almost all (x,ξ), we use that by Lemma 5.3.12 for all t, $T_F^{-1}Q_+(f,f) \in L^1(\Re^d\times\Re^d_{\text{loc}})$ and $F(\cdot\,,t) \in L^1(\Re^d\times\Re^d_{\text{loc}})$. Explicitly, we see that

$$\int_0^t Q_+(f,f)^\# \exp -(F^\#(t) - F^\#(s))\,ds$$

is in $L^1(\Re^d\times\Re^d_{\text{loc}})$ for all t, and because $F^\#$ is non-negative, *increasing* with respect to t and in $L^1(\Re^d\times\Re^d_{\text{loc}})$ with respect to (x,ξ), it follows that $Q_+(f,f)^\# \in L^1(0,T)$ for almost all (x,ξ). For Q_-, the same assertion follows from (3.49). Now we can use Lemma 5.3.12 to conclude that f is a mild solution of the Boltzmann equation in the sense of step 4.

The only remaining step is the verification of the entropy estimate (3.9) from (3.32). To this end, note that from the proof of Lemma 5.3.11, for all $\delta > 0$,

$$\frac{f^n f^n_*}{1+\delta\int f^n\,d\xi} \underset{w}{\longrightarrow} \frac{f f_*}{1+\delta\int f\,d\xi},$$

$$\frac{f^{n\prime} f^{n\prime}_*}{1+\delta\int f^n\,d\xi} \underset{w}{\longrightarrow} \frac{f' f'_*}{1+\delta\int f\,d\xi},$$

in $L^1\big((0,T)\times\Re^d_x\times\Re^d_\eta\times S^{d-1}\big)$. Now, by using the convexity of the function

$$(x,y) \to (x-y)\ln\frac{x}{y}$$

on $\Re_+\times\Re_+$, we see that for all $T>0$

$$\int_0^T\!\!\iint \frac{e(f)}{1+\delta\int f\,d\xi}\,dx\,d\xi\,dt \le \liminf_{n\to\infty}\int_0^T\!\!\iint \frac{e_n(f^n)}{1+\delta\int f^n\,d\xi}\,dx\,d\xi\,dt.$$

The entropy estimate (3.9) follows from this and the monotone convergence theorem in the limit $\delta\to 0$.

This completes the proof of Theorem 5.3.5.

Problems

1. Verify (3.11).
2. Check that the collision kernel for hard spheres,

$$q(\xi - \xi_*, n) = |(\xi - \xi_*) \cdot n|,$$

satisfies (3.12).
3. Verify the estimate (3.18).
4. Show that the estimate (3.25) with the specified properties on h and w entails weak sequential compactness of $\{f_n\}$.
5. Verify (3.28) in detail.
6. Prove (3.32).
7. Prove 3.35).
8. Show that (3.34) implies $\lim_{\delta \to 0} \|f - g_\delta\|_{L^1} = 0$.
Hint: Use that

$$\|f - g_\delta\|_{L^1} = \sup_{\varphi, \|\varphi\|_{L^\infty} \leq 1} \int (f - g_\delta) \cdot \varphi,$$

and

$$\sup_\varphi \lim_n \ldots \leq \liminf_n \sup_\varphi \ldots$$

5.4 Generalizations and Other Remarks

The result proved in the previous section lends itself to several generalizations. The simplest considers the case when the space variable x in the Boltzmann equation varies on a torus $\mathbf{T}^d$ rather than in $\Re^d$. Actually this case is even simpler than the one treated before, because we can dispense with the last conservation equation, Eq. (3.6), which had the purpose of controlling the behavior of f at space infinity and is no longer needed; for the same reason the terms contaning $|x|^2$ can be suppressed in Eqs. (3.8) and (3.9) and in the statement of Theorem 5.3.5. Since this generalization is rather obvious, we shall not deal with it in detail here. In Chapter 9 we shall deal with another generalization, i.e., the case of a bounded domain with rather general boundary conditions.

Another important aspect of DiPerna and Lions's result is that, if we take a sequence of times $\{t_n\}$ going to infinity with n then the weak stability properties of their solution imply that $f(\cdot, t + t_n)$ converges to another solution $M(\cdot, t)$ of the Boltzmann equation. Di Perna and Lions[6] proved that M satisfies the equation, discussed in Chapter 3, which is characteristic of a Maxwellian distribution. This Maxwellian might be degenerate, i.e., vanish almost everywhere, and this what will generally occur for L^1 solutions in $\Re^d \times \Re^d$; in the case of $\mathbf{T}^d \times \Re^d$, however, the Maxwellian will not be degenerate (it must have a nonzero norm). Use of the results of Chapter 3 actually shows that the weak limit of $f(\cdot, t + t_n)$ is an absolute Maxwellian. The detailed argument will be given in Chapter 9, Theorem 9.5.1, when we deal with the case of general boundary conditions.

Here we restrict ourselves to remarking that the result is not as strong as we would like. And this not only because the convergence to a Maxwellian is weak. In fact, this is a restriction that was eliminated by Arkeryd [1] by using techniques of nonstandard analysis, and subsequently by P. L. Lions [15], who was able to dispense with the latter tool. Even if we have strong convergence, we cannot identify the specific Maxwellian to which the sequence $f(\cdot, t+t_n)$ tends; thus, in particular, a different sequence of times might give a different Maxwellian. The reason for this is that we cannot prove that energy is conserved for the DiPerna-Lions solution (otherwise M would be identified by the conserved quantities).

References

1. L. Arkeryd, "On the strong L^1 trend to equilibrium for the Boltzmann equation," *Studies in Appl. Math.* **87**, 283–288 (1992).
2. H. Babovsky, "Initial and boundary value problems in kinetic theory. II: The Boltzmann equation," *Transport Theory Stat. Phys.* **13 (3&4)**, 475–498 (1984).
3. N. Bellomo and G. Toscani, "On the Cauchy problem for the nonlinear Boltzmann equation. Global existence, uniqueness and asymptotic stability," *J. Math. Phys.* **26**, 334–338 (1985).
4. C. Cercignani, "Existence of homoenergetic affine flows for the Boltzmann equation," *Arch. Rat. Mech. Anal.* **105**, 377–387 (1989).
5. B. Dacorogna, "Weak continuity and weak lower semicontinuity of non-linear functionals," *Lecture Notes in Mathematics* **922**, 1–117, Springer-Verlag (1985).
6. R. DiPerna and P. L. Lions, "Global solutions of Boltzmann's equation and the entropy inequality," *Arch. Rat. Mech. Anal.* **114**, 47–55 (1991).
7. R. J. DiPerna and P. L. Lions, "On the Cauchy problem for Boltzmann equations: Global existence and weak stability," *Annals of Mathematics* (1989).
8. N. Dunford and J. T. Schwartz, *Linear operators, I.* Interscience (1963).
9. P. Gérard, "Solutions globales du problème de Cauchy pour l'équation de Boltzmann," *Seminaire Bourbaki* **699** (1987–88).
10. F. Golse, B. Perthame, P. L. Lions and R. Sentis, "Regularity of the moments of the solution of a transport equation," *J. Funct. Anal.* **76**, 110–125 (1988).
11. K. Hamdache, "Quelques résultats pour l'équation e Boltzmann," *C. R. Acad. Sci. Paris* **299**, Série I, 431–434 (1984).
12. R. Illner and M. Shinbrot, "Blow-up of solutions of the gain-term only Boltzmann equation," *Math. Meth. in the Appl. Sci.* **9**, 251–259 (1987).
13. R. Illner and M. Shinbrot, "The Boltzmann equation: Global Existence for a rare gas in an infinite vacuum," *Commun. Math. Phys.* **95**, 217–226 (1984).
14. S. Kaniel and M. Shinbrot, "The Boltzmann equation. I. Uniqueness and local existence," *Commun. Math. Phys.* **59**, 65–84 (1978).
15. P. L. Lions, "Compactness in Boltzmann's equation via Fourier integral operators and applications. I," *Cahiers de Mathématiques de la décision* no. 9301, CEREMADE (1993).

16. J. Polewczak, "Classical solution of the Boltzmann equation in all $\Re^3$: Asymptotic behavior of solutions," *Journal Stat. Phys.* **50 (3&4)**, 611–632 (1988).
17. G. Toscani, "Global solution of the initial value problem for the Boltzmann equation near a local Maxwellian," *Arch. Rat. Mech. Anal.* **102**, 231–241 (1988).

Appendix 5.A

Proof

(Of Lemma 5.3.4.) We need an elementary theorem on linear transport theory.

(5.A.1) Theorem. *Let f, $h \in L^1_{\text{loc}}(\Re \times \Re^d \times \Re^d)$ and suppose that*

$$Tf = h \quad \text{in} \quad \mathcal{D}'(\Re \times \Re^d \times \Re^d). \tag{A.1}$$

Then $f^{\#}(x, \xi, \cdot)$ is for almost all x, ξ absolutely continuous with respect to t,
$h^{\#}(x, \xi, \cdot) \in L^1_{\text{loc}}(\Re)$ and

$$f^{\#}(t_2) - f^{\#}(t_1) = \int_{t_1}^{t_2} h^{\#}(s)\, ds \tag{A.2}$$

for all t_1, $t_2 \in \Re$. If, conversely, f and h are such that for almost all x, ξ $f^{\#}$ is absolutely continuous with respect to t and $h^{\#} \in L^1_{\text{loc}}(\Re)$, then (A.2) entails (A.1).

Proof. Let $\psi \in \mathcal{D}(\Re^d \times \Re^d)$, $\rho \in \mathcal{D}(\Re)$, and multiply (A.2) by $\psi(x - t\xi,\, \xi) \cdot \rho(t)$ to get

$$-\int dt \iint dx\, d\xi \psi(x - t\xi,\, \xi)\rho'(t) f = \int dt \iint dx\, d\xi \psi(x - t\xi,\, \xi)\rho(t) h.$$

The change of variables $(t,\, x,\, \xi) \to (t,\, x + t\xi,\, \xi)$ leads to

$$\iint dx\, d\xi \psi(x,\xi)\left\{\int dt(\rho'(t)f^{\#}+\rho(t)h^{\#})\right\}=0,$$

$$\text{i.e.}\quad \int dt(\rho'(t)f^{\#}+\rho(t)h^{\#})=0$$

for each $\rho \in C^1(\Re)$ with compact support. To complete the proof use a mollifier to approximate the characteristic function of $[t_1, t_2]$ by such ρ. The converse is left to the reader. □

Proof. (Of Lemma 5.3.4.)

i) We assume that $f \in L^1_+(\Re^+_{\text{loc}} \times \Re^d \times \Re^d)$ satisfies

$$\frac{Q\pm(f,f)}{1+f} \in L^1_{\text{loc}}(\Re^+ \times \Re^d \times \Re^d)$$

and

$$T(\ln(1+f)) = \frac{Q(f,f)}{1+f}$$

in the sense of distributions.

Let $\beta_\delta(y) = \frac{1}{\delta}\ln(1+\delta(e^y-1))$ $(y>0)$, then

$$\beta'_\delta(y) = \frac{e^y}{1+\delta(e^y-1)}.$$

By assumption, $T(\ln(1+f))$ and $\ln(1+f)$ are in $L^1_{\text{loc}}(\Re^+ \times \Re^d \times \Re^d)$. Then, by using the theorem and the fact that β_δ is Lipschitz continuous, it follows that $T\beta_\delta(\ln(1+f)) = \beta'_\delta(\ln(1+f))\cdot T(\ln(1+f))$. But $\beta'_\delta(\ln(1+f)) = \frac{1+f}{1+\delta f}$ by direct inspection, so that

$$T\beta_\sigma(\ln(1+f)) = \frac{Q(f,f)}{1+\delta f}$$

in the sense of distributions. As $\beta_\delta(\ln(1+f)) = \frac{1}{\delta}\ln(1+\delta f)$,

$$T\frac{1}{\delta}\ln(1+\delta f) = \frac{Q(f,f)}{1+\delta f} \tag{A.3}$$

in the sense of distributions for all $\delta > 0$.

Let $g_\delta := \frac{1}{\delta}\ln(1+\delta f)$. By (A.3) and the theorem, $g^{\#}_\delta$ is, for almost all $x,\xi \in \Re^N$, and for all $\delta > 0$, absolutely continuous with respect to t, and $\frac{Q^\pm(f,f)^{\#}}{1+\delta f^{\#}} \in L^1_{\text{loc}}[0,\infty)$. Since $f^{\#} = e^{g^{\#}_1} - 1$, $f^{\#}$ is also absolutely continuous with respect to t for almost all x, ξ, and consequently $Q^\pm(f,f)^{\#} \in L^1_{\text{loc}}[0,\infty)$ for almost all x,ξ. Finally, by the theorem for all $t > \delta \geq 0$

$$g^{\#}_\delta(t) - g^{\#}_\delta(s) = \int_s^t \frac{1}{1+\delta f^{\#}} Q(f,f)^{\#}\, d\sigma, \qquad \text{a.e. } x,\xi$$

and as $\delta \to 0$, this becomes

$$f^{\#}(x, \xi, t) - f^{\#}(x, \xi, s) = \int_s^t Q(f, f)^{\#}(x, \xi, \sigma)\, d\sigma.$$

ii) If f is a mild solution then $f^{\#}$ is absolutely continuous with respect to t for almost all $x, \xi \in \Re^d$ and so is $\ln(1 + f^{\#})$. Clearly, $g = \ln(1 + f)$ satisfies $g^{\#}(t) - g^{\#}(s) = \int_s^t \frac{1}{1+f^{\#}} Q(f, f)^{\#}\, d\sigma$. The rest follows from Theorem 5.A.1. □

6

The Initial Value Problem for the Homogeneous Boltzmann Equation

6.1 An Existence Theorem for a Modified Equation

In this chapter we treat the spatially homogeneous Boltzmann equation, i.e., the special case where f does not depend on x. In this case the main difficulty in estimating the collision operator, namely, the pointwise interaction, disappears, and we can develop a rather complete and satisfactory theory. The remaining difficulties are due to large velocities (high energy tails).

We shall use some of the classical arguments already employed in Chapter 5 Section 3. We repeat these arguments here in some detail, in order to make this chapter, which is conceptually easier than the general theory from Section 5.3, as self-contained as possible.

The initial value problem associated with the homogeneous equation is

$$\begin{aligned} \partial_t f &= Q(f,f) \\ f(\cdot,0) &= f_0 \end{aligned} \tag{1.1}$$

where Q is given by (3.1.2). To simplify notation, we shall sometimes just write $Q(f)$ for $Q(f,f)$.

We first show that if we neglect large relative velocities in the collision operator, the initial value problem is easily solvable.

Let us define the symmetrized collision operator with cutoff by:

$$Q^M(f,g) = \frac{1}{2} \int d\xi_1 \int_{n\cdot(\xi-\xi_*)\geq 0} dn\; n\cdot(\xi-\xi_*)\; \chi_M(|\xi-\xi_*|) \quad \{f'g'_* + g'f'_* - fg_* - f_*g\} \tag{1.2}$$

where $\chi_M : \Re^+ \to \Re$ is defined by:

$$\begin{aligned} \chi_M(r) &= 1 \text{ if } r \leq M \\ \chi_M(r) &= 0 \text{ otherwise.} \end{aligned} \tag{1.3}$$

Consider the initial value problem:

$$\begin{aligned} \partial_t f^M &= Q^M(f^M, f^M) \\ f^M(\cdot, 0) &= f_0. \end{aligned} \tag{1.4}$$

The physical meaning of Problem (1.4) is that the molecules of the gas behave like in the Boltzmann dynamics, but if two molecules collide with the modulus of the relative velocity larger than M, they ignore each other. It is easy to verify, at least at a formal level, that the conservation laws and the H-theorem established in Chapter 3 are still valid.

By an obvious application of the measure-preserving property of the collision transformation, we have:

$$\|Q^M(f,f)\|_{L^1} \leq CM\, \|f\|^2_{L^1} \tag{1.5}$$

$$\begin{aligned} &\|Q^M(f,f) - Q^M(g,g)\|_{L^1} \\ =&\|Q^M(f+g, f-g)\|_{L^1} \\ \leq& CM\, \|f+g\|_{L^1}\|f-g\|_{L^1}. \end{aligned} \tag{1.6}$$

C, here and in the sequel, stands for a numerical constant.

The inequalities (1.5-6) allow us to construct a local solution to Problem (1.4): $f^M \in C^1([0,T]; L^1)$ for a sufficiently small time $T > 0$, by means of the standard iteration scheme.

Assuming the initial datum positive and normalized, i.e.:

$$\int d\xi\; f(\xi) = 1 \tag{1.7}$$

by the identity (3.1.15) (mass conservation), we conclude that:

$$\int d\xi\; f^M(\xi, t) = 1 \text{ for all } t \in [0, T]. \tag{1.8}$$

If we could prove that the solution we have found is non-negative we would be in a position to extend the solution to arbitrary times. In fact, by positivity and (1.8) it follows that the L^1- norm of the solution is preserved in time and, since T depends only on the initial L^1-norm of the solution we are allowed to extend the procedure up to time $2T$, and so on. However, the non-negativity of the solution is not obvious, although it is suggested by physical considerations.

We use the following trick to prove non-negativity. Consider the initial value problem

$$\partial_t g + \mu g = \Gamma^M(g) \qquad (1.9)$$
$$g(\cdot, 0) = f_0$$

where:

$$\Gamma^M(g) = Q^M(g) + \mu g \int d\xi \; g(\xi) \qquad (1.10)$$

and $\mu > 0$ will be chosen (sufficiently large) later. Notice that we have already found a local solution f^M to Problem (1.9) because of (1.8). Therefore, if we prove the existence of a positive solution to Problem (1.9) we have also proved that f^M is positive, because Γ^M is Lipschitz continuous, and hence the initial value problem (1.9) has a unique solution. Thus, we focus on proving the existence of a non-negative solution to Problem (1.9).

To this end, let:

$$g^n = e^{-\mu t} f_0 + \int_0^t ds \; e^{-\mu(t-s)} \, \Gamma^M(g^{n-1}) \qquad (1.11)$$
$$g^0 = 0.$$

We observe that for sufficiently large μ, Γ^M is a positive monotone operator in the sense that

$$\Gamma^M(f) \geq \Gamma^M(g) \text{ if } f \geq g \geq 0. \qquad (1.12)$$

In fact, denoting by Q_+^M and Q_-^M the gain and loss part of the collision operator Q^M,

(1.13a)
$$Q_+^M(f,g) = \tfrac{1}{2} \int d\xi_* \int_{n\cdot(\xi-\xi_*)\geq 0} n \cdot (\xi - \xi_*) \geq 0 \; \chi_M(|\xi - \xi_*|)\{f' g'_* + g' f'_*\}$$

(1.13b)
$$Q_-^M(f) = \int d\xi_* \int_{n\cdot(\xi-\xi_*)\geq 0} dn \; n \cdot (\xi - \xi_*) \geq 0 \; \chi_M(|\xi - \xi_*|) \; f(\xi_*) f(\xi)$$
$$=_{\text{def}} R^M(f) f,$$

we have, by using symmetry properties:

$$\begin{aligned} \Gamma^M(f) - \Gamma^M(g) = {} & Q_+^M(f+g, f-g) \\ & - \tfrac{1}{2}\{(f+g)R^M(f-g) + (f-g)R^M(f+g)\} \\ & + \tfrac{1}{2}\mu\{(f+g)\textstyle\int(f-g) + (f-g)\int(f+g)\}. \end{aligned} \qquad (1.14)$$

Hence the positivity property follows by the obvious inequality

$$R^M(f) \le CM \int d\xi \; f(\xi),$$

which is valid for positive f. Consequently $g^n \ge g^{n-1} \cdots$ is a monotonically increasing sequence.

Moreover:

$$\int g^n = e^{-\mu t} + \mu \int_0^t ds \, e^{-\mu(t-s)} \left(\int g^{n-1} \right)^2, \tag{1.15}$$

which implies that if $\int g^{n-1} \le 1$ then $\int g^n \le 1$. By Beppo Levi's theorem, the limit

$$g = \lim_{n \to \infty} g^n \tag{1.16}$$

exists, is non-negative, and satisfies:

$$\int g \le 1. \tag{1.17}$$

It is now straightforward to prove that g solves the initial value problem (1.9) (see Problem 2) so that $g = f^M$. In particular the equality sign holds in (1.17). Thus we have proven the first part of the following theorem.

(6.1.1) Theorem. *There exists a unique positive solution $f^M \in C^1([0,T];L^1)$ to the initial value problem (1.4) for arbitrary times $T \ge 0$, provided that $f_0 \ge 0$ and $\int f_0 = 1$.*

Suppose in addition that $E(f_0) = \frac{1}{2}\int \xi^2 f_0(\xi)$ and $H(f_0) = \int f_0 \ln f_0$ (energy and entropy) are initially finite. Then

$$E(f_0) = E\left(f^M(t)\right) \tag{1.18}$$

$$H\left(f^M(t)\right) \le H(f_0). \tag{1.19}$$

To complete the proof of the theorem, we only need to prove Identity (1.18) and Inequality (1.19). The energy conservation follows easily by general arguments [see (3.1.15)]. The proof of (1.19) is a little technical and requires some knowledge of the L^∞- theory presented later. We outline it in Appendix 6.A.

Problems

1. Consider the initial value problem (1.4) for an arbitary $f_0 \in L^1$ (not necessarily non-negative) and give an estimate of the time T of existence of the local solution.
2. Prove that g defined by (1.16) satisfies the initial value problem (1.9).

6.2 Removing the Cutoff: The L^1-Theory for the Full Equation

We now study the behavior of the solutions in the limit $M \to \infty$. Our target is to obtain a solution to the original initial value problem (1.1). The shortest way to obtain a limit for the squence f^M is to use the Dunford-Pettis compactness criterion which we already used in Chapter 5 (see step 8 of Section 5.3). For the reader's convenience, we now adapt some of the estimates from Chapter 5 to the current situation. By energy conservation:

$$\int_{|\xi|>R} f^M(\xi,t)\,d\xi \le \frac{1}{R^2}\int \xi^2 f^M(\xi,t)\,d\xi \le \frac{2E}{R^2} \tag{2.1}$$

where $E = E\,(f_0) = \frac{1}{2}\int \xi^2\, f_0(\xi)$ is the conserved energy. Moreover, if A is a set such that $\int_A d\xi \le \epsilon$, we have from the H-theorem and the energy inequality, for $a > 1$:
(2.2)

$$\begin{aligned}
&\int_A f^M(\xi,t)d\xi \\
&\le \frac{1}{\ln a}\int_A f^M(\xi,t)\ln f^M(\xi,t)d\xi\ \chi_{\{f^M(t)>a\}} + a\epsilon \\
&\le \frac{1}{\ln a}\int f^M(\xi,t)\ln f^M(\xi,t)\ \chi_{\{f^M(t)>1\}} + a\epsilon \\
&\le \frac{H(f_0)}{\ln a} + a\epsilon - \frac{1}{\ln a}\int f^M(\xi,t)\ln f^M(\xi,t)\,d\xi\ \chi_{\{f^M(t)<1\}} \\
&\le \frac{H(f_0)}{\ln a} + a\epsilon - \frac{1}{\ln a}\int f^M(\xi,t)\ln f^M(\xi,t)\,d\xi\ \chi_{\{\exp(-\xi^2)<f^M(t)<1\}} \\
&\qquad - \frac{1}{\ln a}\int f^M(\xi,t)\ln f^M(\xi,t)\,d\xi\ \chi_{\{f^M(t)\le\exp(-\xi^2)\}} \\
&\le \frac{H(f_0)}{\ln a} + a\epsilon + \frac{1}{\ln a}\left[\int f^M(\xi,t)\xi^2\,d\xi + C\int \exp-(\xi^2/2)\,d\xi)\right] \\
&\quad \text{(because } x^{\frac{1}{2}}|\ln x| \text{ is bounded if } x<1, \\
&\quad \text{and after choosing } \ a=\epsilon^{-\frac{1}{2}}) \\
&\le -\frac{2H(f_0)}{\ln\epsilon} + \sqrt{\epsilon} - C(1+E)/\ln\epsilon.
\end{aligned}$$

Thus we can apply the Dunford-Pettis theorem to the set $\{f^M\}$ in $L^1([0,T)\times\Re^3)$ to extract a weakly convergent subsequence with limit $f \in L^1([0,T]\times\Re^3)$. Obviously $f \ge 0$. Moreover, f satisfies the energy inequality:

$$\int \xi^2 f(\xi,t)\,d\xi \le \int \xi^2 f_0(\xi)d\xi \text{ for a.a. } t\in[0,T]. \tag{2.3}$$

In fact, by definition of weak convergence:

$$\int_0^T \phi(t) \int \chi_H(\xi)\, \xi^2 f^n(\xi)\, d\xi\, dt \to \int_0^T \phi(t) \int \chi_H(\xi)\, \xi^2 f(\xi)\, d\xi\ dt \tag{2.4}$$

where:

$$\chi_H(r) = \begin{cases} r \text{ if } |r| \le H \\ 0 \text{ otherwise} \end{cases} \tag{2.5}$$

and $\phi \in C([0,T])$.

Therefore, for positive ϕ :

$$\frac{1}{2} \int_0^T \phi(t) \int \xi^2\ f(\xi)\, d\xi\ dt \le \int_0^T \phi(t)\, E, \tag{2.6}$$

implying (2.3).

To prove that f is a solution of the Boltzmann equation we have to prove the convergence of the collision operators:

$$Q^n(f^n, f^n) \equiv Q^{M_n}(f^n, f^n) \to Q(f,f).$$

Here M_n denotes the sequence $M_n \to \infty$ for which we have the weak convergence of the sequence f^n.

Before approaching this problem, we need to establish an inequality that will turn out to be very useful in the sequel. We introduce the following family of norms:

$$\|f\|_{1,s} = \int (1+\xi^2)^{s/2} |f(\xi)|\, d\xi \tag{2.7}$$

and the associated family of Banach spaces

$$L^1_s = \{f; \|f\|_{1,s} < \infty\}. \tag{2.8}$$

Then we have the following lemma.

(6.2.1) Lemma. (Povzner[7]) *Suppose that* $s \ge 2$, $f, g \in L^1_s$, *and* $f \ge 0$, $g \ge 0$. *Then the following inequality is true:*

$$\int (1+\xi^2)^{s/2} Q(f,g)|d\xi \le C(s)\{\|f\|_{1,s}\, \|g\|_{1,2} + \|g\|_{1,s}\, \|f\|_{1,2}\}. \tag{2.9}$$

For the proof see Appendix 6.B.

Remark. We can replace Q by Q^M in (2.9), with $C(s)$ independent of M.

The Povzner inequality generalizes, in a sense, the conservation of the energy, taking into account compensation between the gain and the loss terms. If one wants to bound the gain and loss terms separately in the s-norm (2.7), one would need to introduce an $s+1-$ norm of f. However, the positivity of the solution and Inequality (2.9) together allow us to control the solution in the $s-$ norm. In fact, by the use of (2.9) we easily obtain:

$$\|f^M(t)\|_{1,s} \leq \|f(0)\|_{1,s} + C(s)\int_0^t d\tau \|f^M(\tau)\|_{1,s}\|f^M(\tau)\|_{1,2}. \tag{2.10}$$

Thus, by the energy bound (2.3) and Gronwall's inequality,

$$\|f^M(t)\|_{1,s} \leq \|f(0)\|_{1,s} \exp\,(C(s)\,E\,t). \tag{2.11}$$

As we shall see in the sequel, the bound (2.11) may be considerably improved by a time-independent estimate.

We return to the problem of the convergence of the collision operator. By the trivial inequality

$$R^M(f)(\xi) \leq C\int |\xi-\xi_*|f(\xi_*)d\xi_* \leq C(|\xi|\,\|f\|_1 + \|f\|_{1,2}) \tag{2.12}$$

(here, and from now on in this chapter, we abbreviate $\|f\|_1 = \|f\|_{L^1} = \|f\|_{1,0}$) we obtain, by using the Lebesgue measure invariance of the collision transformation:

$$\begin{aligned}
&\|Q_+^M(f)\|_{1,2}\\
=&\|f\,R^M(f)\|_{1,2}\\
\leq& C\int (1+\xi^2)\, f(\xi)(|\xi|\,\|f\|_1 + \|f\|_{1,2})d\xi\\
\leq& C\{\|f\|_{1,4}\,\|f\|_1 + \|f\|_{1,2}^2\}.
\end{aligned} \tag{2.13}$$

Therefore, by (2.13) and (2.11), provided that $f_0 \in L^1_4$,

$$\|f^M(t) - f^M(r)\|_{1,2} \leq \int_r^t d\tau\,\|Q^M\,(f^M(\tau))\,\|_{1,2} \leq C(T)\,|t-r|. \tag{2.14}$$

From a standard equicontinuity argument it follows that

$$f \in C([0,T]; L^1_2)$$

and that we can extract a subsequence, again denoted by $f^n(t)$, which converges weakly in $L^1(\Re^3)$ to $f(t)$, for all $t \in [0,T]$. Therefore:

$$\int d\xi\, d\xi_*\, \phi(\xi,\xi_*)\, f^n(\xi)\, f^n(\xi_*) \to \int d\xi\, d\xi_*\, \phi(\xi,\xi_*)\, f(\xi)\, f(\xi_*) \tag{2.15}$$

for $t \in [0,T]$, $\phi \in L_\infty(\Re^3 \times \Re^3)$.

Finally, for $\phi \in L_\infty(\Re^3)$:

$$
\begin{aligned}
&\left| \int [Q_+^n(f^n) - Q_+(f)](\xi)\phi(\xi)\, d\xi \right| \\
&\leq \left| \int d\xi d\xi_* \int_{n\cdot(\xi-\xi_*)\geq 0} dn\ n \cdot (\xi - \xi_*) \right. \\
&\quad \left. \times\ \chi_A(|\xi - \xi_*|)\phi(\xi')\left\{f^n(\xi)f^n(\xi_*) - f(\xi)f(\xi_*)\right\} \right| \\
&\quad + C\|\phi\|_\infty \left(\left| \int_{|\xi|>A/2} d\xi\ d\xi_*(|\xi| + |\xi_*|)\ f^n(\xi)f^n(\xi_*) \right| \right. \\
&\quad \left. + \left| \int_{|\xi|>A/2} d\xi\ d\xi_*(|\xi| + |\xi_*|)\ f(\xi)f(\xi_*) \right| \right)
\end{aligned}
\tag{2.16}
$$

where A is fixed. The first term on the right-hand side of (2.16) vanishes as $n \to \infty$. The last two terms can be made arbitrarily small, as A is large, because of the energy inequality and mass conservation.

The weak convergence of the loss term follows the same lines. This is enough to conclude that f solves the initial value problem for the Boltzmann equation in integral form. We finally remark that, by virtue of Estimate (2.13), we can integrate this integral equation against ξ^2 to obtain the energy conservation. Summarizing, we have the following.

(6.2.2) Lemma. *Let $f_0 \geq 0$ be an initial value with finite entropy and such that $f_0 \in L^1_4$. Then, there exists $f \in C([0,T]; L^1_2)$ satisfying:*

$$
f(t) = f_0 + \int_0^t ds\ Q\left(f(s), f(s)\right). \tag{2.17}
$$

In addition $f(t) \in L^1_4$ and estimate (2.11) holds. Finally $\|f(t)\|_1 = 1$, and the energy is constant in time.

The reader may feel unsatisfied by a nonconstructive argument in finding a possibly nonunique solution, a priori. We shall discuss other more constructive approaches later. For now we shall prove a uniqueness theorem and some regularity properties.

(6.2.3) Theorem. *Let $f_0 \geq 0$ be an initial datum with finite entropy such that $f_0 \in L^1_4$. Then there exists a unique $f \in C^1([0,T]; L^1)$ satisfying:*

$$
\begin{aligned}
\partial_t f &= Q(f, f) \\
f(\cdot, 0) &= f_0
\end{aligned}
\tag{2.18}
$$

(here the time derivative is understood in the L^1-sense).

Moreover $f(t) \in L^1_4$, Estimate (2.13), and the H-theorem in the form

$$
H\left(f(t)\right) \leq H(f_0) \tag{2.19}
$$

hold.

Proof. The following inequalities follow from trivial computations:

$$\|Q_+(f,g)\|_{1,s} \le C(s)\{\|f\|_{1,s+1}\|g\|_1 + \|g\|_{1,s+1}\|f\|_1\} \tag{2.20}$$

$$\|fR(g) + gR(f)\|_{1,s} \le C(s)\{\|f\|_{1,s+1}\|g\|_1 + \|g\|_{1,s+1}\|f\|_1\}. \tag{2.21}$$

Therefore, by (2.20), (2.21), and (2.11):

$$\|f(t) - f(r)\|_{1,s} \le \int_r^t d\tau \|Q\left(f(\tau), f(\tau)\right)\|_{1,s} \le C(s,T)|t-r| \tag{2.22}$$

provided that $f_0 \in L_{1,s+1}$. Furthermore,

$$\begin{aligned} &\left\| \frac{f(t+h) - f(t)}{h} - Q\left(f(t), f(t)\right) \right\|_{1,2} \\ \le & h^{-1} \int_t^{t+h} d\tau \|Q\left(f(\tau) + f(t), f(\tau) - f(t)\right)\|_{1,2} \\ \le & Ch^{-1} \int_t^{t+h} d\tau \|f(\tau) + f(t)\|_{1,3}\|f(\tau) - f(t)\|_{1,3} \\ \le & Ch. \end{aligned} \tag{2.23}$$

This proves the strong differentiability of the solution in L^1_2. We are now in a position to prove the uniqueness of the solution. To this end, suppose that f and g are two solutions for the same initial value. By the strong differentiability in $L_{1,2}$ we get (see Problem 5):

$$\begin{aligned} &\frac{d}{dt}\|(f(t) - g(t))\|_{1,2} \\ = & \int d\xi (1+\xi^2) \operatorname{sgn}\left(f(t) - g(t)\right) Q(f(t) + g(t)), (f(t) - g(t)). \end{aligned} \tag{2.24}$$

The right-hand side of (2.24) can be expanded in the following way:

$$\begin{aligned} &\int d\xi \int d\xi' \int dn\, n\cdot(\xi - \xi_*)(1+\xi^2) \operatorname{sgn}\left(f-g\right)(\xi) \\ &\quad \{(f+g)(\xi'_*)(f-g)(\xi') + (f+g)(\xi')(f-g)(\xi'_*) \\ &\qquad - (f+g)(\xi_*)(f-g)(\xi) - (f+g)(\xi)(f-g)(\xi_*)\} \\ &\qquad\quad \le \int d\xi \int d\xi' \int dn\, n\cdot(\xi-\xi_*)(1+\xi^2) \\ &\qquad\qquad \{(f+g)(\xi'_*)|f-g|(\xi') + (f+g)(\xi')|f-g|(\xi'_*) \\ &\qquad\qquad\quad - (f+g)(\xi_*)|f-g|(\xi) + (f+g)(\xi)(f-g)(\xi_*)\} \end{aligned} \tag{2.25}$$

(using the identity $\operatorname{sgn} x \cdot x = |x|$).

Since the first and third terms on the right-hand side of (2.24) compensate exactly, we have:

$$\frac{d}{dt}\|f(t) - g(t)\|_{1,2} \leq C\|f(t) - g(t)\|_{1,2}\|f(t) + g(t)\|_{1,3}, \tag{2.26}$$

which implies:

$$\|f(t) - g(t)\|_{1,2} \leq C(T)\|f_0 - g_0\|_{1,2} \tag{2.27}$$

via the Gronwall lemma. Here g_0 denotes the initial datum of $g(t)$.

From (2.27) we deduce the Lipschitz continuity of the solution with respect to the initial datum and the uniqueness of the solution we have found so far.

The entropy inequality follows from the following argument. We established the inequality:

$$H(f^M(t)) \leq H(f_0). \tag{2.28}$$

Since the entropy is a nonlinear functional, we cannot simply obtain the entropy inequality by taking the limit $M \to \infty$, because f^M converges only weakly. However, it is a known fact that if f_n converges weakly to f, and H is a convex functional (as the entropy is), then:

$$H(f) \leq \liminf_{n\to\infty} H(f_n). \tag{2.29}$$

This remark concludes the proof. □

Problems

1. After applying the Dunford–Pettis theorem, prove that $R(f^M)$ converges to $R(f)$ as $M \to \infty$, uniformly in ξ on compact sets.
2. For f_0 and g_0 positive and normalized, consider $f^M(t)$ and $g^M(t)$ as solutions of the initial value problem (1.4) with initial data f_0 and g_0. Prove that
$$\|f^M(t) - g^M(t)\|_1 \leq C(t)\|f_0 - g_0\|_1.$$
3. Prove the lower bound
$$f^M(t) \geq f_0 \exp(-CMt).$$
4. For f_0 such that $\|f_0\|_\infty \leq A$ prove the existence of a small time T such that, for $t < T$:
$$\|f(t)\|_\infty \leq 2A.$$

Hint: Prove the estimate $|Q_+(f)| \leq C(M)\|f\|_\infty$.

5. Prove Formula (2.24). Hint: For any smooth γ prove that
$$\frac{d}{dt}\int d\xi\, \gamma(f(t))(1+\xi^2)^{s/2} = \int d\xi\, \gamma'(f(t))Q(f(t))(1+\xi^2)^{s/2}.$$

Let γ_ϵ be a mollified version of $|\cdot|$ such that $\gamma_\epsilon(x) \to |x|$ as $\epsilon \to 0$. Write the identity in integral form and take the limit $\epsilon \to 0$.
Realize that

$$\int d\xi \ \mathrm{sgn}\ (f(t))Q(f(t))(1+\xi^2)^{s/2}$$

is integrable in t to prove the claim.

6.3 The L^∞-Theory and Classical Solutions

A very natural question arising after constructing a solution for the spatially homogeneous Boltzmann initial value problem is whether the solution preserves some smoothness properties of the initial datum. A preliminary question is to give L^∞-bounds on the solution, assuming that f_0 is essentially bounded. To do this, we need a more appropriate description of the gain part of the collision operator. The main difficulty in handling the gain operator is that it is expressed in terms of the outgoing velocities, which are linearly dependent on the incoming velocities. A more direct representation formula for Q_+ can be given as follows. Let us recall the collision equations:

$$\begin{aligned} \xi' &= \xi - n\cdot(\xi-\xi_*)n \\ \xi'_* &= \xi_* + n\cdot(\xi-\xi_*)n. \end{aligned} \tag{3.1}$$

Consider the sphere of diameter $|\xi - \xi_*|$ centered at $\xi_0 = \frac{\xi+\xi_*}{2}$ and denote it by $k(\xi,\xi_*)$ (see Fig. 26).

The intersection point of the sphere k with the line passing through ξ in the direction of n represents ξ'. Hence, by direct inspection:

$$(\xi'-\xi)\cdot(\xi'_*-\xi) = 0. \tag{3.2}$$

Thus, for ξ and ξ' fixed, ξ'_* lies in the plane $E_{\xi\xi'}$ orthogonal to $\xi'-\xi$. For fixed ξ, we can treat ξ' and ξ'_* as independent variables in the definition of Q_+, and write:

$$Q_+(f)(\xi) = \int d\xi' \frac{f(\xi')}{|\xi'-\xi|} \int_{E_{\xi\xi'}} d\xi'_*\ f(\xi'_*) \tag{3.3}$$

where the integration on $d\xi'_*$ is restricted to the plane $E_{\xi\xi'}$. This formula can be derived either by a direct computation or as a consequence of the representation formula,

$$\int d\xi\ \phi(\xi)Q_+(f)(\xi) = \int d\xi \int d\xi_* \frac{f(\xi)f(\xi_*)}{|\xi-\xi_*|} \int_{k(\xi,\xi_*)} \phi(\xi')\,d\sigma(\xi'), \tag{3.4}$$

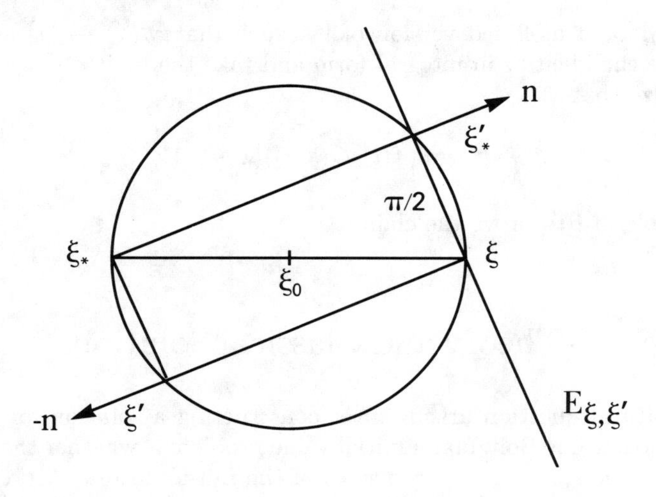

FIGURE 26.

which is valid for any continuous ϕ. Here, $d\sigma$ denotes the surface element in $k(\xi, \xi_*)$. We suggest the proof of (3.4) as an exercise (Problem 1). However, both formulas are relevant for proving the desired L^∞-bounds on the solution and will be proved in Appendix 6.C.

We set:

$$I(E) = \int_E d\xi \; f(\xi) \tag{3.5}$$

where E is a given plane and

$$I(\xi) = \int d\xi_* \frac{f(\xi_*)}{|\xi - \xi_*|}. \tag{3.6}$$

It is clear that we can control the time derivatives of $I(E)$ and $I(\xi)$, respectively, by means of

$$\int_E d\xi \; Q_+(f)(\xi) \tag{3.7}$$

for a given plane E and

$$\int d\xi_* \frac{Q_+(f)(\xi_*)}{|\xi - \xi_*|} \tag{3.8}$$

for $I(\xi)$. Denoting by $d = d(\xi, E)$ the distance between the plane E and ξ, the expression in (3.7) can be written as

$$
\begin{aligned}
&\lim_{\epsilon\to 0} \int Q_+(f)(\xi)\frac{\exp(-d^2)/\epsilon}{(\epsilon\pi)^{1/2}}\, d\xi \\
(3.9)\qquad &= \lim_{\epsilon\to 0} \int d\xi \int d\xi_* \frac{f(\xi)f(\xi_*)}{|\xi-\xi_*|} \int_{k(\xi,\xi_*)} d\sigma(\xi') \frac{\exp -(d^2(\xi',E))/\epsilon}{(\epsilon\pi)^{1/2}} \\
&= \pi \int d\xi\, d\xi_*\, \theta f(\xi) f(\xi_*)
\end{aligned}
$$

where $\theta = 1$ if $k \cap E \neq \emptyset$ and $\theta = 0$ otherwise. We have also used that $|\xi - \xi_*|$ is invariant under the collision transformation. Therefore:

$$(3.7) \leq \pi. \tag{3.10}$$

Moreover:

$$(3.8) = \int d\xi_1 \int d\xi_2 \frac{f(\xi_1)f(\xi_2)}{|\xi_1-\xi_2|} \int_{k(\xi_1,\xi_2)} \int \frac{1}{|\xi-\xi'|}\, d\sigma(\xi') \leq 2\pi. \tag{3.11}$$

The last step follows by the following inequality

$$\int_{k(\xi_1,\xi_2)} \frac{1}{|\xi-\xi'|}\, d\sigma(\xi') \leq \int_{k(\xi_1,\xi_2)} \frac{1}{|\xi_0-\xi'|}\, d\sigma(\xi') \tag{3.12}$$

where $\xi_0 = \frac{(\xi+\xi_1)}{2}$ is the center of the sphere $k(\xi, \xi_1)$ of diameter $|\xi - \xi_1|$. The inequality says that the potential generated by a spherically symmetric distribution of superficial charges takes its maximum at the center of the sphere (see Problem 2). By (3.3), (3.10), and (3.11) we can conclude that:

$$\frac{d}{dt} Q_+(f(t)) \leq 4\pi^2 \tag{3.13}$$

from which:

$$f(t,\xi) \leq f_0(\xi) + Q_+(f_0)(\xi)t + 2\pi^2 t^2. \tag{3.14}$$

This argument is not completely rigorous. The solution we constructed in the previous theorems exists in L^1 so that its integral on a plane or in a sphere is not even defined. However, $I(\xi)$ and $I(E_{\xi\xi'})$ make sense for almost all ξ in $\Re^3$ and almost all ξ and ξ' in $\Re^3 \times \Re^3$. Everything can be made rigorous without any significant effort to obtain an L^∞-bound of $f(t)$ in terms of $\|f_0\|_\infty$ and $\|Q_+(f_0)\|_\infty$.

The estimate (3.13) can be improved toward a time-independent estimate. Actually, one can prove

$$\sup_{t\in\Re} \|f(t)\|_\infty \leq C(f_0) \tag{3.15}$$

by establishing a lower bound for the loss term:

(6.3.1) Lemma. *Suppose that $f_0 \geq 0$ with finite energy and entropy. Then:*

$$R(f(t)) \geq C(f_0)(1 + |\xi|), \tag{3.16}$$

where $C(f_0)$ depends only on $\|f_0\|_{1,2}$ and $H(f_0)$.

Proof.

$$\begin{aligned}
&R(f)(\xi) \\
&\geq \int |\xi - \xi_*| f(\xi_*)\, d\xi_* \\
&\geq \int_{|\xi-\xi_*|>A} |\xi - \xi_*| f(\xi_*)\, d\xi_* 1 \\
&\geq A\Big(\int f - \int_{|\xi-\xi_*|\leq A} f\Big) \qquad (3.17) \\
&\geq A\Big(\|f\|_1 - \frac{1}{\ln B} \int_{|\xi-\xi_1|\leq A} f \ln f \chi_{(f>B)} - \int_{|\xi-\xi_*|\leq A} f\chi_{(f\leq B)}\Big) \\
&\geq A\left(\|f\|_1 - \frac{H(f)}{\ln B} - B\pi A^3\right).
\end{aligned}$$

Here $\chi_{\{\}}$ denotes the characteristic function of the set $\{\}$. Now choosing A and B suitably, we can find $C > 0$ [(depending only on $\|f\|_1$ and $H(f)$] for which

$$R(f)(\xi) \geq C. \tag{3.18}$$

Finally, for $\frac{|\xi|}{2} > \frac{\|f\|_{1,2}}{\|f\|_1}$

$$R(f)(\xi) \geq \int |\xi - \xi_*| f(\xi_*)\, d\xi_* \geq |\xi|\, \|f\|_1 - \|f\|_{1,2} \geq \frac{1}{2}|\xi|\, \|f\|_1. \tag{3.19}$$

Combining (3.19) with (3.18) completes the proof. □

By virtue of Lemma 6.3.1, the earlier estimates can be improved in the following way:

$$\frac{dI(E)}{dt} + CI(E) \leq \pi \tag{3.20}$$

$$\frac{dI(\xi)}{dt} + CI(\xi) \leq 4\pi^2 \tag{3.21}$$

yielding bounds on $\|Q_+(f(t))\|_\infty$ uniform in time. Also, from the estimate

$$f(t) \le e^{-Ct} f_0 + \int_0^t ds\, e^{-C(t-s)} Q_+(f(s)) \tag{3.22}$$

we finally obtain bounds on $\|f(t)\|_\infty$ that are uniform in time.

The continuity in time of the solution $f(t,\xi)$ follows from the fact that $|Q|$ is pointwise bounded, as follows from the bounds on Q_+ and the obvious inequality:

$$R(f)(\xi) \le C(1 + |\xi|). \tag{3.23}$$

This also implies the continuity in time of $Q(f(t))(\xi)$ and hence the differentiability of $f(t,\xi)$ in time. Summarizing, we have the following theorem.

(6.3.2) Theorem. *Suppose that $f(\xi) \le \frac{C}{(1+\xi^2)^{s/2}}$ with $s > 6$. Then the solution of the Boltzmann equation $f(t,\xi)$ satisfies:*

$$\sup_{t \in \Re^+} \|f(t)\|_\infty < C, \tag{3.24}$$

where C depends only on f_0. In addition, for almost all $\xi \in \Re^3$ and all $t \in \Re^+$, $f(t,\xi)$ is differentiable in time and

$$\partial_t f = Q(f) \qquad \text{pointwise.} \tag{3.25}$$

Remark. Strictly speaking, if we want $\|f_0\|_{1,4} < +\infty$, we need $s > 7$. However, it is enough to require $\|f_0\|_{1,3} < +\infty$ if we do not insist on strong differentiability.

Problems

1. Prove the identity (3.4).
2. Show that the potential generated by two identical charges takes the maximum in the middle point of the segment joining the two charges. Use this fact to prove the inequality (3.12).
3. Suppose that $f_0 \ge e^{-\alpha\xi^2}$. Prove that

   $$f(\xi, t) \ge c e^{-1/2\alpha\xi^2}.$$

   Hint: Use the bound (3.23) and observe that the inequality does not make any use of the H-theorem.

6.4 Long Time Behavior

Physics and the formal version of the H-theorem suggest, as discussed in Chapter 3 Section 4, that the solution of the Boltzmann equation $f(t)$ should converge to a Maxwellian

$$M(\xi) = A \exp(-\beta|\xi - v|^2) \tag{4.1}$$

where the parameters A, β, v are determined by the conserved quantities mass, energy, and momentum of the intial density f_0. We know already (see Chapter III) that the H-functional, when restricted to positive L^1-functions with assigned mass, energy, and momentum, takes its minimum value on the corresponding Maxwellian distribution. We also know that the H-functional is decreasing in time. Combining these facts with a careful study of the collision terms, we can prove the convergence of the solution to a Maxwellian.

(6.4.1) Theorem. *Under the hypotheses of Theorem 6.3.2, $f(t)$ converges weakly to a Maxwellian, i.e., for all $g \in L_\infty(\Re^3)$*

$$\lim_{t\to\infty} \int d\xi\; g(\xi) f(\xi, t) = \int d\xi\; g(\xi) M(\xi). \tag{4.2}$$

Proof. $H(f(t))$ is decreasing. By monotonicity $H'(f(t))$ exists almost everywhere. The negative function:

$$\frac{1}{4}(f'f'_* - ff_*) \ln \frac{ff_*}{f'f'_*} \tag{4.3}$$

is therefore integrable (with respect to $d\xi\, d\xi_*\, dn\; n \cdot (\xi - \xi_*) \chi_{\{n\cdot(\xi-\xi_*)\geq 0\}}$ for almost all $t \geq 0$.

Since H is bounded from below we have that

$$\int_0^\infty H'(f(s))\, ds < +\infty \tag{4.4}$$

and so we can find a sequence $t_n, t_n \to \infty$, such that

$$\lim_{t\to\infty} H'(f(t_n)) = 0. \tag{4.5}$$

By the Dunford–Pettis theorem, we can extract a weakly convergent subsequence (still denoted by $f(t_n) \equiv f_n$) to some ϕ. To prove the theorem, it is enough to prove that the collision equation

$$\phi'\phi'_* - \phi\phi_* = 0 \tag{4.6}$$

is satisfied. Indeed, for $D = \sup f(\xi, t)$, we have

$$\int |\phi'\phi'_* - \phi\phi_*|^2 \le \liminf \int |f'_n f'_{n*} - f_n f_{n*}|^2 \qquad \text{(by 2.29)}$$

$$\le D^2 \liminf \int |f'_n f'_{n*} - f_n f_{n*}| \left| \ln \frac{f_n f_{n*}}{f'_n f'_{n*}} \right|$$

$$\left(\text{because } \left| \ln \frac{f_n f_{n*}}{f'_n f'_{n*}} \right| \ge D^{-2} |f'_n f'_{n*} - f_n f_{n*}|; \text{see Problem 1} \right)$$

$$(4.7) \qquad = D^2 \liminf \int (f'_n f'_{n*} - f_n f_{n*}) \ln \frac{f_n f_{n*}}{f'_n f'_{n1}} = 0.$$

□

Problem

1. Verify (4.7) by proving that if $0 < x, y, x_*, y_* \le D$, then

$$\frac{1}{D^2} |xy - x_* y_*| \le \left| \ln \frac{xy}{x_* y_*} \right|.$$

Hint: Reduce it to the inequality $z - 1 \ge \ln z$, valid for $0 < z$.

6.5 Further Developments and Comments

The literature concerning the homogeneous Boltzmann equation is rather extensive. Here, we limit ourselves to commenting on the references we believe to be of some relevance in connection with the theory as developed in this chapter.

The first result on the theory of the existence of the solutions of the homogeneous Boltzmann equation was obtained by T. Carleman in 1932[4]. In this paper, he proved the existence of a unique solution for the homogeneous initial value problem in the particular case in which f depends only on $|\xi|$ and t.

The treatise Ref. 3, which appeared posthumously, develops the classical theory of the homogeneous equation for continuous bounded initial data f_0 such that:

$$(5.1) \qquad \sup_{\xi} f_0(\xi)(1 + \xi^2)^{s/2} < +\infty \qquad s \ge 6.$$

Starting from the representation formulas for the collision operator and the analysis presented in Section 4, Carleman constructed a unique solution satisfying the uniform estimate:

$$(5.2) \qquad \sup_{\xi;t} f(\xi, t)(1 + \xi^2)^{s/2} < C.$$

He also established the approach to equilibrium in the uniform norm, using an equicontinuity property to apply the Ascoli–Arzelà theorem instead of the Dunford–Pettis criterion.

This monograph is fundamental in the history of the kinetic theory of gases, because it presented far–reaching new mathematical ideas and constituted, from a methodological point of view, the first rigorous and modern approach to the Boltzmann equation.

Many years later, L. Arkeryd [1] developed an L^1-theory (for a very general class of interaction kernels, including the case of hard spheres) along the same lines presented in the earlier section of this chapter. The monotonicity argument for the cutoff interaction is related to a paper by Morgenstern [8], who developed an L^1-theory for the Maxwell molecules.

In Ref. 1 the cutoff M is removed by compactness, but also by monotonicity. The argument is subtle and interesting. We sketch it here.

Given f and f^M as in Section 1, define g^M to be the solution of the initial value problem

$$\partial_t g^M + hg^M = Q_+^M(g^M) - g^M R(g^M) + G(g^M), \tag{5.3}$$

where:

$$h(\xi) = \mu(1+\xi^2)\|f_0\|_{1,2} \tag{5.4}$$

$$G(g^M) = \mu(1+\xi^2)\|g^M\|_{1,2}\, g^M(\xi). \tag{5.5}$$

For μ sufficiently large it follows that

$$0 \le g^M \le g^{M'} \le f \qquad (\text{if} \quad M \le M') \tag{5.6}$$

and

$$g^M \le f^M. \tag{5.7}$$

Therefore, the difference $f - f^M$ can be estimated in terms of the differences

$$\Delta M_1 = f^M - g^M \qquad \text{and} \qquad \Delta M_2 = f - g^M. \tag{5.8}$$

By virtue of the positivity of ΔM_1 and ΔM_2, we can easily obtain the estimate:

$$\frac{d}{dt}\|\Delta M_1\|_{1,2} \le \omega(M) + C\|f^M\|_{1,4}\,\|\Delta M_1\|_{1,2} \tag{5.9}$$

where $\omega(M) \to 0$ as $M \to \infty$.

In Ref. 1 it is also shown how to prove uniform estimates like:

$$\sup_{t\in\Re+}\|f(t)\|_{1,s} < C \tag{5.10}$$

where C depends only on f_0. The weak convergence to a Maxwellian is also proved for general L^1 data.

The L^∞-theory [3] and the L^1-theory [1] can be combined to give an L^p-theory, as discussed by Gustafsson [7].

The strong differentiability of the solution in L^1_s was investigated by DiBlasio [5]. This property allows us to establish the differentiability of the solution with respect to the initial datum.

Regarding the asymptotic behavior of the solution, we remark that the attractivity of the Maxwellian in the weak topology proved above (but, as mentioned, also valid in the uniform topology) does not imply the stability of the equilibrium in the sense of Liapunov. In the theory of differential equations, one can find examples of critical points that are attractive but not stable. Concerning stability results for the global equilibrium, we mention the work by Arkeryd et al. [2]

Finally, we quote a result due to Elmroth [6] in which the L^1-convergence to the Maxwellian follows from the weak convergence. To get the result, one combines the content of our Theorem 6.4.1 with the discussion in Chapter 3, Section 4.

References

1. L. Arkeryd, "On the Boltzmann equation." *Arch. Rat. Mech. Anal.* **45**, 1–34 (1972).
2. L. Arkeryd, R. Esposito, and M. Pulvirenti, "The Boltzmann equation for weakly inhomogeneous data," *Commun. Math. Phys.* **111**, 393–407 (1987).
3. T. Carleman, *Problèmes Mathématiques dans la Théorie Cinétique des Gaz,* Almqvist & Wiksell, Uppsala (1957).
4. T. Carleman, "Sur la théorie de l'équation intégro–differentielle de Boltzmann," *Acta Mathematica* **60**, 91–146 (1933).
5. G. DiBlasio, "Differentiability of spatially homogeneous solutions of the Boltzmann equation," *Commun. Math. Phys.* **38**, 331–340 (1974).
6. T. Elmroth, "On the H-function and convergence towards equilibrium for a space-homogeneous molecular density," *SIAM J. Appl. Math.* **44**, 150–159 (1984).
7. T. Gustafsson, *L^p- estimates for the nonlinear spatially homogeneous Boltzmann equation,* Arch. Rat. Mech. Anal. **92**, 23–57 (1986).
8. D. Morgenstern, "Analytical studies related to the Maxwell–Boltzmann equation," *J. Rat. Mech. Anal.* **4**, 533–555 (1955).
9. A. Ya. Povzner, "The Boltzmann equation in the kinetic theory of gases," *Amer. Math. Soc. Transl.* **ser. 2, 47**, 193–216 (1962).

Appendix 6.A

Proof of the H-Theorem

The proof of the H-theorem in the form of Inequality (1.19) can be organized in five steps:

1) Fix an arbitrary time T, then if $\|f_0\|_\infty \le C$ then $\sup_{t \le T} \|f^M(t)\|_\infty < C'$.
2) Replace f_0 by $f_0^n = \min\left[n, \max\left(f_0, \frac{\exp(-\xi^2)}{n}\right)\right]$ and denote by $f^{n,M}(t)$ the solution of the corresponding initial value problem. For these $f^{n,M}(t)$ the usual arguments yielding the "formal" H-theorem can be used to prove:

$$H(f^{n,M}(t)) \le H(f_0^n). \tag{A.1}$$

3) Prove the inequality:

$$H(f_0^n) \le H(f_0). \tag{A.2}$$

4) Prove that:

$$\|f^{n,M}(t) - f^M(t)\|_1 \to 0 \quad \text{as} \quad n \to \infty. \tag{A.3}$$

5) Extract a subsequence (still denoted by $f^{n,M}(t)$) converging to $f^M(t)$ and prove the claim:

$$H(f^M(t)) \le H(f_0). \tag{A.4}$$

Proof.

Step 1. This was proved in connection with the L^∞-theory. Notice that we make use of the H-theorem (which we are going to prove) only to prove estimates uniform in time, thereby avoiding a loop!

Step 2. One can prove the inequality:

$$C(n,t)\exp(-\xi^2) \le f^{n,M}(\xi,t) \le C \tag{A.5}$$

(see Section 3, Problem 3). Then the usual arguments used in the formal proof of the H-theorem work.

Step 3. This follows by direct inspection.

Step 4. The Lipschitz continuity of $f^M(t)$ with respect to the initial condition (see Section 2, Problem 2) yields

$$\|f^{n,M}(t) - f^M(t)\|_1 \le \|f_0^n - f_0\|_1 \to 0 \quad \text{as} \quad n \to \infty. \tag{A.6}$$

Step 5.

$$\int d\xi\ f^M(t)\ln f^M(t)\chi_{(|\xi|<R)} = \lim \int d\xi\ f^{n,M}(t)\ln f^{n,M}(t)\chi_{(|\xi|<R)}$$

(by the dominated convergence theorem)

$$H(f_0) \ge \lim \int d\xi\ f^{n,M}(t)\ln f^{n,M}(t)\chi_{(|\xi|\ge R)} \times \{\chi_{(f^M(t)<\exp(-\xi^2))} + \chi_{(f\ge\exp(-\xi^2))}\}$$

(by steps 2 and 3)

$$\begin{aligned} H(f_0) &\ge \lim \int d\xi \sqrt{f^{n,M}}\ln f^{n,M}(t)\chi(|\xi| \ge R)\exp(-\xi^2)\chi(f^M(t) < 1) \\ &\quad + \lim \int d\xi\ f^{n,M}|\xi|\chi(|\xi| \ge R) \\ &\le H(f_0) - C\int d\xi\ \chi(|\xi| \ge R)\exp(-\xi^2) + R^{-1}\int d\xi\ f_0(\xi)\xi^2. \end{aligned} \tag{A.7}$$

(Because of the energy conservation and the fact that $\sqrt{x}\ln x$ is bounded for $x < 1$.) The claim follows by taking the limit $R \to \infty$. □

Appendix 6.B

Proof of Lemma 6.2.1

For a and b positive numbers, $b > a$ and any $\gamma \in [0,1]$:

$$(B.1) \qquad (a+b)^s \leq a^s + b^s + 2^{s-1}b^{s-1}a \leq a^s + b^s + 2^{s-1}b^{s-\gamma}a^\gamma,$$

and hence for positive a and b :

$$(B.2) \qquad a^s + b^s \leq (a+b)^s \leq a^s + b^s + C(b^{s-\gamma}a^\gamma + a^{s-\gamma}b^\gamma).$$

As a consequence of (B.2) and the energy conservation for $\gamma \in [0,2]$:

$$\begin{aligned} &(1+\xi'^2)^{s/2} + (1+\xi_*'^2)^{s/2} - (1+\xi^2)^{s/2} - (1+\xi_*^2)^{s/2} \\ (B.3) \qquad &\leq C_s[(1+\xi^2)^{\gamma/2}(1+\xi_*^2)^{(s-\gamma)/2} + (1+\xi^2)^{(s-\gamma)/2}(1+\xi_*^2)^{\gamma/2}]. \end{aligned}$$

Now, the right-hand side of (2.9) can be written as:

$$\begin{aligned} &\frac{1}{2} \int\limits_{n\cdot(\xi-\xi_*)\geq 0} d\xi\, d\xi_*\, dn\, n\cdot(\xi-\xi_*)[(1+\xi'^2)^{s/2} + (1+\xi_*'^2)^{s/2} \\ (B.4) \qquad &\qquad - (1+\xi^2)^{s/2} - (1+\xi_*^2)^{s/2}]f(\xi)f(\xi). \end{aligned}$$

By using the obvious inequality:

$$(B.5) \qquad |\xi - \xi_*| \leq (1+\xi^2)^{1/2}(1+\xi_*^2)^{1/2}$$

and choosing $\gamma = 1$ in (B.3), we easily get inequality (2.9).

Appendix 6.C

Proof of Formula (3.4)

By the invariance of the Lebesgue measure with respect to the collision transformation, we have:

(C.1)

$$\int d\xi\; \phi(\xi) Q_+(f)(\xi) = \int d\xi \int d\xi_1\; f(\xi) f(\xi_*) \int\limits_{n\cdot(\xi-\xi_1)\geq 0} n\cdot(\xi-\xi_1)\phi(\xi').$$

Recalling the collision laws, one can replace the integration over the unit sphere by the integration over the sphere k (see Fig. 26). A simple analysis of the Jacobian of the transformation yields:

$$\int\limits_{n\cdot(\xi-\xi_*)\geq 0} n\cdot(\xi-\xi_*)\phi(\xi') = \int\limits_{k(\xi,\xi_*)} \phi(\xi') \frac{d\sigma(\xi')}{|\xi-\xi_*|}. \tag{C.2}$$

Proof of Formula (3.3)

For a fixed ξ, choose $\phi(\xi') = \frac{1}{(\epsilon\pi)^{3/2}} \exp\left(\frac{(\xi'-\xi)^2}{\epsilon}\right)$. Then:

$$Q_+(f)(\xi) = \lim_{\epsilon\to 0}\int d\xi_1\ \phi(\xi_1)Q_+(f)(\xi_1)$$

$$\text{(C.3)} \qquad = \lim_{\epsilon\to 0}\int d\xi_1 \int d\xi_2\ \frac{f(\xi_1)f(\xi_2)}{|\xi_1-\xi_2|}I(\xi)$$

where:

$$(C.4) \qquad I(\xi) = \int_{k(\xi_1,\xi_2)} \frac{1}{(\epsilon\pi)^{3/2}} \exp\left(-\frac{(\xi'-\xi)^2}{\epsilon}\right) d\sigma(\xi').$$

Introducing the middle point $\xi_0 = \frac{(\xi_1+\xi_2)}{2}$ and using the obvious identity:

$$(C.5) \qquad (\xi'-\xi)^2 = (\xi'-\xi_0)^2 + (\xi_0+\xi)^2 + 2(\xi'-\xi_0)\cdot(\xi_0+\xi)$$

we can perform explicitly the integral (C.4) to obtain:
$(C.6)$

$$I(\xi) = \frac{|\xi-\xi_1|}{2(\epsilon\pi)^{1/2}|\xi-\xi_0|}$$
$$\times\left\{\exp -\frac{1}{\epsilon}\left(|\xi-\xi_0| - \frac{|\xi_1-\xi_2|}{2}\right)^2 - \exp\frac{1}{\epsilon}\left(|\xi-\xi_0| + \frac{|\xi_1-\xi_2|}{2}\right)^2\right\}.$$

Inserting this expression in (C.3), we realize that the contribution of the second term in the right-hand side of (C.6) vanishes in the limit. The first one describes the effect of a one-dimensional δ function concentrated on the set $|\xi-\xi_0| = |\xi_1-\xi_2|/2$, which is equivalent to the plane $(\xi-\xi_1)\cdot(\xi_2-\xi)$. This completes the proof.

7

Perturbations of Equilibria and Space Homogeneous Solutions

7.1. The Linearized Collision Operator

Our first aim in this chapter will be to find a global solution $f = f(x, \xi, t)$ of the Cauchy problem for the Boltzmann equation for hard spheres:

$$\frac{\partial f}{\partial t} + \xi \cdot \frac{\partial f}{\partial x} = Q(f, f) \tag{1.1}$$

when f is close to some absolute Maxwellian M, which without loss of generality (thanks to possible scalings and choice of a suitable reference frame) can be assumed to be $(2\pi)^{-3/2} \exp(-\xi^2/2)$. To this end we introduce a new unknown h related to the distribution function f by

$$f = M + M^{1/2} h. \tag{1.2}$$

The Boltzmann equation (1.1) takes on the form:

$$\frac{\partial h}{\partial t} + \xi \cdot \frac{\partial h}{\partial x} = Lh + \Gamma(h, h) \tag{1.3}$$

where L is the linearized collision operator defined by:

$$Lh = 2M^{-1/2} Q(M^{1/2} h, M) \tag{1.4}$$

[here Q is the bilinear operator defined in Eq. (3.1.3)] and $\Gamma(h,h)$ is the nonlinear part, which should be small compared to the linear part, and is given by:

$$\Gamma(g,h) = M^{-1/2}Q(M^{1/2}g, M^{1/2}h) \tag{1.5}$$

with $g = h$.

A more explicit expression of Lh reads as follows

$$Lh = \int_{\Re^3}\int_{S^2_+}(h'R'_* + R'h'_* - Rh_* - hR_*)R_*|V\cdot n|d\xi_* dn \tag{1.6}$$

where, for convenience, R denotes $M^{1/2}$ and we took into account that $M'M'_* = MM_*$. Because of Eq. (3.1.10) (with Rh in place of f, M in place of g and g/R in place of ϕ), we have the identity:

$$\begin{aligned}\int_{\Re^3} gLhd\xi = -\frac{1}{4}\int_{\Re^3}\int_{\Re^3}\int_{S^2_+}&(h'R'_* + R'h'_* - hR_* - Rh_*)\\ &\times(g'R'_* + R'g'_* - gR_* - Rg_*)|V\cdot n|d\xi_* d\xi dn.\end{aligned} \tag{1.7}$$

This relation expresses a basic property of the linearized collision term. In order to make it clear, let us introduce the Hilbert space of square summable functions of ξ endowed with the scalar product

$$(g,h) = \int_{\Re^3}\overline{g}hd\xi \tag{1.8}$$

where the bar denotes complex conjugation. Then Eq. (1.7) (with $\overline{g}$ in place of g) gives (thanks to the symmetry of the expression on the right-hand side (1.7) with respect to the interchange $g \Leftrightarrow h$):

$$(g, Lh) = (Lg, h). \tag{1.9}$$

Further:

$$(h, Lh) \le 0, \tag{1.10}$$

and the equality sign holds if and only if

$$h'/R' + h'_*/R'_* - h/R - h_*/R_* = 0, \tag{1.11}$$

i.e., unless h/R is a collision invariant.

Eqs. (1.9) and (1.10) indicate that the operator L (provided it is taken with its maximal domain in L^2) is self-adjoint and nonpositive in L^2. In order to exploit these and other good properties of L, we shall introduce the operator

$$B = L - \xi \cdot \frac{\partial}{\partial x} \tag{1.12}$$

and write Eq. (1.3) in the following integral form:

$$h(t) = T(t)h(0) + \int_0^t ds T(t-s)\Gamma(h(s), h(s)) \tag{1.13}$$

where $T(t)$ is the semigroup generated by B. We can hope that in some norm (see Lemma 2.6):

$$\| \Gamma(h,h) \| \leq C \| h \|^2 \quad ; \tag{1.14}$$

then if we could prove (and this would be the crucial estimate) that, for some positive c,

$$\| T(t) \| \leq Ce^{-ct} \tag{1.15}$$

we would have

$$\| h(t) \| \leq Ce^{-ct} \| h(0) \| + \int_0^t ds\, e^{-c(t-s)} \| h(s) \|^2 . \tag{1.16}$$

If we let

$$\hat{h} = \sup_t \| h(t)e^{ct} \| \tag{1.17}$$

we arrive at

$$\hat{h} \leq C \| h(0) \| + C\hat{h}^2 \tag{1.18}$$

which implies that $\hat{h}$ is bounded whenever $\| h(0) \|$ is small.

Unfortunately this strategy cannot be followed so easily. In fact since the dissipative part of B is contained in L, one would like to prove Eq. (1.15) for the semigroup generated by L itself. L, however, has five linearly independent eigenfunctions corresponding to the zero eigenvalue; these are the functions h such that h/R is a collision invariant, because then Eq. (1.11) holds and, according to Eq. (1.6), Lh vanishes. Then $T(t)h = h$ for any linear combination of these eigenfunctions and the desired property does not hold for the semigroup generated by L. The operator $\xi \cdot \partial/\partial x$, however, generates a semigroup, which, although norm-preserving, has the tendency to spread the molecular distribution in a uniform way; this will help in obtaining the desired estimate in $\Re^3$. In fact to prove this estimate will be our major aim in the following sections. This will require a preliminary study of the spectral properties of L and of the Fourier transform of $B, B(k)$.

7.2 The Basic Properties of the Linearized Collision Operator

In order to study the linearized collision operator L, given by Eq. (1.6), we start by remarking that we can split L as $K - \nu(|\xi|)I$, where $K = K_2 - K_1$ is an integral operator, I the identity, and ν a function bounded from below by a constant ν_0 and from above by a linear function. Specifically

$$\nu(|\xi|) = \int_{\Re^3}\int_{S^2_+} M_*|V \cdot n| d\xi_* dn$$

$$K_1 h = R(|\xi|)\int_{\Re^3}\int_{S^2_+} h_* R_* |V \cdot n| d\xi_* dn \tag{2.1}$$

$$K_2 h = \int_{\Re^3}\int_{S^2_+} (h' R'_* + R' f'_*) R_* |V \cdot n| d\xi_* dn$$

It is easy to pick out the kernel of K_1, $k_1(\xi, \xi_*)$ by inspection:

$$k_1(\xi, \xi_*) = \pi |\xi - \xi_*| R(|\xi|) R(|\xi_*|). \tag{2.2}$$

To find the kernel of K_2 requires a little more work. To reduce it a little bit we use a trick. Let us consider a unit vector m that lies in the plane of V and n and is orthogonal to n. We can then write $V = n(n \cdot V) + m(m \cdot V)$, which implies $\xi - n(n \cdot V) = \xi_* + m(m \cdot V)$, $\xi_* + n(n \cdot V) = \xi - m(m \cdot V)$; thus if we use m in place of n in the second part of the integral appearing in (2.1), it becomes identical to the first, except for the fact that θ is replaced by $\pi/2 - \theta$ (and ϕ by $\phi \pm \pi$). But even this difference disappears, because $|V \cdot n| dn = |V| \cos\theta \sin\theta d\theta d\phi = |V \cdot m \mid dm$. Thus

$$K_2 h = 2\int_{\Re^3}\int_{S^2_+} h' R'_* R_* |V \cdot n| d\xi_* dn = \int_{\Re^3}\int_{S^2} h' R'_* R_* |V_* \cdot n| dV_* dn \tag{2.3}$$

where in the last integral n runs over the entire sphere S^2 and $V_* = \xi_* - \xi$ is used as integration variable in place of ξ_* (a unit Jacobian transformation). Next consider the components of V_* parallel and perpendicular to n:

(2.4) $$V_* = v + w; \qquad v = n(n \cdot V_*); \qquad w = V_* - n(n \cdot V_*).$$

We now perform the integral in Eq. (2.3) in the following order: first w (on a plane Π perpendicular to n), then v, then n. With n fixed, the replacement of V_* by v and w is just a choice of coordinates. After integrating with respect to w, we combine the one-dimensional v-integration in the direction n with the integral with respect to n over the unit sphere to give a three-dimensional integration over the vector $v = |v|n$; here we must introduce a factor 2 because v describes $\Re^3$ twice (for a given V_*, n and $-n$ give the same point). Thus since the Jacobian from dv to $dnd|v|$ is $|v|^2$, we have:

(2.5) $$dndV_* = 2|v|^{-2}dvdw$$

and Eq. (2.3) becomes

(2.6) $$K_2h = 2\int_{\Re^3}\int_{\Pi} h(\xi + v)R(\xi + v + w)R(\xi + w)|v|^{-1}dwdv$$

where the integral with respect to w over Π (the plane through the origin perpendicular to v) has to be performed first. The kernel of the integral operator K_2 is now clear. Introducing the new variable $\xi_* = v + \xi$, the kernel is:

(2.7) $$k_2(\xi, \eta) = 2|\xi_* - \xi|^{-1}\int_{\Pi} R(\xi_* + w)R(\xi + w)dw$$

where Π is now perpendicular to $\xi - \xi_*$. Since:

(2.8)
$$|\xi_*+w|^2+|\xi+w|^2 = \xi_*^2+\xi^2+2(\xi_*+\xi)\cdot w+2w^2 = 2|w+\frac{1}{2}(\xi_*+\xi)|^2+\frac{1}{2}|\xi_*-\xi|^2$$

(2.9) $$R(\xi_* + w)R(\xi + w) = R(2^{1/2}w + 2^{-1/2}(\xi_* + \xi))R(2^{-1/2}(\xi_* - \xi)).$$

The vector $\frac{1}{2}(\xi_* + \xi)$ has a part in the plane Π, say ζ, which can be eliminated by letting $z = w + \zeta$ (a translation in Π); the remaining part is the projection on the direction of $\xi_* - \xi$, i.e.,

(2.10) $$\frac{1}{2}(\xi_* + \xi) \cdot \frac{\xi_* - \xi}{|\xi_* - \xi|} = \frac{1}{2}\frac{|\xi_*|^2 - |\xi|^2}{|\xi_* - \xi|}.$$

By means of Eqs. (2.9) and (2.10), Eq. (2.7) becomes:

$$k_2(\xi,\xi_*) = 2(2\pi)^{3/4}|\xi_* - \xi|^{-1} R(2^{-1/2}\frac{|\xi_*|^2 - |\xi|^2}{|\xi_* - \xi|}) \\ R(2^{-1/2}(\xi_* - \xi)) \int_{\Pi} R(2^{1/2}z)dz. \tag{2.11}$$

The integral is now easily performed with the result $(2\pi)^{-3/4}2\pi = (2\pi)^{-1/4}$ and Eq. (2.11) becomes

$$k_2(\xi,\xi_*) = 4\pi|\xi_* - \xi|^{-1} R(2^{-1/2}\frac{|\xi_*|^2 - |\xi|^2}{|\xi_* - \xi|}) R(2^{-1/2}(\xi_* - \xi)). \tag{2.12}$$

Finally, we can make more explicit the expression for the collision frequency

$$\begin{aligned}
&\nu(|\xi|) \\
=&\int_{\Re^3}\int_{S^2} M_*|V\cdot n|d\xi_* dn \\
=&\pi\int_{\Re^3} M(\xi_*)|\xi_* - \xi|d\xi_* = \int_{\Re^3} M(\xi + v)|v|dv \\
=&(2\pi)^{-1/2}\int_0^\infty\int_0^\pi \exp(-|\xi|^2/2 - |v|^2/2 - |\xi||v|\cos\theta)|v|^3 \sin\theta d|v|d\theta \\
=&(2\pi)^{-1/2}|\xi|^{-1}[\int_0^\infty \exp(-|\xi|^2/2 - t^2/2 + |\xi|t)t^2 dt \\
&- \int_0^\infty \exp(-|\xi|^2/2 - t^2/2 - |\xi|t)t^2 dt] \\
=&(2\pi)^{-1/2}|\xi|^{-1}[\int_{-|\xi|}^\infty \exp(-u^2/2)(u + |\xi|)^2 du \\
&- \int_{|\xi|}^\infty exp(-u^2/2)(u - |\xi|)^2 du] \\
=&(2\pi)^{-1/2}|\xi|^{-1}[2\int_0^{|\xi|} exp(-u^2/2)(u^2 + |\xi|^2)du + 4|\xi|\exp(-|\xi|^2/2)] \\
=&(2\pi)^{-1/2}[2(|\xi| + |\xi|^{-1})\int_0^{|\xi|} exp(-u^2/2)du + 2\exp(-|\xi|^2/2)]
\end{aligned} \tag{2.13}$$

where we first performed the trivial integration with respect to n, then changed the integration variable from ξ_* to $v = \xi_* - \xi$ and transformed the resulting integral from Cartesian to polar coordinates in velocity space; then we performed the integration with respect to the angle variables and changed the name of the remaining integration variable from $|v|$ to t for convenience; the resulting two integrals in t are first transformed by letting $t = u + |\xi|$ and $t = u - |\xi|$, respectively, and then, after expanding the squares and symplifying, the last step has been performed with a partial integration. (We remark that the expression for $\nu(|\xi|)$ given by Grad[7] is wrong.)

Thus we have proved the following

(7.2.1) Theorem. *The linearized collision operator (defined on the functions $h(.)$ of L^2 such that $[\nu(|.|)]^{1/2}h(.)$ is also in L^2) is self-adjoint and non-positive in L^2, with a fivefold null eigenspace spanned by $M^{1/2}\psi_\alpha$, where ψ_α $(\alpha = 0,1,2,3,4,5)$ are the collision invariants. It can be decomposed in the difference*

$$L = K - \nu(|\xi|)I \tag{2.14}$$

where $\nu(|\xi|)$is given by (2.13) and satisfies the bound (see Problem 1),

$$0 < \nu_0 \leq \nu(|\xi|) \leq \nu_1(1+|\xi|^2)^{1/2} \tag{2.15}$$

with ν_0 and ν_1 positive numbers, while I is the identity operator and K is an integral operator with a real measurable symmetric kernel $k(\xi,\xi_*)$ given by:

$$\begin{aligned} k(\xi,\xi_*) =& (2\pi)^{-1/2} 2|\xi_* - \xi|^{-1} \exp(-\frac{1}{8}\frac{(|\xi_*|^2 - |\xi|^2)^2}{|\xi_* - \xi|^2} - \frac{1}{8}|\xi_* - \xi|^2) \\ & - \frac{1}{2}|\xi - \xi_*| \exp[-(|\xi|^2 + |\xi_*|^2)/4]. \end{aligned} \tag{2.16}$$

For later purposes we shall need estimates of this kernel. It is trivial to prove that

$$k(\xi,\xi_*) \leq (c_1|\xi_* - \xi|^{-1} + c_2|\xi - \xi_*| \exp(-\frac{1}{8}|\xi_* - \xi|^2). \tag{2.17}$$

This estimate has the following consequence.

(7.2.2) Theorem. *The kernel of the operator K is integrable and square integrable with respect to ξ_*. The integrals are bounded by a constant, independent of ξ.*

Proof. The proof is trivial, thanks to (2.17), because of the exponential decay of the kernel and the fact that $|\xi_* - \xi|^{-1}$ and $|\xi_* - \xi|^{-2}$ are integrable in a neighborhood of $\Re^3$ about ξ. The fact that the bounds are independent of ξ follows from the fact that the right hand side of Eq. (2.17) is translation invariant. □

We now prove the following theorem.

(7.2.3) Theorem. *The kernel $k(\xi,\xi_*)$ of the operator K is such that for any $r \geq 0$ we have*

$$\int k(\xi,\xi_*)(1+|\xi_*|^2)^{-r}d\xi_* \leq k_0(1+|\xi|^2)^{-r-1/2}. \tag{2.18}$$

Proof. If we had just $-r$ as the exponent on the right-hand side it would be enough to use estimate (2.17) (see Problem 2). To gain the additional $-1/2$ requires a longer proof. If we look at the explicit expression (2.16) we realize that the part after the minus sign (arising from k_2) is easy to deal with, because of the exponential $\exp(-|\xi|^2/4)$ (see Problem 3). We thus have only to prove that

$$I = (1+|\xi|^2)^{r+1/2}\int(1+|\xi+v|^2)^{-r}|v|^{-1}\exp\left(-\frac{1}{8}\frac{\left(2\xi\cdot v+|v|^2\right)^2}{|v|^2}-\frac{1}{8}|v|^2\right)dv \tag{2.19}$$

is bounded uniformly in ξ. We split the integration into I_1+I_2, where the former refers to $|v|>|\xi|/4$ and the latter to $|v|<|\xi|/4$. We have

$$I_1 < (1+|\xi|^2)^{r+1/2}\int_{|v|>\frac{|\xi|}{4}}|v|^{-1}\exp\left(-\frac{1}{8}|v|^2\right)dv, \tag{2.20}$$

which is obviously bounded. For I_2 we can restrict ourselves to $|\xi| \geq 1$, because otherwise the result is clear. For $|\xi| \geq 1$ we use $1+|\xi+v|^2 > 1+9|\xi|^2/16$ and polar coordinates to obtain:

$$\begin{aligned} I_2 =&(1+|\xi|^2)^{r+1/2}(1+9|\xi|^2/16)^{-r} \\ &\times 2\pi\int_0^\infty\int_0^\pi \exp\left(-\frac{1}{8}(2|\xi|\cos\theta+|v|)^2-\frac{1}{8}|v|^2\right)|v|\sin\theta d|v|d\theta \\ <&(1+|\xi|^2)^{r+1/2}(1+9|\xi|^2/16)^{-r} \\ &\times (2\pi)^{1/2}|\xi|^{-1}\int_0^\infty \exp\left(-\frac{1}{8}|v|^2\right)|v|d|v| \\ =&4(1+|\xi|^2)^{r+1/2}(1+9|\xi|^2/16)^{-r}(2\pi)^{1/2}|\xi|^{-1} \end{aligned} \tag{2.21}$$

which for $|\xi| \geq 1$ is clearly bounded. In an intermediate step here we have used the inequality:

$$\begin{aligned} &\int_0^\pi \exp\left(-\frac{1}{8}(2|\xi|\cos\theta+|v|)^2\right)\sin\theta d\theta \\ &< \int_{-\infty}^\infty \exp\left(-\frac{1}{2}|\xi|^2t^2\right)dt = (2\pi)^{1/2}|\xi|^{-1}. \end{aligned} \tag{2.22}$$

Inequality (2.13) is thus proved. □

In the following we shall denote by $\mathbf{B}(\mathcal{X},\mathcal{Y})$ the set of all linear bounded operators from a Banach space $\mathcal{X}$ into a Banach space $\mathcal{Y}$, and by $\mathbf{C}(\mathcal{X},\mathcal{Y})$ its subset consisting of compact operators. When $\mathcal{Y}=\mathcal{X}$, we simply write $\mathbf{B}(\mathcal{X})$ and $\mathbf{C}(\mathcal{X})$. We also denote by L^∞_β the Banach space of the functions h such that $(1+|\xi|^2)^{\beta/2}h$ is in $L^\infty(\Re^3)$. Then we can prove the following.

(7.2.4) Theorem. *The integral operator K is in $\mathbf{B}(L^2)\cap\mathbf{B}(L^\infty_\beta, L^\infty_{\beta+1}), \beta\geq 0$. It is also in $\mathbf{B}(L^2, L^\infty_0)$ and in $\mathbf{C}(L^2)$.*

Proof. The fact that $K\in\mathbf{B}(L^2,L^\infty_0)$ follows from Theorem 7.2.2. We now prove that $K\in\mathbf{C}(L^2)$. Let χ_R be the characteristic function of $\{\xi : |\xi|\leq R\}$. Then in $\mathbf{B}(L^2)$ we have (thanks to Theorem 7.2.3):

$$\begin{aligned} &\| (1-\chi_R)K \| < C(1+R)^{-1}\to 0 \\ &\| K(1-\chi_R) \| < C(1+R)^{-1}\to 0 \qquad (R\to\infty). \end{aligned} \tag{2.23}$$

In order to prove these results we apply the Schwarz inequality in a suitable way; we just indicate how to prove the first of these relations (for the second, see Problem 4):

$$\| (1-\chi_R)K \|^2 = \sup_{\|h\|=1}\int d\xi(1-\chi_R(\xi))\left[\int d\eta k(\xi,\eta)h(\eta)\right]^2$$

$$\leq \sup_{\|h\|=1}\int d\xi(1-\chi_R(\xi))\left[\int d\eta k(\xi,\eta)\right]\left[\int d\eta k(\xi,\eta)(h(\eta))^2\right]$$

$$\leq k_0 \sup_{\|h\|=1}\int d\xi(1-\chi_R(\xi))(1+|\xi|)^{-1}\left[d\eta k(\xi,\eta)(h(\eta))^2\right]$$

$$\leq k_0(1+|R)^{-1}\sup_{\|h\|=1}\int d\xi\left[\int d\eta k(\xi,\eta)(h(\eta))^2\right]$$

$$\leq k_0(1+|R)^{-1}\sup_{\|h\|=1}\int d\eta(h(\eta))^2 = k_0(1+|R)^{-1}.$$

In addition, $\chi_R K$, because of Theorem 7.2.2, is a Hilbert-Schmidt operator. Then, thanks to Eq. (2.23), K is compact, because the set of compact operators is a closed linear manifold in $\mathbf{B}(L^2)$. Finally the fact that $K\in\mathbf{B}(L^2)\cap\mathbf{B}(L^\infty_\beta,L^\infty_{\beta+1})$, for $\beta\geq 0$, follows from Theorem 7.2.3 and $K\in\mathbf{C}(L^2)$. □

We next consider the spectrum of L. Since in this section and the following ones we shall use several standard theorems on the perturbation of linear operators, for the sake of the reader we state them here and refer to the book of Kato[8] for the proofs. We denote by $K(\ldots)$ the theorem $(\ldots)$ in Kato's book. In the following statements (where only the numbers of the equations and some symbols have been modified with respect to Ref. 2) $\mathcal{C}(\mathcal{X})$ means the set of closed operators from $\mathcal{X}$ to $\mathcal{X}$ and a holomorphic family of operators $T(\kappa)$ of type (A) is [8] such that $D(T(\kappa)) = D$ (independent of κ) and $T(\kappa)u$ is holomorphic in κ for every $u \in D$. We also recall the notion of relative compactness used in the following statements. Let T and a be operators with the same domain space X (but not necessarily with the same range space). Assume that $D(T) \subset D(A)$ and for any sequence $u_n \in D(T))$ with both u_n and Tu_n bounded, Au_n contains a convergent subsequence. Then A is said to be relatively compact with respect to T or simply T-compact.

(K.IV.5.35) Theorem. *The essential spectrum is conserved under a relatively compact perturbation. More precisely, let $T \in \mathcal{C}(\mathcal{X})$ and let A be T-compact. Then T and $T + A$ have the same essential spectrum.*

(K.VII.2.6) Theorem. *Let T be a closable operator from $\mathcal{X}$ to $\mathcal{Y}$ with $D(T) = D$. Let $T^{(n)}, n = 1, 2, \ldots$ be operators from $\mathcal{X}$ to $\mathcal{Y}$ with domains containing D and let there be constants $a, b, c \geq 0$ such that*

$$(2.24) \qquad \| T^{(n)}u \| \leq c^{n-1}(a \| u \| + b \| Tu \|), \qquad (u \in D,\ n = 1, 2, \ldots).$$

Then the series

$$(2.25) \qquad T(\kappa)u = Tu + \kappa T^{(1)}u + \kappa^2 T^{(2)}u + \ldots \qquad (u \in D)$$

defines an operator $T(\kappa)$ with domain D for $|\kappa| < 1/c$. If $|\kappa| < (b + c)^{-1}, T(\kappa)$ is closable and the closures $\tilde{T}(\kappa)$ for such κ form a holomorphic family of type (A).

(K.VII.2.7) *Remark.* The form of the condition (2.24) is chosen so as to be particularly convenient when $T^{(2)} = T^{(3)} = \ldots = 0$. In this case we can choose $c = 0$ if

$$(2.26) \qquad \| T^{(1)}u \| \leq a \| u \| + b \| Tu \|, (u \in D) \ldots.$$

(K.VII.1.8) Theorem. *If $T(\kappa)$ is holomorphic in κ near $\kappa = 0$, any finite systems of eigenvalues $\lambda_h(\kappa)$ of $T(\kappa)$ consists of branches of one or several analytic functions that have at most algebraic singularities near $\kappa = 0$. The same is true of the corresponding eigenprojections and eigennilpotents $Q_h(\kappa)$.*

We now prove the following theorem.

(7.2.5) Theorem. *The spectrum $\sigma(L)$ of the operator L is made up of a discrete and an essential spectrum: the former is contained in the interval $(-\nu_0, 0]$, where $\nu_0 = \nu(0) = 4(2\pi)^{-1/2}$, while the latter coincides with $(-\infty, -\nu_0]$.*

Proof. This result follows from a particular case of Theorem K.IV. 5.35, i.e., Weyl's theorem on the perturbation of a self-adjoint operator[8] [the multiplication by $-\nu(|\xi|)$) by a compact operator (K)], and Theorem 2.1. □

Before ending this section we prove a result on the nonlinear term $\Gamma(h,h)$ or the correponding bilinear operator $\Gamma(g,h)$ defined in Eq. (1.5).

(7.2.6) Lemma. *The projection of $\Gamma(g,h)$ on the null space of L vanishes and there exists a constant $c \geq 0$ such that*

$$\| [\nu(\xi)]^{-1}\Gamma(h,g) \| \leq C \| h \| \| g \| \tag{2.27}$$

in L^∞_β for any $\beta \geq 0$.

Proof. The first part of the statement is obvious (because of the properties of $Q(f,g)$; the second part follows from the fact that $|g| \leq \| g \| (1+|\xi|^2)^{-\beta/2}$and hence for any piece Γ_i $(i = 1-4)$ in which we can split $\Gamma(= \Gamma_1 + \Gamma_2 - \Gamma_3 - \Gamma_4)$ we have:

$$\begin{aligned} &\| [\nu(\xi)]^{-1}\Gamma_i(g,h) \| \\ &\leq \| R^{-1}[\nu(\xi)]^{-1}Q(R(1+|\xi|^2)^{-\beta/2}, R(1+|\xi|^2)^{-\beta/2}) \| \| h \| \| g \| \\ &\leq \| (1+|\xi|^2)^{-\beta/2} \| \| h \| \| g \| = C \| h \| \| g \| \end{aligned} \tag{2.28}$$

where we have noted that, e.g.,

$$\begin{aligned} (1+|\xi'|^2)^{-\beta/2}(1+|\xi'_*|^2)^{-\beta/2} &\leq (1+|\xi'|^2+|\xi'_*|^2)^{-\beta/2} \\ &\leq (1+|\xi|^2+|\xi^2_*)^{-\beta/2} \leq (1+|\xi|^2)^{-\beta/2} \end{aligned} \tag{2.29},$$

and this concludes the proof. □

Problems

1. Prove (2.15) and find explicit values for ν_0 and ν_1.
2. Prove that for any positive r, $j(\xi,v) = \exp(-\ \frac{1}{16}|v|^2)(1+|\xi|^2)^r(1+|\xi+v|^2)^{-r} \leq$ constant. (Note that if $|\xi+v| \leq |v|$ then $|\xi| \leq 2|v|$ and if $|\xi+v| \geq |v|$ then $|\xi| \leq 2|\xi+v|$. In the first case one can easily prove that $j(\xi,v) \leq (64r)^r$; in the second case that $j(\xi,v) \leq (4)^r$).

3. Prove that $(1+|\xi|^2)\int\int(1+|\xi_*|^2)^{-r}|\xi-\xi_*|\exp[-(|\xi|^2+|\xi_*|^2)/4]$ is uniformly bounded in ξ, for any $s, r \geq 0$.
4. Prove that $\| K(1-\chi_R) \| < C(1+R)^{-1} \to 0 \ (R\to\infty)$.

7.3 Spectral Properties of the Fourier-Transformed, Linearized Boltzmann Equation

We want to look for a solution of the Cauchy problem of Eq. (1.3) in $\Re^3$ or in a periodic box. As a preliminary step we consider the *linearized Boltzmann equation*, obtained by neglecting the nonlinear term in Eq. (1.3):

$$\frac{\partial h}{\partial t} + \xi \cdot \frac{\partial h}{\partial x} = Lh. \tag{3.1}$$

We first consider the case of $\Re^3$ and use the Fourier transform in x:

$$\hat{h}(k,\xi,t) = (2\pi)^{-3/2}\int h(x,\xi,t)e^{-ik.x}d^3x. \tag{3.2}$$

Then $\hat{h}$ satisfies

$$\frac{\partial \hat{h}}{\partial t} + i\xi \cdot k\hat{h} = L\hat{h} \tag{3.3}$$

or, for short:

$$\frac{\partial \hat{h}}{\partial t} = B(k)\hat{h} \tag{3.4}$$

where

$$B(k) = L - i\xi \cdot kI = K - \sigma(\xi;k)I. \tag{3.5}$$

Here $\sigma(\xi;k)$ is a function given by

$$\sigma(\xi;k) = \nu(|\,\xi\,|) + ik\cdot\xi. \tag{3.6}$$

Let us consider k as a parameter so that we deal with $L^2(\Re^3)$ for the moment. We want to study the semigroup $T(t;k)$ generated by $B(k)$. The first result is the following.

(7.3.1) Lemma. *The operator $B(k)$ with domain $D(B(k)) = \{f(\xi) : f \in L^2, |\xi| f \in L^2\}$ is an unbounded operator with domain dense in L^2, generating a strongly continuous semigroup $T(t;k)$ with*

$$||T(t;k)|| \leq 1. \tag{3.7}$$

Proof. The multiplication operator $S(k) = -\sigma(\xi;k)I$ generates the strongly continuous semigroup

$$U(t;k) = \exp[-\sigma(\xi;k)t]I. \tag{3.8}$$

$B(k)$, being a compact perturbation of $S(k)$, generates a strongly continuous semigroup $T(t;k)$ in L^2. Estimate (3.7) follows because L is self-adjoint, nonpositive, and $-ik \cdot vI$ antisymmetric. □

We note that this theorem establishes the existence of a unique solution of the Cauchy problem for the linearized Boltzmann equation in L^2; in fact if the initial condition is $h(0) = h_0, then h(t) = T(t)h_0$ where $T(t)h_0$ is the inverse Fourier transform of $T(t;k)\hat{h}_0$, where $\hat{h}_0$ is the Fourier transform of h_0.

We shall now study the asymptotic behavior of $T(t;k)$ when $t \to \infty$. To this end it is useful to recall the representation of $T(t;k)$ in terms of its Laplace transform $R(\lambda;k)$, which equals the resolvent of $B(k)$:

$$R(\lambda;k) = (\lambda I - B(k))^{-1}. \tag{3.9}$$

The mentioned representation reads as follows:

$$T(t;k)h = \frac{1}{2\pi i} \text{ s-} \lim_{\delta \to \infty} \int_{\gamma - i\delta}^{\gamma + i\delta} \exp(\lambda t) R(\lambda;k) h d\lambda \tag{3.10}$$

$(t, \gamma > 0, h \in D(B(k))$. This is a formal relation that we shall presently justify. To this end, we need a few results concerning the operator $R(\lambda;k)$. We shall write $\text{Re}\lambda$ and $\text{Im}\lambda$ for the real and imaginary parts of a complex number λ. The first result is given in the following lemma.

(7.3.2) Lemma. *For any fixed k, the operator $R(\lambda;k)$ is an analytic function of λ in the half-plane $\text{Re}\lambda \geq -\nu_0 + \epsilon$ $(\epsilon > 0)$ with the exception of a finite number of poles of finite multiplicity $\{\lambda_j(k)\}$. These poles satisfy the following conditions:*

1) $\text{Re}\lambda_j(k) \leq 0$ *and* $\text{Re}\lambda_j(k) = 0$ *implies* $\lambda = k = 0$.
2) $|\text{Im}\lambda_j(k)| \leq c(\epsilon)$.

Proof. $R(\lambda; k)$ can be expressed as follows

$$R(\lambda; k) = (\lambda I - B(k))^{-1} = (\lambda I - S(k) - K)^{-1}$$
$$(3.11) \qquad = (I - \overline{R}(\lambda; k)K)^{-1}\overline{R}(\lambda; k)$$

where $\overline{R}(\lambda; k) = (\lambda I - S(k))^{-1} = (\lambda + \sigma(\xi; k))^{-1} I$ is the multiplication by a function, analytic in λ for $\lambda > -\nu$. Since K is compact, $(I - \overline{R}(\lambda; k)K)^{-1}$ exists as a bounded operator with the exception of countably many isolated points. Eq. (3.11) shows that $R(\lambda; k)$ has the same property. In addition the points where $R(\lambda; k)$ is unbounded are those for which there is a function $\psi \neq 0$ such that $\overline{R}(\lambda; k)K\psi = \psi$ or

$$(3.12) \qquad B(k)\psi = \lambda\psi.$$

Condition 1) follows from direct calculation of Reλ from this equation. Further the compactness of K implies that the eigenvalues can only accumulate near the line Re$\lambda = -\nu_0$; this implies that for Re$\lambda \geq -\nu_0 + \epsilon$ ($\epsilon > 0$) there is only a finite number of eigenvalues and condition 2) holds. □

Further information on $\overline{R}(\lambda; k)$ is provided by

(7.3.3) Lemma. *For any* $\epsilon > 0$ $\| K\overline{R}(\lambda; k) \| \to 0$ *for* $|k| \to \infty$, *uniformly for* Re$\lambda \geq -\nu_0 + \epsilon$ *and* $\| K\overline{R}(\lambda; k) \| \to 0$ ($|\text{Im}\lambda| \to \infty$) *uniformly for* Re$\lambda \geq -\nu_0 + \epsilon$ *and* k *such that* $|k| < k_0$, *for any fixed* $k_0 > 0$.

Proof. In fact, if $\chi_R(\xi)$ is the characteristic function of the ball $|\xi| \leq R$, then the square-integrability of the kernel of K implies (Schwarz's inequality):

$$(3.13) \qquad \| K\chi_R\overline{R}(\lambda; k) \| \leq C\left(\int_{|\xi|<R} |\lambda + ik \cdot \xi + \nu(\xi)|^{-2} d\xi\right)^{1/2}.$$

The last integral can be subdivided into two contributions, one extended to the subset $|\text{Im}\lambda + k \cdot \xi| \leq |k|\delta$ ($\delta > 0$) and the other to the complement of this subset in $|\xi| < R$. The first subset is not larger than a parallelepipedon with two edges of length R and the third of length 2δ, so that its measure is less than $2R^2\delta$ and in it the integrand is less than ϵ^{-2}, while the second set has measure less than $4\pi R^3/3$ and the integrand is certainly not larger than $(|k|\delta)^{-2}$. Hence

$$(3.14) \qquad \| K\chi_R\overline{R}(\lambda; k) \| \leq CR[\epsilon^{-2}\delta + R(|k|\delta)^{-2}]^{1/2}.$$

If we choose $\delta = (R/|k|)^{2/3}$ we have

$$(3.15) \qquad \| K\chi_R\overline{R}(\lambda; k) \| \leq C(\epsilon)R^{4/3}|k|^{-1/3}.$$

On the other hand, Eq. (2.23) gives (thanks to $\| \overline{R}(\lambda; k) \| \leq \epsilon^{-1}$):

$$\| K(1 - \chi_R)\overline{R}(\lambda; k) \| \leq C\epsilon^{-1}(1 + R^2)^{-1/2}. \tag{3.16}$$

If we put together the two estimates and choose $R = |k|^{1/6}$, we obtain the first statement of the lemma. To prove the second one, let $|\mathrm{Im}\lambda| \geq 2k_0\mathrm{R}$. Then $|\mathrm{Im}\lambda + k \cdot \xi| \geq |\mathrm{Im}\lambda|/2$ whenever $|k| \leq k_0$ and $|\xi| \leq R$, so that for $\mathrm{Re}\lambda > -\nu_0 + \epsilon$,

$$\| K\chi_R\overline{R}(\lambda; k) \| \leq C(4\pi R^3/3)^{1/2}(\epsilon^2 + |\mathrm{Im}\lambda|^2/4)^{1/2}. \tag{3.17}$$

We can now choose here and in (3.16) $R = |\mathrm{Im}\lambda|^{2/5}$, which is possible for $|\mathrm{Im}\lambda| \geq (2k_0)^{5/3}$, since we have chosen $|\mathrm{Im}\lambda| \geq 2k_0R$. □

To proceed further, we need this lemma.

(7.3.4) Lemma. *For any* $\gamma = \mathrm{Re}\lambda > -\nu_0$, *we have*

$$\int_{\gamma - i\delta}^{\gamma + i\delta} \| \overline{R}(\lambda; k)h \|^2 \, d\lambda \leq \pi(\gamma + \nu_0)^{-1} \| h \|^2 . \tag{3.18}$$

Proof. If we recall the representation

$$\overline{R}(\lambda; k) = \int_0^\infty \exp(-\lambda t)U(t; k)dt \qquad (\mathrm{Re}\lambda > -\nu_0) \tag{3.19}$$

and denote by $\chi_+(t)$ the Heaviside step function, we have

$$\overline{R}(\lambda; k) = (2\pi)^{-1/2} \int_0^\infty \exp(-\lambda t)[(2\pi)^{1/2}U(t; k)\chi_+(t)]dt. \tag{3.20}$$

$(\mathrm{Re}\lambda > -\nu_0)$ which shows that, for a fixed value of $\mathrm{Re}\lambda$, $\overline{R}(\lambda; k)$ is the Fourier transform (in the variable $\mathrm{Im}\lambda$) of the function of t

$$(2\pi)^{1/2}U(t; k)\chi_+(t) \exp[-(\mathrm{Re}\lambda)t].$$

Then Parseval's equality gives

$$\int_{\gamma - i\delta}^{\gamma + i\delta} \| \overline{R}(\lambda; k)h \|^2 \, d\lambda = 2\pi \int_0^\infty \| U(t; k)h \|^2 \exp(-2\gamma t)dt. \tag{3.21}$$

Since $\| U(t; k)h \| \leq \| h \| \exp(-\nu_0 t)$, the lemma follows. □

According to Lemma 7.3.2, at the right of the line $\mathrm{Re}\lambda = -\nu_0 + \epsilon$ ($\epsilon > 0$) there is only a finite number of eigenvalues λ_j and they can be numbered as

$$\mathrm{Re}\lambda_1 \geq \mathrm{Re}\lambda_2 \geq \mathrm{Re}\lambda_3 \geq \ldots \mathrm{Re}\lambda_r \geq -\nu_0 + \epsilon. \tag{3.22}$$

We shall denote by P_j the projector on the eigenspace of $B(k)$ corresponding to the eigenvalue λ_j. If the eigenvalue is not simple, we denote by m_j its multiplicity. In this case we only know that $B(k) - \lambda_j I$ is nilpotent of rank m_j on the functions $P_j f$ obtained by projecting on the associated subspace of dimension m_j, but, in general, the functions of this subspace are not eigenfunctions of $B(k)$, because the associated matrix will in general be a "Jordan block." We remark that, in principle, we should write $\lambda_j(k), P_j(k), \ldots$ in place of $\lambda_j, P_j, \ldots$, but we shall do this only after the proof of Theorem 7.3.5, when we shall discuss the dependence of these quantities upon k.

Let us denote by P the projector on the subspace spanned by all the P_j, i.e.,

$$P = \sum_{j=1}^{r} P_j. \tag{3.23}$$

We can prove the following.

(7.3.5) Theorem. *Assume (for a given ϵ) that $\mathrm{Re}\lambda_j \neq -\nu_0 + \epsilon$. Then*

$$T(t;k)P = \sum_{j=1}^{r} (\exp(\lambda_j t)(P_j + \sum_{k=1}^{m_j} \frac{t^k}{k!} Q_j^k) \tag{3.24}$$

$$\| T(t;k)(I - P) \| \leq C \exp[(-\nu_0 + \epsilon)t]. \tag{3.25}$$

Here the Q_j are nilpotent operators associated with the Jordan block corresponding to the eigenvalue λ_j and C is a constant independent of k. More precisely (see Kato[8], p. 181) they are the residue of $-(\lambda - \lambda_j)R(\lambda;k)$ at a multiple pole of $R(\lambda;k)$.

Proof. The inverse Laplace transform for $T(t;k)$ is given by Eq. (3.10), which is no longer formal because of the estimate (2.18) . Further, we have

$$\begin{aligned} R(\lambda;k) &= (\lambda I - B(k))^{-1} \\ &= (\lambda I - S(k) - K)^{-1} = \overline{R}(\lambda;k)(I - K\overline{R}(\lambda;k))^{-1} \end{aligned} \tag{3.26}$$

and hence

$$R(\lambda;k) = \overline{R}(\lambda;k) + R(\lambda;k)K\overline{R}(\lambda;k) \tag{3.27}$$

and using again Eq. (3.26) on the right-hand side of Eq. (3.27):

$$R(\lambda;k) = \overline{R}(\lambda;k) + \overline{R}(\lambda;k)[I - K\overline{R}(\lambda;k)]^{-1}K\overline{R}(\lambda;k). \tag{3.28}$$

Then

$$\begin{aligned} T(t;k)h =& U(t;k)h + \frac{1}{2\pi i}\text{s-}\lim_{\delta\to\infty}\int_{\gamma-i\delta}^{\gamma+i\delta} \exp(\lambda t)\overline{R}(\lambda;k) \\ & \times [I - K\overline{R}(\lambda;k)]^{-1}K\overline{R}(\lambda;k)h d\lambda \end{aligned} \tag{3.29}$$

($\gamma > 0$). We now shift the integration line from $\text{Re}\lambda > 0$ to $\text{Re}\lambda = -\nu_0 + \epsilon$. Since the integrand has only poles in $\text{Re}\lambda \geq -\nu_0 + \epsilon$, we have

$$\int_{\gamma-i\delta}^{\gamma+i\delta} \exp(\lambda t)\overline{R}(\lambda;k)[I - K\overline{R}(\lambda;k)]^{-1}K\overline{R}(\lambda;k)h d\lambda$$

$$\begin{aligned} &= 2\pi i \sum_{j=1}^{r} \operatorname*{Res}_{\lambda=\lambda_j} \exp(\lambda t)\overline{R}(\lambda;k)[I - K\overline{R}(\lambda;k)]^{-1}K\overline{R}(\lambda;k)h d\lambda \\ &+ \int_{-\nu_0+\epsilon-i\delta}^{-\nu_0+\epsilon+i\delta} \exp(\lambda t)\overline{R}(\lambda;k)[I - K\overline{R}(\lambda;k)]^{-1}K\overline{R}(\lambda;k)h d\lambda + Z \end{aligned} \tag{3.30}$$

where

$$Z = \left(\int_{-\nu_0+\epsilon-i\delta}^{\gamma+i\delta} - \int_{\nu_(0)+\epsilon+i\delta}^{\gamma-i\delta}\right) exp(\lambda t)\overline{R}(\lambda;k)[I - K\overline{R}(\lambda;k)]^{-1}K\overline{R}(\lambda;k)h d\lambda. \tag{3.31}$$

In order to evaluate the residues in Eq. (2.30), we first remark that, because of (3.28), they are the same as the residues of $R(\lambda;k)$ (since λ_j is in the resolvent set of S). Then, because of the expansion of the resolvent (Kato[8], p.181), the following result follows:

$$\begin{aligned} &\operatorname*{Res}_{\lambda=\lambda_j} \exp(\lambda t)\overline{R}(\lambda;k)[I - K\overline{R}(\lambda;k)]^{-1}K\overline{R}(\lambda;k) \\ &= \operatorname*{Res}_{\lambda=\lambda_j} \exp(\lambda t)R(\lambda;k) = (\exp(\lambda_j t)(P_j + \sum_{k=1}^{m_j}\frac{t^k}{k!}Q_j^k). \end{aligned} \tag{3.32}$$

Next, because of Lemma 3.3, which also implies the boundedness of $[I - K\overline{R}(\lambda;k)]^{-1}$ for $\text{Im}\lambda$ sufficiently large, and since for $\text{Re}\lambda > -\nu_0$:

$$\| \overline{R}(\lambda;k) \| = \| (\lambda + ik\cdot\xi + \nu(\xi))^{-1}I \| \leq (\text{Re}\lambda + \nu_0)^{-1} \tag{3.33}$$

$\| Z \| \to 0$ when $\delta \to \infty$. Finally, since (by assumption) there are no eigenvalues on $\mathrm{Re}\lambda = -\nu_0 + \epsilon$, $\| [I - K\overline{R}(\lambda; k)]^{-1} \| \leq C$ on any compact set of that line and, because of the second statement of Lemma 3.3, on the entire line as well, with C independent of k. Then, for any $h, g \in L^2$:

(3.34)

$$|(\int_{-\nu 0+\epsilon-i\delta}^{-\nu_0+\epsilon+i\delta} exp(\lambda t)\overline{R}(\lambda; k)[I - K\overline{R}(\lambda; k)]^{-1} \times K\overline{R}(\lambda; k)h d\lambda, g)|$$

$$\leq \exp[-(\nu_0 - \epsilon)t] \int_{-\delta}^{\delta} |(\overline{R}(-\nu_0 + \epsilon + i\tau; k)$$

$$\times [I - K(-\nu_0 + \epsilon + i\tau; k)]^{-1} K\overline{R}(-\nu_0 + \epsilon + i\tau; k) d\tau, g)|$$

$$\leq C \| K \| \exp[-(\nu_0 - \epsilon)t] \int_{-\delta}^{\delta} \| \overline{R}(-\nu_0 + \epsilon + i\tau; k)h \|$$

$$\times \| \overline{R}(-\nu_0 + \epsilon - i\tau; k)g \| d\tau|.$$

Then, because of Eq. (3.18) the last integral is majorized by $\pi(\gamma + \nu_0)^{-1} \| h \| \| g \|$. This implies not only the convergence (for $\delta \to \infty$) of the operator $\int_{-\nu 0+\epsilon-i\delta}^{-\nu_0+\epsilon+i\delta} \exp(\lambda t)\overline{R}(\lambda; k)[I - K\overline{R}(\lambda; k)]^{-1} K\overline{R}(\lambda; k) d\lambda$ in the weak operator topology, but also that its limit satisfies:

$$\| \int_{-\nu_0+\epsilon-i\infty}^{-\nu_0+\epsilon+i\infty} \exp(\lambda t)\overline{R}(\lambda; k)[I - K\overline{R}(\lambda; k)]^{-1} K\overline{R}(\lambda; k) \| \leq C \exp[-(\nu_0 - \epsilon)t] \quad (t \in \Re_+). \tag{3.35}$$

If we combine the estimates we obtain the theorem. □

This important result was first obtained by Ukai[14] and is a key result for the treatment of the study of the asymptotic behavior of the linearized Boltzmann equation and of the existence theory for the weakly nonlinear Boltzmann equation.

The next step is due to Ellis and Pinsky[5] (see also McLennan[10] and Arsen'ev[3]):

(7.3.6) Theorem. *One can find positive numbers k_0 and σ_0 ($< \nu_0$) and functions $\mu_j(|k|) \in C^\infty([-k_0, k_0])$, $j = 0, 1, \ldots, 4$ such that*
a) for any $k \in \Re^3$ with $|k| \leq k_0$, there are five eigenvalues λ_j given by $\lambda_j(k) = \mu_j(|k|)$ where

$$\mu_j(|k|) = i\mu_j^{(1)}|k| - \mu_j^{(2)}|k|^2 + O(|k|^3) \qquad (k| \to 0) \tag{3.36}$$

and $\mu_j^{(1)} \in \Re$ and $\mu_j^{(2)} \in \Re_+$. In addition

$$P_j(k) = P_j^{(0)}(k/|k|) + |k|P_j^{(1)}(k/|k|)$$

$$Q_j(k) = 0, \tag{3.37}$$

for $j = 1, 2, \ldots, 5$, where $P_j^{(0)}$ are orthogonal projectors and

$$P_0 = \sum_{j=1}^{5} P_j^{(0)}(k/|k|) \tag{3.38}$$

does not depend on $k/|k|$ and is the projector on the five-dimensional eigenspace of the collision invariants.
b) for any $k \in \Re^3$ with $|k| > k_0$, there are no eigenvalues with $\mathrm{Re}\lambda \leq -\sigma_0$.

Proof. The fact that the eigenvalues only depend upon $|k|$ follows from the fact that the linearized collision operator commutes with any rotation $\mathcal{R}$ of $\Re^3$ and $\mathcal{R}k \cdot \mathcal{R}\xi = k \cdot \xi$; hence if $\varphi(\xi; k)$ is an eigenfunction corresponding to an eigenvalue λ, then $\varphi(\mathcal{R}\xi; \mathcal{R}k)$ is an eigenfunction corresponding to the same eigenvalue, which thus can only depend on $|k|$. Please note that the eigenfunction itself and the corresponding projector do not generally depend on $|k|$ alone, contrary to what is stated sometimes. We can now replace k by κe (where $e = k/\kappa$ is a unit vector and κ is $\pm|k|$) and look for a solution depending analytically on κ, according to Theorem K.VI.1.8 on the analytic perturbations of linear operators[8] in L^2, which, according to Theorem K.VII.2.6 and Remark K.VII.2.7, applies here, because there is a constant M, such that $\| e \cdot \xi h \| \leq M(\| h \| + \| Lh \|)$; thus the eigenvalues are analytic functions of κ. In particular, since for $\kappa = 0$ there are five eigenfunctions corresponding to the zero eigenvalue, for a sufficiently small κ there will be five eigenvalues (which may be distinct or not) whose expression will be given by Eq. (3.36). In order to show that $\mu_j^{(1)} \in \Re$ we remark that the operator $B(k)$ is invariant with respect to the product $\mathcal{CP}$ of the operations $\mathcal{P}$ of changing κ into $-\kappa$ and $\mathcal{C}$ of taking the complex conjugate; the same invariance applies to the eigenvalues, and this proves that $\mu_j^{(1)} \in \Re$. To prove that $\mu_j^{(2)} \in \Re_+$, we remark that if ψ_j are the normalized eigenfuctions, then $\psi_j(\kappa) = \psi_j(0) + \kappa\psi_j'(0) + \kappa^2\psi_j"(0)/2 + \ldots$ and hence:

$$\begin{aligned}\lambda_j &= (\psi_j, B(k)\psi_j) = (\psi_j, L\psi_j) - i\kappa(\psi_j, e \cdot \xi\psi_j) \\ &= (\psi_j(0), L\psi_j(0)) + \kappa(\psi_j'(0), L\psi_j(0)) + \kappa(\psi_j(0), L\psi_j'(0)) \\ &\quad + \kappa^2(\psi_j'(0), L\psi_j'(0)) + (\kappa^2/2)(\psi_j(0), L\psi_j"(0)) \\ &\quad + (\kappa^2/2)(\psi_j"(0), L\psi_j(0)) - i\kappa(\psi_j(0), e \cdot \xi\psi_j(0)) \\ &\quad - i\kappa^2(\psi_j'(0), e \cdot \xi\psi_j(0)) - i\kappa^2(\psi_j(0), e \cdot \xi\psi_j'(0)) + O(\kappa^3).\end{aligned} \tag{3.39}$$

Since, however, the functions $\psi_j(0)$ are collision invariants, many of the scalar products above are zero and we are left with:

$$\begin{aligned}\lambda_j =& - i\kappa(\psi_j(0), e \cdot \xi\psi_j(0)) + \kappa^2(\psi_j'(0), L\psi_j'(0)) \\ &- i\kappa^2[(\psi_j'(0), e \cdot \xi\psi_j(0)) + (\psi_j(0), e \cdot \xi\psi_j'(0))] + O(\kappa^3).\end{aligned} \tag{3.40}$$

We now remark that the term in square brackets is real (since the factor $e\cdot\xi$ is self-adjoint, the two terms in the brackets are complex conjugate of each other); but this would imply that λ_j is not invariant with respect to $\mathcal{CP}$, as it must be by the above argument, unless the term in square brackets vanishes. Our final expression for λ_j will thus be

$$\lambda_j = -i\kappa(\psi_j(0), e\cdot\xi\psi_j(0)) + \kappa^2(\psi_j'(0), L\psi_j'(0)) + O(\kappa^3). \tag{3.41}$$

This now coincides with Eq. (3.36) and indeed the coefficient of κ^2 is negative because of the properties of the linearized collision operator. (Note that $\psi_j'(0)$ cannot be a collision invariant because $L\psi_j'(0) = i[e\cdot\xi - (\psi_j(0), e\cdot\xi\psi_j(0))]\psi_j(0) \neq 0$).

Finally since the spectrum of $B(k)$ is discrete and depends analytically on κ, we can obtain (3.37) and (3.38); the latter is obvious and the former follows from the fact that analyticity allows us to take a purely imaginary κ and obtain a self-adjoint operator $B(\kappa e)$, for which diagonalization is possible (without Jordan blocks).

In order to prove *b*), we first prove that for any $\delta > 0$, there exists $k_0 = k_0(\delta)$ such that whenever $|k| < k_0$ and λ is in the discrete spectrum of $B(k)$, the following holds:

$$\mathrm{Re}\lambda \geq -\sigma_1 \qquad \text{implies} \qquad |\mathrm{Im}\lambda| \leq \delta \tag{3.42}$$

$$\mathrm{Re}\lambda \geq -\mu/2 \qquad \text{implies} \qquad |\lambda| \leq \delta \tag{3.43}$$

where $-\mu$ is, among the nonzero eigenvalues of L, the closest to the origin, while σ_1 is any real number between μ and $\nu(0)$.

In fact if (3.42) is violated, then, for some $\delta > 0$, there exists a sequence of real numbers $\{k_n\}$ converging to zero, a corresponding sequence of eigenvalues $\{\lambda_n\}$, and a sequence $\{h_n\}$, of L^2-functions (with unit norm) such that:

$$B(k_n)h_n = \lambda_n h_n \tag{3.44}$$

$$|\mathrm{Im}\lambda_n| > \delta, \qquad\qquad \mathrm{Re}\lambda_n \geq -\sigma_1. \tag{3.45}$$

Let us show that $\overline{\lim}|\mathrm{Im}\lambda_n| < +\infty$. In fact, Eq. (3.44) states that

$$Kh_n = (\lambda_n + \nu(\xi) + ik_n\cdot\xi)h_n. \tag{3.46}$$

Since K is compact, we may assume that $Kh_n \to g$ in L^2, by choosing a subsequence, if necessary. Now if $|\mathrm{Im}\lambda_n|$ converged to $+\infty$, we would have, from (3.46), that h_n should converge to zero, in contradiction to $\| h_n \| = 1$. Hence for some C, there are infinitely many indices n such that $\delta < |\mathrm{Im}\lambda_n| \leq$C. Since $0 \geq \mathrm{Re}\lambda_n \geq -\sigma_1$, we may extract a convergent subsequence $\{\lambda_n\}$ with limit λ such that $\mathrm{Im}\lambda \neq 0$. Taking the limit in Eq. (3.46), we obtain that h_n has a nonzero limit in L^2, which satisfies

$$Kh = (\lambda + \nu(\xi))h \tag{3.47}$$

and this, because of the self-adjointness of L implies $\mathrm{Im}\lambda = 0$, a contradiction that proves (3.42). To prove (3.43), we proceed in a similar way: if (3.43) is violated, then, for some $\delta > 0$, there exists a sequence of real numbers $\{k_n\}$ converging to zero, a corresponding sequence of eigenvalues $\{\lambda_n\}$ and a sequence $\{h_n\}$ of L^2-functions (with unit norm) such that Eq. (3.44) holds with

$$-\mu/2 \leq \mathrm{Re}\lambda_n \leq -\delta. \tag{3.48}$$

Because of (3.42), which we have just proved, $|\mathrm{Im}\lambda_n| \leq \delta$ and we can extract a subsequence $\{\lambda_n\}$ converging to a real λ with $-\mu/2 \leq \lambda \leq -\delta$. Taking the limit in Eq. (3.46), we obtain, as above, Eq. (3.47). Since $\lambda \neq 0$ and L is self-adjoint, h must be orthogonal to the null space of L; but this would imply, by the definition of $\mu, \lambda < \mu$, *a* contradiction that proves (3.43).

In order to finish the proof, let us show that there is a neighborhood $\mathcal{N}_1 \times \mathcal{N}_2$ of the origin in $\Re \times \mathbf{C}$, such that if $\lambda = \lambda(\kappa e)$ ($e \in S^2$) with $\kappa \in \mathcal{N}_1$ and $\lambda \in \mathcal{N}_2$, then λ is one of the five eigenvalues discussed in part a) of the theorem. Let us define $H = \nu^{-1}K - \nu^{-1/2}P_\nu\nu^{1/2}$, where P_ν projects on the subspace spanned by the eigenfunctions of $\nu^{-1/2}K\nu^{-1/2}$ corresponding to a unit eigenvalue. Clearly, H is compact. Also H cannot have $\lambda = 1$ as eigenvalue. Otherwise there would be a function h such that $Hh = h$ or

$$Lh = \nu^{1/2}P_\nu\nu^{1/2}\mathrm{h}. \tag{3.49}$$

Then projecting upon the null space of L we obtain that $\nu^{1/2}h$ is orthogonal to the range of P_ν (Problem 1). Since P_ν is a projector this implies that $P_\nu(\nu^{1/2}h) = 0$ and because of Eq. (3.49) h must be in the null space of L; this together with $P_\nu(\nu^{1/2}h) = 0$ implies $h = 0$ (Problem 2) and $\lambda = 1$ is not an eigenvalue of H.

Let us now prove that there exists a neighborhood $\mathcal{N}_1 \times \mathcal{N}_2$ of the origin in $\Re\times\mathbf{C}$ such that for $(\kappa, \lambda) \in \mathcal{N}_1 \times \mathcal{N}_2$, the operator $I-(\nu+ik\cdot\xi+\lambda)^{-1}\nu H$ is invertible. The operator $(\nu+ik\cdot\xi+\lambda)^{-1}\nu H$, being the product of a bounded operator by a compact operator, is compact. By the Fredholm alternative it is then sufficient to prove that there is no function h such that

$$h = (\nu + ik \cdot \xi + \lambda)^{-1}\nu Hh \tag{3.50}$$

for k and λ sufficiently small. Assume the contrary, i.e. that there exist sequences $\{k_n\}, \{\lambda_n\}$ converging to zero and $\{h_n\}$ ($\| h_n \| = 1$) satisfying (3.50); since H is compact and the factor multiplying it in (3.50) is bounded uniformly and converges strongly to 1 when k and λ converge to zero, we have that h_n converges to some g (with $\| g \| = 1$) that satisfies the limiting equation $Hg = g$. We have shown, however, that $Hg = g$ implies $g = 0$ and the invertibility of $I - (\nu + ik \cdot \xi + \lambda)^{-1} \nu H$ is proved.

We are now ready to attack the original problem $B(k)h = \lambda h$ by rewriting it in terms of H

$$\nu H h + \nu^{1/2} P_\nu \nu^{1/2} h = (\nu + ik \cdot \xi + \lambda) h \tag{3.51}$$

or

$$h = [I - (\nu + ik \cdot \xi + \lambda)^{-1} \nu H]^{-1} (\nu + ik \cdot \xi + \lambda)^{-1} \nu^{1/2} P_\nu \nu^{1/2} h. \tag{3.52}$$

This form of the problem gives h once $P_\nu \nu^{1/2} h$ is known; and since P_ν has a finite range, we can compute $P_\nu \nu^{1/2} h$ by solving a system of five linear algebraic equations in five unknowns, by simply projecting Eq. (3.52). The determinant of the system will be some analytic function of λ and κ, $D(\lambda, \kappa)$. For $\kappa = 0$, Eq. (3.51) is equivalent to $Lh = \lambda h$ and hence in a neighborhood of $\lambda = 0$ there will be only a fivefold degenerate zero of $D(\lambda, 0)$; by continuity, for a sufficiently small κ, there will be just five zeroes of $D(\lambda, \kappa)$ in a neighborhood of the origin in $\Re \times \mathbf{C}$, as was to be shown. This, when combined with (3.42) and (3.43) gives part *b*) of Theorem 7.3.6. □

From this theorem and the previous one, we obtain the following.

(7.3.7) Corollary. *There is a constant $C \geq 0$ such that*
a) for any $k \in \Re^3$ with $|k| \leq k_0$ (where k_0 is the same as in Theorem 7.3.6):

$$T(t; k) = \sum_{j=0}^{n+1} (\exp(\mu_j(|k|)t) P_j(K) + V(t, k) \tag{3.53}$$

$$\| V(t; k) \| \leq C \exp[(-\nu_0 + \epsilon)t] \qquad (t \geq 0) \tag{3.54}$$

b) for any $k \in \Re^3$ with $|k| > k_0$:

$$\| T(t; k) \| \leq C \exp[(-\nu_0 + \epsilon)t] \quad (t \geq 0). \tag{3.55}$$

The constant C in Eqs. (3.54) and (3.55) is independent of k because this is guaranteed by Theorem 7.3.5.

Problems

1. Prove that if P_ν projects on the eigenfunctions of $\nu^{-1/2}K\nu^{-1/2}$ corresponding to a unit eigenvalue. and $\nu^{1/2}P_\nu\nu^{1/2}h$ is orthogonal to the null eigenspace of L then $\nu^{1/2}h$ is orthogonal to the range of P_ν. (Hint: let $\chi_\alpha = R\psi_\alpha$ where ψ_α is a collision invariant; then P_ν projects on the space spanned by...; then $P_\nu(\chi_\alpha\nu^{1/2}) = \ldots$ and $(\chi_\alpha\nu^{1/2}, \nu^{1/2}h) = \ldots$).
2. Prove that if P_ν is as in Problem 1 then $P_\nu(\chi\nu^{1/2}) = 0$ implies $\chi = 0$ for any χ in the null space of L. (Hint: use the details of Problem 1).

7.4 The Asymptotic Behavior of the Solution of the Cauchy Problem for the Linearized Boltzmann Equation

We can now establish a decay estimate for $T(t)$, the semigroup generated by the operator $B = L - \xi\cdot\partial/\partial x$ in ordinary space. To this end we introduce the L^2-Sobolev space $H^s(\Re^3{}_x)$ and define

$$H_s = L^2(\Re^3{}_\xi, H^s(\Re^3{}_x)), \qquad L^{q,2} = L^2(\Re^3{}_\xi; L^q(\Re^3{}_x)) \tag{4.1}$$

and prove the following theorem.

(7.4.1) Theorem. *For any $s \in \Re$ and $q \in [1, 2]$, there is a constant $C \geq 0$ such that*

$$\| T(t)h \|_{H_s} \leq C(1+t)^{-m} \| h \|_{H_s \cap L^{q,2}} \tag{4.2}$$

$$\| T(t)(I - P_0)h \|_{H_s} \leq C(1+t)^{-m-1/2} \| h \|_{H_s \cap L^{q,2}} \tag{4.3}$$

where $m = 3(2-q)/4q$.

Proof. By Parseval's equality for Fourier transforms we have

$$\| T(t)h \|^2_{H_s} = \int (1 + |k|^{2s}) \| T(t;k)\hat{h} \|^2_{L^2} d^3k. \tag{4.4}$$

Recalling Corollary 7.3.7, we split the integral into two contributions I_1 and I_2, referring to $|k| \leq k_0$ and $|k| \geq k_0$, respectively. Then I_2 is bounded by $\exp[-2(\nu_0 - \epsilon)t] \| h \|^2_{H_s}$ and

$$I_1 \leq C\Big(\sum_{j=0}^{n+1} I_{1,j} + \exp[-2(\nu_0 - \epsilon)t] \| h \|^2_{H_s}\Big) \tag{4.5}$$

where

$$(4.6) \qquad I_{j,1} = \int_{|k|`lek_0} \exp(2\mathrm{Re}\mu_j(|k|)t) \parallel \hat{h}(k,.) \parallel^2_{L^2} .$$

By Theorem 7.3.6, there is a positive constant σ, such that

$$(4.7) \qquad \mathrm{Re}\mu_j(|k|) \leq -\sigma k^2 \qquad (j = 0, 1, \ldots, n+1, |k| \leq k_0).$$

Then by Hölder's inequality

$$(4.8) \qquad I_{j,1} \leq \Big(\int_{|k|\leq k_0} \exp(-\sigma q'|k|^2 t)dk)^{1/q'} \parallel \hat{h}(k,.) \parallel^2_{L^{2p'}(\Re^3_k; L^2_\xi(r^3))}\Big)$$

where $(1/p' + 1/q' = 1)$. The integral is majorized by

$$(4.9) \qquad \begin{aligned} &\int_{|k|\leq k_0} \exp(-\sigma q'|k|^2 t)dk \\ &\leq \exp[\sigma q'|k_0|^2] \int \exp[-\sigma q'|k|^2 (1+t)]dk \leq C(1+t)^{-3/2} \end{aligned}$$

while the norm in (4.8) is majorized, thanks to a well-known interpolation inequality for the Fourier transform[12], by $(2\pi)^{2-2/p'} \parallel\parallel h \parallel_{L^2} \parallel^2_{L^q} \leq (2\pi)^{2-2/p'} \parallel\parallel h \parallel_{L^q} \parallel^2_{L^2} = (2\pi)^{2-2/p'} \parallel h \parallel^2_{L^{q,2}}$, with $q = 2p'/(2p'-1)$, which proves (4.2). (Here we have used the fact, that, by convexity,

$$\parallel\parallel h \parallel_{L^2} \parallel_{L^q} \leq \parallel\parallel h \parallel_{L^q} \parallel_{L^2},$$

for $q \leq 2$). To prove (4.3) we proceed in the same way, but now we take into account that we get an extra factor $|k|^{q'}$ in the integral estimated in Eq. (4.9), thanks to Theorem 7.3.6; this leads to an exponent $3/2 + q'/2$ in place of $3/2$ and hence to an exponent $3/(2q') + 1/2$ in the final estimate (in place of $3/(2q')$). □

(7.4.2) *Remark.* The exponent $m = 3(2-q)/4q$ in the previous theorem is larger than $1/2$ if $q \in [1, 6/5)$ and takes the maximum value $3/4$ for $q = 1$.

The result that we have just proved indicates that the solution decays in time and that the component orthogonal to the collision invariants decays just a little bit faster. Since we want to use this result as a tool for attacking the weakly nonlinear problem, we must face the problem that $\Gamma(h,h)$ is not well defined in H_s, but it is, as will be shown in the next section, in the space $H_{s,\beta}$ defined by:

$$(4.10) \qquad \begin{aligned} &h \in H_{s,\beta} \Rightarrow h \in L^\infty_{\mathrm{loc}}(\Re^3_\xi, H^s(\Re^3{}_x)), \\ &\parallel h \parallel_{s,\beta} = \sup_\xi (1+|\xi|^2)^{\beta/2} \parallel h(\cdot,\xi) \parallel_{H^s(R^3)} < \infty. \end{aligned}$$

Hence, before proceeding further, we translate the decay estimate, which we have just found, into $H_{s,\beta}$, as first suggested by Grad[6]. Let us set

$$|h|_{m,s,\beta} = \sup_{t\geq 0}(1+t)^m \parallel h(t) \parallel_{s,\beta} \tag{4.11}$$

and prove the following.

(7.4.3) Theorem. *Let $q \in [1,2], s \in \Re, \beta \geq 0$ and $m = 3(2-q)/4q$ and $h = h(x,\xi)$ a function of $H_s \cap L^{q,2} \cap H_{s,\beta}$. Then there is a constant $C \geq 0$ such that, for any h:*

$$|T(t)(I-P_0)^n h|_{m+n/2,s,\beta} < C \parallel h \parallel_{H_s \cap L^{q,2} \cap H_{s,\beta}} . \tag{4.12}$$

Proof. This result can be obtained from the circumstance that the semigroup $U(t)$ generated by the operator $A = -\xi \cdot \partial/\partial x - \nu(\xi)I$ is related to the semigroup $T(t)$ generated by the full Boltzmann operator $B = A + K$, through

$$T(t)g = U(t)g + \int_0^t U(t-s)KT(s)gds. \tag{4.13}$$

Here g ("the initial data") is a function of x, ξ belonging to some Banach space, such as those used before. Eq. (4.13) is nothing other than the integral form of Eq. (3.1), obtained by rewriting the latter as

$$\frac{\partial h}{\partial t} + \xi \cdot \frac{\partial h}{\partial x} + \nu(\xi)h = Kh \tag{4.14}$$

and integrating along the characteristic lines of the left-hand side. Let us put

$$|h|_{m,X} = \sup_{t\geq 0}(1+t)^m \parallel h(t) \parallel_X, \tag{4.15}$$

a special case of which is Eq. (4.11). We are now going to exploit Theorem 7.2.4 and the fact that $\parallel U(t) \parallel \leq \exp(-\nu_0 t)$ in both H_s and $H_{s,\beta}$. Then Eq. (4.13) readily gives (see Problem 1):

$$|T(t)g|_{m,X} \leq C \parallel g \parallel_X + |T(t)g|_{m,Y} \tag{4.16}$$

for the pairs $X = H_{s,0}, Y = H_s$, and $X = H_{s,\beta+1}, Y = H_{s,\beta}, \beta \geq 0$. This permits an iterative use of this formula with respect to β to show that it also holds for $X = H_{s,\beta}, Y = H_s$ $(\beta \geq 0)$. The proof is now complete, because Theorem 7.4.1 gives an estimate of $\| T(t)g \|_{m,H_s}$. □

A problem we have to face when dealing with the nonlinear problem is that $\Gamma(h,h)$ is not bounded in $H_{s,\beta}$. In order to circumvent this difficulty, we shall need the smoothing properties of time integration, in the form of the following.

(7.4.4) Theorem. *Let* $0 \leq m' \neq 1/2, s \in \Re, \beta \geq 0$, *and* $0 \leq m < \min(2m', 5/4, 2m' + 1/4)$. *Then*

$$|Gh|_{m,s,\beta} < C(|h_{2m',s,\beta} + |\nu h|_{2m',H_s \cap L^{1,2}}) \tag{4.17}$$

where

$$Gh = \int_0^t T(t-s)(I-P_0)\nu h(s)ds. \tag{4.18}$$

Proof. For any $\alpha \geq 0$ we have

$$|G_n h|_{m,s,\beta} \leq C|h|_{m,s,\beta} \qquad (n = 0,1) \tag{4.19}$$

where

$$G_n h = \int_0^t U(t-s)(I-P_0)^n \nu h(s)ds \qquad (n = 0,1). \tag{4.20}$$

In fact, we have, taking the norm in $H^s(\Re^3_x)$,

$$\begin{aligned} &\| G_n h \|_{H^s(\Re^3_x)} \\ &\leq \int_0^t \exp[-\nu(\xi)(t-s)] \\ &\quad (I-P_0)^n \nu(\xi) \| h(\cdot,\xi,s) \|_{H^s(\Re^3_x)} \, ds \\ &\leq C \int_0^t \exp[-\nu(\xi)(t-s)]\nu(\xi)(1+s)^{-m}(1+|\xi|^2)^{-\beta/2}|h|_{m,s,\beta} ds \\ &\leq C(1+t)^{-m}|h|_{m,s,\beta}(1+|\xi|^2)^{-\beta/2} \qquad (n=0,1) \end{aligned} \tag{4.21}$$

where we have used the fact that $I - P_0$ is a bounded operator. Eq. (4.19) now easily follows. In order to obtain (4.17), we remark (see Problem 1) that

$$Gh = G_1 h + G_0([\nu(.)]^{-1} KGh) \tag{4.22}$$

and, proceeding as in the proof of (4.16), we obtain (for any $m \geq 0$):

$$|Gh|_{m,s,\beta} \leq C(|G_1 h|_{m,s,\beta} + |Gh|_{m,H^s}). \tag{4.23}$$

Combining this with Theorem 7.4.1 (for $q = 1$) yields (4.17). □

Problems

1. Prove (4.16).
2. Prove (4.22) (first differentiate, use the relation between the generators of T and U, and then integrate again).
3. Complete the proof of Theorem 4.4 by showing that Eq. (4.23) and Theorem 4.1 yield Eq. (4.17).

7.5 The Global Existence Theorem for the Nonlinear Equation

We now have all the preliminary results to be used to solve Eq. 1.3. Using the operator G defined in Eq. 4.18, we can write the corresponding integral equation in the following form:

$$h(t) = T(t)h_0 + G([\nu(.)]^{-1}\Gamma(h,h))(t) \equiv N(h)(t) \tag{5.1}$$

where Lemma 7.2.6 was taken into account. Eq. (5.1) shows that we must find a fixed point of the nonlinear mapping N. To this end, we first need the following lemma.

(7.5.1) Lemma. *Let $m \geq 0, s > 3/2$, and $\beta > 2$. Then there is a constant $C \geq 0$ such that*

$$|[\nu(.)]^{-1}\Gamma(h,g)|_{2m,s,\beta} + |\Gamma(h,h)|_{2m,H_s\cap L^{1,2}} \leq C|h|_{m,s,\beta}|g|_{m,s,\beta}. \tag{5.2}$$

Proof. The theorem is a consequence of the following three facts: (i) $H_{s,\beta}$ is a Banach algebra (i.e., a Banach space closed with respect to multiplication) for $s > 3/2$; (ii) H_s is continuously imbedded in $H_{1,\beta}$ if $\beta > 3/2$; (iii) if $h, g \in L^2$, then $uv \in L^1$ (this is applied to the dependence on x). Consequently the lemma follows from Lemma 7.2.6. □

Let us take $q \in [1,2]$ and set $m = m' = 3(2-q)/(4q)$ so that the conditions of Theorem 7.4.4 are satisfied. Combining Theorems 7.4.3 and 7.4.4 with Lemma 7.5.1, we see that the operator N appearing in Eq. (5.1) satisfies:

$$
(5.3) \qquad \begin{aligned} |N(h)|_{m,s,\beta} &\leq C_0 \parallel h_0 \parallel_{H_{s,\beta}\cap L^{q,2}} + C_1|h|^2_{m,s,\beta} \\ |N(h) - N(g)|_{m,s,\beta} &\leq C_1(|h|_{m,s,\beta} + |g|_{m,s,\beta}|h - g|_{m,s,\beta}). \end{aligned}
$$

This shows that N is contractive if h_0 is sufficiently small. We have thus proved the following theorem.

(7.5.2) Theorem. *Let $q \in [1,2], s > 3/2,$ and $\beta > 2$. Then there are positive constants c_0 and c_1, such that for any h_0 with*

$$
(5.4) \qquad \parallel h_0 \parallel_{H_{s,\beta}\cap L^{q,2}} \leq c_0
$$

Eq. (5.1) has a unique global solution $h \in L^\infty([0,\infty); H_{s,\beta})$ satisfying

$$
(5.5) \qquad |h|_{m,s,\beta} \leq c_1 \parallel h_0 \parallel_{H_{s,\beta}\cap L}\ q,2 \qquad (m = 3(2-q)/(4q).
$$

This result was proved independently by N. B. Maslova and A. N. Firsov[9], Nishida and Imai[11], and Ukai[13], after the paper by Ukai[14] had given the basic results on the weakly nonlinear Boltzmann equation. In the latter paper Ukai had actually given a deeper result proving that the solution is more regular than proved in Theorem 7.5.2 and is actually a classical solution of the Boltzmann equation. In order to discuss this result, we define

$$
(5.6) \qquad \dot{H}_{s,\beta} = \{h \in H_{s,\beta};\parallel [1 - \chi_R(|\xi| + |k|)]\hat{h} \parallel_{s,\beta} \to 0 \text{ as } R \to \infty\}
$$

where k is the variable conjugate to x in the Fourier transform. Ukai and Asano[16] proved the following facts: (i) $U(t)$ and hence $T(t)$ are, for any s and β, C_0-semigroups on $\dot{H}_{s,\beta}$, although not in $H_{s,\beta}$, with the domains of the generators related by:

$$
(5.7) \qquad D(A) = D(B) \supset \dot{H}_{s+1,\beta+1}.
$$

Also

$$
(5.8) \qquad [\nu(.)]^{-1}\Gamma(.,.) \quad \text{maps} \quad \dot{H}_{s,\beta} \times \dot{H}_{s,\beta} \quad \text{into} \quad \dot{H}_{s,\beta} \quad \text{if} \quad s > 3/2, \beta \geq 0.
$$

It follows that $N(h) \in C^0([0,\infty); \dot{H}_{s,\beta})$ if $h \in C^0([0,\infty); \dot{H}_{s,\beta})$ and if $h_0 \in \dot{H}_{s,\beta}$. Eq. (5.7) with $s-1, \beta-1$ in place of s, β then leads to Theorem 7.5.3.

(7.5.3) Theorem. *Let h, h_0 be as in Theorem 5.2. If, in addition, $h_0 \in \dot{H}_{s,\beta}$, then $h \in C^0([0,\infty); \dot{H}_{s,\beta}) \cap C^1([0,\infty); \dot{H}_{s-1,\beta-1})$ and is a classical solution of Eq. (1.3) with initial value h_0 and hence $f = M + M^{1/2}h$ is a classical solution of Eq. (1.1).*

Since $H_{s,\beta} \in \dot{H}_{s-\epsilon,\beta-\epsilon}$ for any $\epsilon > 0$, the previous two theorems lead to the following.

(7.5.4) Theorem. *Let h, h_0 be as in Theorem 7.5.2. Then*

$$h \in L^\infty([0,\infty); H_{s,\beta}) \cap C^0([0,\infty); H_{s-\epsilon,\beta-\epsilon}) \cap C^1([0,\infty); H_{s-1-\epsilon,\beta-1-\epsilon})$$

($\epsilon > 0$) and is a classical solution of Eq. (1.3), and hence $f = M + M^{1/2}h$ is a classical solution of Eq. (1.1).

This is the theorem originally given by Ukai[13,14]; the conciseness of his papers and the fact that many readers did not appreciate the meaning of the "$-\epsilon$" in the subscripts of Theorem 7.5.4 generated the rumor, unfortunately echoed by some of the books on kinetic theory, that the statement of Ukai[11,14] was not completely correct, but this, as we have seen, is not the case.

7.6 Extensions: The Periodic Case and Problems in One and Two Dimensions

It is easy to see that the previous arguments also provide the global existence for the Cauchy problem for Eq. (1.3) when the solution is looked for in a box with periodicity boundary conditions. This result has a physical meaning because the solution of the problem in a box with specular reflection can be reduced[7] to that with periodicity conditions by considering 2^3 contiguous boxes, each of which is the mirror image of the neighboring ones (Problem 1).

In the periodic case, it is natural to use the Fourier series instead of the Fourier integral. The proof is actually simpler because k is never close to the origin (unless $k = 0$). Then Theorem 7.4.1 simplifies because the projection onto the subspace spanned by the collision invariants does not decay in time and the remaining part decays exponentially. We remark that the reason for the decay is different in the two cases. In a bounded domain, the dissipativity of L has a crucial role together with the fact that the natural basis for representing the space dependence of the solution is discrete (a Fourier series replaces the Fourier integral). In the case of $\Re^3$ the dispersion properties of the free-streaming operator ensure a decay (although not exponential) in time.

In fact Theorem 7.4.1 is now replaced by the following.

(7.6.1) Theorem. *For any $s \in \Re$, there is a constant $\sigma_0 > 0$ such that*

$$\| T(t)h \|_{H_s} \leq C \| h \|_{H_s} \tag{6.1}$$

(6.2) $$\| T(t)(I - P_0)h \|_{H_s} \leq C \exp(-\sigma_0 t) \| h \|_{H_s}$$

where $C \geq 0$ is independent of u and $t \geq 0$.

Here, of course, $H_s = L^2(\Re^3{}_\xi; H^s(\mathbf{T}^3_x))$, where $\mathbf{T}^3_x$ is a three-dimensional torus and (4.4) is, e.g, replaced by

(6.3) $$\| T(t)h \|^2_{H_s} = \sum_{k \in \mathbf{Z}^3} (1 + | k|^{2s}) \| T(t;k)\hat{h}(k,.) \|^2_{L^2} .$$

Similarly, the other results for the case of $\Re^3$ can be translated into theorems for $\mathbf{T}^3$ to arrive at the global existence result for the periodic case.

(7.6.2) Theorem. *Let $s > 3/2$ and $\beta > 2$. Then there are positive constants c_0 and c_1, such that for any h_0 with*

(6.4) $$\| h_0 \|_{H_{s,\beta}} \leq c_0.$$

Eq. (1.1) associated with periodicity boundary conditions has a unique global solution $h \in L^\infty([0,\infty); H_{s,\beta})$, which, if, in addition, $P_0 h_0 = 0$, then $P_0 u(t) = 0$ for all $t \geq 0$ and

(6.5) $$\sup_{T \geq 0} \exp(\sigma_0 t) |h|_{m,s,\beta} \leq c_1.$$

For the sake of clarity, let us remark that the projection P_0 is taken in the Hilbert space $L^2(\Re^3{}_\xi \times \mathbf{T}^3_x)$ and thus the restriction $P_0 h_0 = 0$ is not so important, because it can always be satisfied by an appropriate choice of the parameters in the Maxwellian M. This choice is, of course, not available in the case of $\Re^3{}_\xi \times \Re^3{}_x$, because the Maxwellian is constant in x and hence not integrable in $\Re^3{}_x$.

Theorem 6.2 is due to S. Ukai[14] and was the first global existence theorem concerning the Cauchy problem for the space-inhomogeneous Boltzmann equation.

Another important remark concerns the solution of the Boltzmann equation when the data, and hence the solution, depends on just one or two space variables. The existence theorems in bounded domains apply without any difficulty (see Problems 2 and 3), because one has only to restrict k to belong to $\mathbf{Z}$ and $\mathbf{Z}^2$ rather than to $\mathbf{Z}^3$. The matter is more delicate in one and two dimensions, because of the role played by the space dimension in the estimates of Theorem (4.7). As remarked by Ukai[15], however, the results remain valid in this case as well (see Problems 4 and 5).

Problems

1. Show that the initial value problem in a box with specular reflection reduces to the problem with periodicity boundary conditions (see Ref. 6).
2. Extend the theorem on the torus $\mathbf{T}_x^3 \times \Re_\xi^3$ to $\mathbf{T}_x^2 \times \Re_\xi^3$.
3. Extend the theorem on the torus $\mathbf{T}_x^3 \times \Re_\xi^3$ to $\mathbf{T}_x \times \Re_\xi^3$.
4. Extend the existence theorem from $\Re_x^3 \times \Re_\xi^3$ to $\Re_x^2 \times \Re_\xi^3$.
5. Extend the existence theorem from $\Re_x^3 \times \Re_\xi^3$ to $\Re_x \times \Re_\xi^3$.

7.7 A Further Extension: Solutions Close to a Space Homogeneous Solution

The constructive existence theory developed so far in this book essentially concerns data that are space homogeneous, small perturbations of a vacuum, or small perturbations of equilibrium. It is natural to try to handle another situation, i.e., the case when the initial data are sufficiently close to a space homogeneous distribution, or, in other words, the case when the space gradients are small.

To fix the ideas we shall consider the periodic gas, i.e., a gas in a three-dimensional flat torus $\mathbf{T}^3$. Let $f_0(x, \xi)$ be the initial value for the distribution function. Then we define

$$g_0(\xi) = \int_{\mathbf{T}^3} f_0(x, \xi) dx \tag{7.1}$$

and let

$$u_0(x, \xi) = f_0(x, \xi) - g_0(\xi). \tag{7.2}$$

Then one can hope to prove a global existence theorem for solutions of the Boltzmann equation with initial data f_0, which satisfy a smallness assumption on u_0, on the basis of the following steps:

Step 1 (Local theorem). Let g be the solution of the homogeneous problem with initial data g_0:

$$\frac{\partial g}{\partial t} = Q(g, g); \qquad g(0) = g_0. \tag{7.3}$$

If we set

$$u = f - g \tag{7.4}$$

where f is assumed to solve the Boltzmann equation with initial data f_0:

$$\frac{\partial f}{\partial t} + \xi \cdot \frac{\partial f}{\partial x} = Q(f,f); \qquad f(0) = f_0 \tag{7.5}$$

then u satisfies

$$\frac{\partial u}{\partial t} + \xi \cdot \frac{\partial u}{\partial x} = 2Q(g,u) + Q(u,u); \qquad u(0) = u_0. \tag{7.6}$$

For this problem it is conceivable to establish a local existence theorem in $[0, t_1]$, where t_1 is as large as we want, provided u_0 is small enough.

Step 2 (Approach to equilibrium). We know (from the study of the space homogeneous problem) that $g(t)$ approaches a Maxwellian M as $t \to \infty$.

Step 3 (Perturbation of equilibrium). Thanks to step 2 and choosing t_1 in step 1 sufficiently large, $g(t_1)$ is simultaneously close to M and $f(t_1)$. We can then use this circumstance to try to exploit the theory developed in the previous sections for solutions close to a Maxwellian in order to extend the solution to (t_1, ∞).

Unfortunately the above strategy cannot be carried out so easily, because the words "close" and "small," which we have repeatedly used, refer to different topologies.

Let us analyze step 2. We know the following proposition from the theory of space homogeneous solutions (see Chapter 6).

(7.7.1) Theorem. *Let $g_0 \in B_s (s < s_0)$ with s_0 sufficiently large. Then*

$$\lim_{t \to \infty} \| g(t) - M \|_r = 0 \qquad (r < s) \tag{7.7}$$

where M is the Maxwellian with the same conserved moments as g_0.

The theorem (where B_s is the Banach space with norm $\| g \| = \sup_\xi (1 + |\xi|^2)^{s/2} |g(\xi)|)$ is due to Carleman[4] and strengthens our analysis from Chapter 6.

We remark, however, that even assuming that g is bounded by a Maxwellian, we do not know that the same is true for $g(t)$. Actually, we have a control on the solution in spaces (such as B_s) with polynomial weights only. This fact is either an unexpected feature or a gap in the rather complete theory for the space homogeneous Boltzmann equation. As a matter of fact, because of this circumstance, Steps 1 and 3 of the above strategy become problematic. In fact, the theory of small deviations from equilibrium, as discussed in the present chapter, requires a Maxwellian bound at $t = 0$, and this makes it impossible to apply this theory to the problem under consideration. There is, however, a suitable decomposition of velocity space, which, when combined with the theory discussed so far and new estimates of the collision operator, allows us to prove the following theorem, due to Arkeryd, Esposito, and Pulvirenti[2], which extends the analysis of the solutions close to equilibrium to the case of polynomially bounded perturbations.

(7.7.2) Theorem. *For fixed* s_0 *and* l_0 *sufficiently large and any* $s > s_0$ *and* $l > l_0$, *we can find* $b, b', \gamma > 0$, *such that, if* $u_o \in H_{l,s}$ *with*

$$\| u_0 \|_{l,s} < B, \tag{7.8}$$

then there exists a unique classical solution of the Boltzmann equation $f = M + u$ *(where* M *is the Maxwellian associated with* f_0*), where*

$$u \in L^\infty([0,T], H_{l,s}) \cap C^l([0,T], H_{l-1-\epsilon,s-1-\epsilon}) \tag{7.9}$$

satisfying the bound:

$$\| u(t) \|_{l,s} < b' e^{-\gamma t}. \tag{7.10}$$

The proof of this theorem is rather technical and will be omitted here. For the proof see the paper by Arkeryd, Esposito, and Pulvirenti[2].

The next step in our program is the proof of a local existence theorem in $H_{l,s}$.

(7.7.3) Theorem. *Under the same assumptions as in Theorem 7.7.2, let* g_0 *and* u_0 *be as in Eqs (7.1) and (7.2), with* $g_0 \in B_{s+1}, u_0 \in H_{l,s}$ *and*

$$\| u_0 \|_{l,s} < \frac{1}{24C} \exp(-2t_1 C \log 4) \tag{7.11}$$

where $t_1 > 0$ *is given and* C *is large enough. Then there is a unique classical solution* $f = g + u$ *of the Boltzmann equation with initial datum* f_0 *up to time* t_1. *Moreover:*

$$\| u \|_{l,s} < \| u_0 \|_{l,s} \exp(2t_1 C \log 4). \tag{7.12}$$

The proof of this theorem is also rather technical: it is based on sharp estimates of the collision operator and will be omitted here. For the proof see the paper by Arkeryd, Esposito, and Pulvirenti[2].

With these results at our disposal, we are now in a position to prove the following theorem.

(7.7.4) Theorem. *Let g_0 and u_0 be as in Eqs (7.1) and (7.2), with $g_0 \in B_{s+1}, u_0 \in H_{l,s}$ with $s > s_0, l > l_0$ and s_0, l_0 sufficiently large.*
Then there is a unique positive global classical solution $f = g + u$ of the Boltzmann equation with initial datum f_0, where g solves the homogeneous equation and

$$(7.13) \qquad u \in L^\infty([0,\infty), H_{l,s}) \cap C^l([0,\infty), H_{l-1-\epsilon,s-1-\epsilon})$$

provided $\| u_0 \|_{l,s}$ is sufficiently small. Moreover, $f(t) \to M$ (where M is the Maxwellian associated with f_0) in $H_{l,s}$, as $t \to \infty$.

Proof. The proof can be easily obtained thanks to Theorems 7.7.2 and 7.7.3. In fact, for fixed s and l, by Theorem 7.7.1, for any $b > 0$

$$(7.14) \qquad \| g(t) - M \|_s < b/2 \qquad \text{for} \qquad t > t_*$$

provided t_* is large enough. By Theorem 7.3

$$(7.15) \qquad \| f(t^*) - g(t^*) \|_{l,s} = \| u(t^*) \|_{l,s} \leq b/2$$

provided $\| u_0 \|_{l,s}$ is small enough. Finally

$$(7.16) \qquad \| f(t^*) - M \|_{l,s} \leq \| u(t^*) \|_{l,s} + \| g(t^*) - M \|_{l,s} < b$$

so that we can extend the solution to arbitrarily long times by Theorem 7.7.2. □

As mentioned before, the arguments in this section are taken from a paper by Arkeryd, Esposito, and Pulvirenti[2]. Subsequent developments are due to Arkeryd[1] and Wennberg[17].

References

1. L. Arkeryd, “Stability in L^1 for the spatially homogeneous Boltzmann equation,” *Arch. Rat. Mech. Anal.* **103**, 151–167 (1988).
2. L. Arkeryd, R. Esposito, and M. Pulvirenti, “The Boltzmann equation for weakly inhomogeneous data,” *Commun. Math. Phys.*, **111**, 393 (1988).
3. A. A. Arsen'ev, “The Cauchy problem for the linearized Boltzmann equation,” *USSR Comput. Math. and Math. Phys.* **5**, 110–136 (1965).
4. T. Carleman, *Problèmes mathématiques dans la théorie cinétique des gaz*, Almqvist & Wiksells, Uppsala, 1957.
5. R. S. Ellis and M. A. Pinsky, “The first and second fluid approximations to the linearized Boltzmann equation,” *J. Math. Pures et Appl.* **54**, 125–156 (1972).
6. H. Grad, “Asymptotic equivalence of the Navier-Stokes and nonlinear Boltzmann equations,” *Proc. Symp. Appl. Math.* **17**, R. Finn, ed., 154–183, AMS, Providence (1965).
7. H. Grad, “Asymptotic theory of the Boltzmann equation, II” in *Rarefied Gas Dynamics*, J. A. Laurmann, ed., Vol. I, 26–59 (1963).

8. T. Kato, *Perturbation Theory of Linear Operators,* Springer, New York (1966).
9. N. B. Maslova and A. N. Firsov, "Solution of the Cauchy problem for the Boltzmann equation" (in Russian), Vestnik Leningrad Univ. **19**, 83–85 (1975).
10. J. A. McLennan, "Convergence of the Chapman-Enskog expansion for the linearized Boltzmann equation," *Phys. Fluids* **8**, 1580–1584 (1965).
11. T. Nishida and K. Imai, "Global solutions to the initial value problem for the non-linear Boltzmann equation," *Publ. R.I.M.S. Kyoto Univ.* **12**, 229–239 (1976).
12. E. C. Titchmarsh, *Introduction to the Theory of Fourier Integral*, Oxford University Press (1948).
13. S. Ukai, "Les solutions globales de l'équation de Boltzmann dans l'espace tout entier et dans le demi-espace," *C. R. Acad. Sci., Paris,* **282A**, 317–320 (1976).
14. S. Ukai, "On the existence of global solutions of a mixed problem for the nonlinear Boltzmann equation," *Proc. Japan Acad.* **50**, 179–184 (1974).
15. S. Ukai, "Solutions of the Boltzmann equation," in *Patterns and Waves-Qualitative Analysis of Nonlinear Differential Equations, Studies in Mathematics and Its Applications* **18**, 37–96 (1986).
16. S. Ukai and Asano, "On the Cauchy problem of the Boltzmann equation with a soft potential," *Publ. R.I.M.S. Kyoto Univ.* **18**, 477–519 (1982).
17. B. Wennberg, "Stability and exponential convergence in L^p for the spatially homogeneous Boltzmann equation," submitted to *J. Nonlinear Anal. Th. & A.*

8
Boundary Conditions

8.1 Introduction

If we want to describe a physical situation where a gas flows past a solid body or is contained in a region bounded by one or more solid bodies, the Boltzmann equation must be accompanied by boundary conditions, which describe the interaction of the gas molecules with the solid walls. It is to this interaction that one can trace the origin of the drag and lift exerted by the gas on the body and the heat transfer between the gas and the solid boundary. Hence, in order to write down the correct boundary conditions for the Boltzmann equation we need information that stems from a discipline that may be regarded as a bridge between the kinetic theory of gases and solid-state physics.

The difficulties of a theoretical investigation are due, mainly, to our lack of knowledge of the structure of surface layers of solid bodies and hence of the effective interaction potential of the gas molecules with the wall. When a molecule impinges upon a surface, it is adsorbed and may form chemical bonds, dissociate, become ionized, or displace surface molecules. Its interaction with the solid surface depends on the surface finish, the cleanliness of the surface and its temperature. It may also vary with time because of outgassing from the surface. Preliminary heating of a surface also promotes purification of the surface through emission of adsorbed molecules. In general, adsorbed layers may be present; in this case, the interaction of a given molecule with the surface may also depend on the distribution of molecules impinging on a surface element.

The first observations of the interaction of gases with solid surfaces (apart from early studies, which arrived at the conclusion that the gas does not slip on the wall under standard conditions) are due to Kundt and Warburg[18]. They noted that the flow rates through tubes at very low pressures are appreciably higher than predicted by the familiar Poiseuille formula and attributed that effect to slip at the boundary. Maxwell[23] suggested that slip would be a consequence of kinetic theory and computed the amount of slip for a particular model of the gas-surface interaction, still very popular and known under his name. Smoluchowski[29] described temperature jump in a similar way. A more systematic effort started with Knudsen[17]. Full-scale research, however, only began about thirty years ago, under the impetus of space flight and thanks to the developments in high vacuum technology.

We cannot deal here with the details of physical models and computations that aim at simulating the complex phenomena to which we have briefly alluded. We refer the interested reader to the surveys of Kuščer[19] and Cercignani[7] as well as to the book y Cercignani[3].

8.2 The Scattering Kernel

In general, a molecule striking a surface with a velocity ξ' reemerges from it with a velocity ξ that is strictly determined only if the path of the molecule within the wall can be computed exactly. This computation is very hard, because it depends on a great number of details, such as the locations and velocities of all the molecules of the wall and an accurate knowledge of the interaction potential (see Sections 4 and 6). In fact an exact, or sufficiently accurate, calculation should be able to predict all the phenomena that we mentioned in the previous section.

Hence it is more convenient to think in terms of a probability density $R(\xi' \to \xi; x, t; \tau)$ that a molecule striking the surface with velocity between ξ' and $\xi' + d\xi'$ at the point x and time t will reemerge (see Fig. 27) at practically the same point with velocity between ξ and $\xi + d\xi$ after a time interval τ (adsorption or sitting time). If R is known, then we can easily write down the boundary condition for the distribution function $f(x, \xi, t)$. To simplify the discussion, the gas will be presently assumed to be monatomic. In addition the surface will be assumed to be at rest. We remark that the probability density is in general a distribution and not an ordinary function; this requires a suitable interpretation of the equations, which we shall write. This aspect of the matter will not be discussed explicitly, given the heuristic character of the present chapter.

The mass (or number, depending on normalization) of molecules emerging with velocity between ξ and $\xi + d\xi$ from a surface element dA about x in the time interval between t and $t + dt$ is

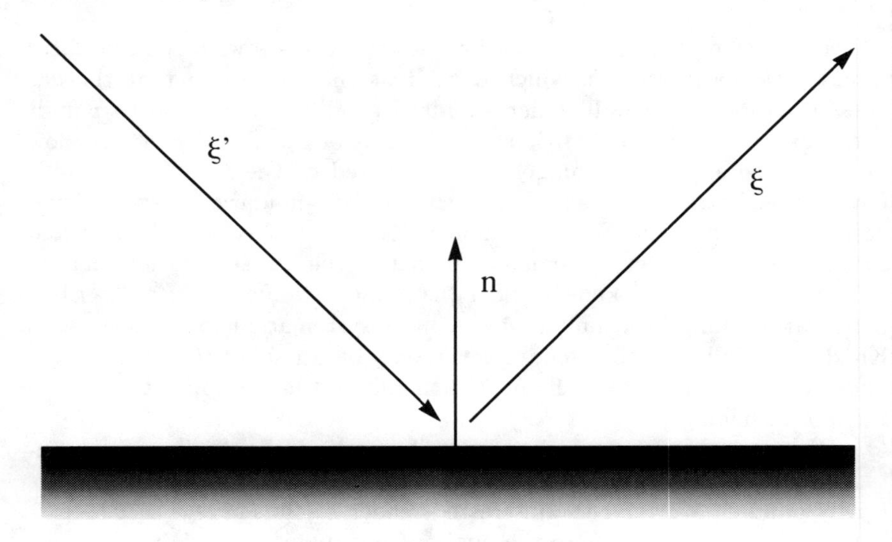

FIGURE 27.

$$d^*\mathcal{M} = f(x, \xi, t)|\xi \cdot n| dt dA d\xi \qquad (x \in \partial\Omega, \xi \cdot n > 0) \tag{2.1}$$

where n is the unit vector normal to the surface $\partial\Omega$ at x and directed from the wall into the gas. Analogously, the probability that a molecule impinges upon the same surface element with velocity between ξ' and $\xi' + d\xi'$ in the time interval between $t - \tau$ and $t - \tau + dt$ $(\tau > 0)$ is

$$d^*\mathcal{M}' = f(x, \xi', t - \tau)|\xi' \cdot n| dt dA d\xi' \qquad (x \in \partial\Omega, \xi' \cdot n < 0). \tag{2.2}$$

If we multiply $d^*\mathcal{M}'$ by the probability of a scattering event from velocity ξ' to a velocity between ξ and $\xi + d\xi$ with an adsorption time between τ and $\tau + d\tau$ (i.e. $R(\xi' \to \xi; x, t; \tau) d\xi d\tau$) and integrate over all the possible values of ξ' and τ, we must obtain $d^*\mathcal{M}$ (here we assume that each molecule reemerges from the surface element into which it entered, which is not so realistic when τ is large):

$$d^*\mathcal{M} = d\xi \int_0^\infty d\tau \int_{\xi' \cdot n < 0} R(\xi' \to \xi; x, t; \tau) d^*\mathcal{M}' \qquad (x \in \partial\Omega, \xi \cdot n > 0). \tag{2.3}$$

Equating the expressions in Eqs. (2.1) and (2.3) and cancelling the common factor $dA d\xi dt$, we obtain

(2.4)

$$f(x,\xi,t)|\xi\cdot n| = \int_0^\infty d\xi' \int_{\xi'\cdot n<0} R(\xi'\to\xi;x,t;\tau)f(x,\xi',t-\tau)|\xi'\cdot n|d\xi'$$

$$(x\in\partial\Omega, \xi\cdot n>0).$$

If $\bar{\tau}$ is an average adsorption time, $\bar{\xi}$ the average normal velocity with which the gas molecules impinge upon the surface, and $\bar{n}$ the number density of the gas, $\bar{n}\bar{\xi}dA$ molecules will impinge, per unit time on a surface element of area dA and stay there an average time $\bar{\tau}$; if σ_0 is an effective range of the gas-surface interaction, each molecule will occupy an area of the order of σ_0^2 and the total area occupied by the adsorbed molecules will be $\bar{n}\bar{\xi}\bar{\tau}\sigma_0^2 dA$, i.e. a fraction $\bar{n}\bar{\xi}\bar{\tau}\sigma_0^2$ of the surface will be occupied. This is, of course, just a rough order-of-magnitude argument since the molecules may penetrate somewhat into the solid and not necessarily remain at the surface of it.

If $\bar{n}\bar{\xi}\bar{\tau}\sigma_0^2$ is not close to zero, the nature of the interaction of each incident molecule depends on the total number and energy of the incident molecules. Under conditions of extremely low density (as, for example, in the case of an orbiting satellite) $\bar{n}\bar{\xi}\bar{\tau}\sigma_0^2 << 1$ and each incident molecule interacts with the surface independently of the others. The same independence may show up in the other limiting case $\bar{n}\bar{\xi}\bar{\tau}\sigma_0^2 \cong 1$ (for example, in chemical adsorption when $\bar{\tau}$ may be very large, or in a dense gas when $\bar{n}$ can be extremely large); in this case the impinging molecule interacts directly with the adsorbed layer rather than with the atoms of the solid surface (and the effective adsorption time for that interaction can be much shorter than $\bar{\tau}$).

Whenever the effective adsorption time $\bar{\tau}_e$ and the number density are such that $\bar{n}\bar{\xi}\bar{\tau}\sigma_0^2 << 1$, scattering events can safely be considered instantaneous and statistically independent and we can assume that the kernel $R(\xi'\to\xi;x,t;\tau)$ does not depend on the distribution function $f(x,\xi,t)$; hence we can compute $R(\xi'\to\xi;x,t;\tau)$ under the assumption that one gas molecule of given velocity ξ' impinges upon the wall. We shall always assume that we deal with the situation that we have just described; then the kernel $R(\xi'\to\xi;x,t;\tau)$ provides a complete description of the surface scattering law.

If, in addition $\bar{\tau}_e$ is small compared to any characteristic time of interest in the evolution of f, we can let $\tau=0$ in the argument of f appearing on the right-hand side of Eq. (2.4); in this case the latter becomes:

$$f(x,\xi,t)\mid\xi\cdot n\mid = \int_{\xi'\cdot n<0} R(\xi'\to\xi;x,t)f(x,\xi',t)\mid\xi'\cdot n\mid d\xi' \tag{2.5}$$

$$(x\in\partial\Omega, \quad \xi\cdot n>0)$$

where

(2.6)
$$R(\xi' \to \xi; x, t) = \int_0^\infty d\tau R(\xi' \to \xi; x, t; \tau).$$

Eq. (2.5) is, in particular, valid for steady problems.

Although the idea of a scattering kernel had appeared before, it is only at the end of the 1960s that a systematic study of the properties of this kernel appears[5,10,21]. In particular, the following properties were pointed out[3,5,8,10,11,19–21]:

1. Non-negativeness, i.e. R cannot take negative values:

   (2.7)
   $$R(\xi' \to \xi; x, t; \tau) \geq 0$$

   and, as a consequence:

   (2.8)
   $$R(\xi' \to \xi; x, t) \geq 0.$$

2. Normalization, if permanent adsorption is excluded; i.e. R, as a probability density for the totality of events, must integrate to unity:

   (2.9)
   $$\int_{\xi \cdot n > 0} \int_0^\infty R(\xi' \to \xi; x, t; \tau) d\tau d\xi = 1$$

   and, as a consequence:

   (2.10)
   $$\int_{\xi \cdot n > 0} R(\xi' \to \xi; x, t) d\xi = 1.$$

   From this property it follows that the normal component of the bulk velocity is zero (Problem 1).

3. Reciprocity; this is a subtler property that follows from the circumstance that the microscopic dynamics is assumed to be time reversible and the wall is assumed to be in a local equilibrium state, not significantly disturbed by the impinging molecule. It reads as follows:

   (2.11)
   $$|\xi' \cdot n| M_w(\xi') R(\xi' \to \xi; x, t; \tau) = |\xi \cdot n| M_w(\xi) R(-\xi \to -\xi'; x, t; \tau)$$

   and, as a consequence:

   (2.12)
   $$| \xi' \cdot n | M_w(\xi') R(\xi' \to \xi; x, t) = | \xi \cdot n | M_w(\xi) R(-\xi \to -\xi'; x, t).$$

   Here M_w is a (nondrifting) Maxwellian distribution having the temperature of the wall, which is uniquely identified apart from a factor.

   We remark that the reciprocity normalization relations imply another property:

3′. Preservation of equilibrium, i.e., the Maxwellian M_w must satisfy the boundary condition (2.4):

$$(2.13)\quad M_w(\xi) \mid \xi\cdot n \mid = \int_0^\infty d\tau \int_{\xi'\cdot n<0} R(\xi' \to \xi; x, t; \tau) M_w(\xi')|\xi'\cdot n| d\xi'$$

$$(\xi \cdot n > 0)$$

equivalent to:
(2.14)

$$M_w(\xi) \mid \xi \cdot n \mid = \int_{\xi'\cdot n<0} R(\xi' \to \xi; x, t) M_w(\xi')|\xi' \cdot n| d\xi' \qquad (\xi \cdot n > 0).$$

In order to obtain Eq. (2.13) it is sufficient to integrate Eq. (2.11) with respect to ξ' and τ, taking into account Eq. (2.9) (with $-\xi$ and $-\xi'$ in place of ξ' and ξ, respectively. We remark that frequently, one assumes Eq. (2.13) [or (2.14)], without mentioning Eq. (2.11) [or (2.12)]; although this is enough for many purposes, reciprocity is very important when constructing mathematical models, because it places a strong restriction on the possible choices. A detailed discussion of the physical conditions under which reciprocity holds was recently given by Bärwinkel and Schippers[1].

Problem

1. Show that if Eqs. (2.5) and (2.10) apply, then the normal component of the bulk velocity of the gas, as defined in Chapter 3, vanishes at the wall.

8.3 The Accommodation Coefficients

The scattering kernel is a fundamental concept in gas-surface interaction, by means of which other quantities should be defined. Its use is often avoided by using the so-called accommodation coefficients, with the consequence of lack of clarity, misinterpretation of experiments, bad definitions of terms, and misunderstanding of concepts. The basic information on gas-surface interaction, which should in principle be obtained from a detailed calculation based on a physical model as discussed in the next section, is summarized in a scattering kernel. The further reduction to a small set of accommodation coefficients can be advocated for practical purposes, provided this concept is firmly related to the scattering kernel.

In order to describe the accommodation coefficients in a systematic way, it is convenient to introduce, for any pair of functions ϕ and ψ, the notations:

$$(3.1)\qquad (\psi, \phi)_+ = \int_{\xi\cdot n>0} \psi(\xi)\phi(\xi) M_w(\xi) \mid \xi \cdot n \mid d\xi$$

$$(\psi, \phi)_- = \int_{\xi \cdot n < 0} \psi(\xi)\phi(\xi)M_w(\xi) \mid \xi \cdot n \mid d\xi. \tag{3.2}$$

Now, if we factor M_w out of the distribution function f and write

$$f = M_w \phi \tag{3.3}$$

we can define the accommodation coefficient for the quantity ψ when the distribution function at the wall is $M_w\phi$, in the following way:

$$\alpha(\psi, \phi) = [(\psi, \phi)_- - (\psi, \phi)_+]/[(\psi, \phi)_- - (\psi, \iota)_+] \tag{3.4}$$

where ι denotes a constant, such that

$$(\iota, \iota)_+ = (\iota, \phi)_+. \tag{3.5}$$

Physically the numerator in Eq. (3.4) is the difference between the impinging and emerging flow of the quantity, whose density is ψ, when the distribution is $M_w\phi$; the denominator is the same thing when the restriction of f to $\xi \cdot n > 0$ is replaced by the wall Maxwellian, normalized in such a way as to give the same entering flow rate as f. In particular, if we let $\psi = \xi \cdot n$, we obtain the accommodation coefficient for normal momentum, if we let $\psi = \xi \cdot t$ we obtain the accommodation coefficient for tangential momentum (in the direction of the unit vector t, tangent to the wall); if we let $\psi =\mid \xi \mid^2$, we obtain the accommodation coefficient for energy. It is convenient to restrict the definition in Eq. (3.4) to functions enjoying the property $\psi(\xi) = \psi(\xi - 2n(n \cdot \xi))$, which are even functions of ξ·n. This condition is not satisfied by $\psi = \xi \cdot n$; accordingly, if one wants to define an accommodation coefficient for normal momentum, one has to take $\psi =\mid \xi \cdot n \mid$.

In general, $\alpha(\psi, \phi)$ turns out to depend on the distribution function of the impinging molecules; accordingly the definition (3.4) is not so useful, in general. The notion of accommodation coefficient becomes more useful if we select a particular class of functions ϕ, as will be done later.

Let us first investigate the relation between the accommodation coefficients and the scattering kernel $R(\xi' \to \xi)$ (we omit indicating the space and time arguments). We assume that reciprocity, as expressed by Eq. (2.12), holds and we define the linear operator A by:

$$A\phi = [M_w(\xi)|\xi \cdot n|]^{-1} \int_{\xi' \cdot n > 0} R(-\xi' \to \xi)M_w(\xi')\phi(\xi') \mid \xi' \cdot n \mid d\xi'. \tag{3.6}$$

Because of reciprocity, A is symmetric with respect to the scalar product $(\cdot, \cdot)_+$:

$$(\psi, A\phi)_+ = (A\psi, \phi)_+. \tag{3.7}$$

We now define the reflection or parity operators in velocity space as follows:

$$\begin{aligned} P\phi &= \phi(-\xi) \\ P_n\phi &= \phi(\xi - 2n(n\cdot\xi)) \\ P_t &= \phi(-\xi + 2n(n\cdot\xi)) \end{aligned} \tag{3.8}$$

and observe (Problem 1) that

$$P^2 = P_n^2 = P_t^2 = I, \qquad P_tP_n = P_nP_t = P, \qquad (\psi, P_t\phi) = (P_t\psi, \phi) \tag{3.9}$$

where I is the identity operator. As stated earlier, the function ψ in Eq. (3.4) will be taken to satisfy the restriction

$$P_n\psi = \psi. \tag{3.10}$$

Eq. (3.4) then becomes

$$\alpha(\psi, \phi) = [(P\psi, P\phi)_+ - (\psi, AP\phi)_+]/[(P\psi, P\phi)_+ - (\psi, \iota)_+]. \tag{3.11}$$

We remark that once the gas-surface interaction is specified through the kernel $R(\xi' \to \xi)$, then only the impinging distribution function (the restriction of $f = M_w\phi$ to $\xi \cdot n < 0$) needs to be known in Eq. (3.11), where, because of Eq. (3.9) and (3.10) we can replace P by P_t.

Let us now consider particularly meaningful instances of accommodation coefficients. The first particularization was introduced by one of the authors [5,10] and is mathematically simple but has no direct physical significance for general kernels. In order to introduce it, let us consider the following eigenvalue problem:

$$A\psi = \lambda P_t\psi. \tag{3.12}$$

$\psi = \iota$ is an eigensolution corresponding to $\lambda = 1$; all the other solutions must be associated with eigenvalues such that $|\lambda| < 1$ [unless the kernel $R(\xi' \to \xi)$ is a delta function], thanks to the property (see Section 5):

$$(A\psi, A\psi)_+ \le (\psi, \psi)_+. \tag{3.13}$$

This inequality is a consequence of the properties of the kernel R, and the equality sign holds in only two cases:

a) the kernel is a delta function; b) the kernel is general and ψ is constant. Thus, if we discard the singular cases of kernels, which reduce to a delta function (cases that can be easily discussed in a direct fashion), we have in addition to $\psi_0 = \iota$ a sequence of eigenfunctions ψ_k associated with eigenvalues λ_k $(k = 1, 2, 3, \ldots)$, with $| \lambda_k | < 1$ and ψ_k orthogonal to ι [note that $(\iota, P_t\psi_k)_+ = 0$ because A is self-adjoint, and this implies $(\iota, \psi_k)_+ = 0$ because $P_t\iota = \iota$ (Problem 2)]. Then, for any ϕ, we have the accommodation coefficients

$$\alpha_k = \alpha(\psi_k, \phi) = [(P_t\psi_k, P\phi)_+ - (\psi_k, AP\phi)_+]/(P_t\psi_k, P\phi)_+ = 1 - \lambda_k$$

$$(k = 1, 2, 3, \ldots). \tag{3.14}$$

The obvious advantage of this set of coefficients is that they are independent of the distribution function of the impinging molecules; in addition, giving them (together with the set of eigenfunctions ψ_k) is equivalent to giving the scattering kernel. However, there is also an obvious disadvantage. The eigenfunctions ψ_k may be hard to compute and, in most cases, do not have a simple physical significance.

Another possibility was considered by Shen and Kuščer[19,20,27]. Let us consider a set of physically meaningful quantities $\{\psi_k\}$ and let $\psi = \psi_i, \phi = \iota + \psi_k$, where ι is a constant that does not necessarily satisfy the constraint in Eq. (3.5); if we still adopt the definition in Eq.(3.4), we obtain a matrix of accommodation coefficients:

$$\alpha_{ik} = \alpha(\psi_i, \iota + \psi_k) = [(\psi_i, P_t\psi_k)_- - (\psi_i, A\psi_k)_+]/(\psi_i, P_t\psi_k)_-. \tag{3.15}$$

These coefficients have a clear meaning, but only for the special distribution function $f = M_w(\iota + \psi_k)$. They are more easily computed, in general, than those defined earlier, because they do not require solving Eq. (3.12).

We state now a formula that will be useful later. Let the $\lambda_k s$ form a discrete set and let ψ_k be the corresponding eigenfunctions. Then one can show[3] that (in a distributional sense):

$$R(\xi' \to \xi) = M_w(\xi)|\xi \cdot n| \sum_{k=0}^{\infty} \lambda_k \phi_k(-\xi')\phi_k(\xi) \tag{3.16}$$

where

$$\phi_k(\xi) = P_t[\psi_k(\xi)]. \tag{3.17}$$

We end this section with a comment on the role and significance of the accommodation coefficients. As the name indicates, they describe how much the molecules accommodate to the state of the wall. A complete accommodation is when the molecules are conserved in number but otherwise forget completely their impinging distribution; the emerging distribution is then proportional to M_w, as is clear from Eq. (3.16) ($\lambda_k = 0$ for $k \neq 0$ and $\lambda_0 = 1$). The opposite case is when the gas remembers as much as possible of the impinging distribution; then the kernel is a delta function and we have specular reflection ($R(\xi' \to \xi) = \delta(\xi' - \xi + 2n(n \cdot \xi))$ and all the accommodation coefficients vanish. In order to describe the gas surface interaction, one would need infinitely many coefficients and this is, of course, equivalent to knowing the scattering kernel. In some particular problems (such as the so-called free-molecular flows, where the intermolecular collisions are so rare that one can neglect them), it turns out that one can describe the quantities of interest to engineers in terms of a few accommodation coefficients and this explains why the concept is popular. Another circumstance where the accommodation coefficients are useful arises in connection with particularly simple models for the scattering kernel (see the next section). These models contain a small number of parameters, which can easily be expressed in terms of an equal number of accommodation coefficients, which may be, accordingly, used to parameterize the kernel.

Problems

1. Check that relations (3.9) hold.
2. Prove that if ψ is an eigensolution of Eq. (3.12) and ι a constant function, then $(\iota, \psi)_+ = 0$.
3. Prove Eq. (3.16) (see Ref. 3).

8.4 Mathematical Models

In view of the difficulty of computing the kernel $R(\xi' \to \xi)$ from a physical model of the wall, a different procedure, which is less physical in nature, has been proposed. The idea is to construct a mathematical model in the form of a kernel $R(\xi' \to \xi)$ that satisfies the basic physical requirements expressed by Eqs. (2.8), (2.10), and (2.12) and is not otherwise restricted except by the condition of not being too complicated. We must warn the reader that in this section we are merely describing some of the attempts; he is not expected to verify the technicalities without consulting the original references. In addition, he must be aware of the fact that in the construction of models based on "simplicity with constraint," there is a great amount of arbitrariness; we are in a sense showing how to add new (largely arbitrary) postulates to our theory in order to be able to describe the complicated phenomena of gas-surface interaction in a simple form. From the point of view of physics, these models should be regarded as simplifications subjected to verification by comparison of the results of calculations with experiments.

A possible approach to the construction of such models is through the eigenvalue equation (3.12). The starting point is Eq. (3.16), which is rewritten here for the sake of the reader

$$R(\xi' \to \xi) = M_w(\xi)|\xi \cdot n| \sum_{k=0}^{\infty} \lambda_k \phi_k(-\xi')\phi_k(\xi) \tag{4.1}$$

where

$$\phi_k(\xi) = P_t[\psi_k(\xi)]. \tag{4.2}$$

In order to construct a model, it is necessary to choose the eigenfunctions ψ_k and the eigenvalues λ_k. Of course, it is convenient to make the choice in such a way that the series in Eq. (4.1) can be evaluated in finite terms. One possibility would be to take only a finite number (≥ 1) of eigenvalues different from zero[5,10], but this has the disadvantage that, generally speaking, the positivity requirement, Eq.(2.8), cannot be met. The simplest choice, then, is to take the first eigenvalue to be unity [with a constant eigenfunction, as required by Eq. (2.14)] and the others all equal to the same value $1 - \alpha$ ($0 \leq \alpha \leq 1$). The kernel then turns out to be (Problem 1)

$$R(\xi' \to \xi) = \alpha M_w(\xi) \mid \xi \cdot n \mid + (1-\alpha)\delta(\xi - \xi' + 2n(\xi' \cdot n)) \tag{4.3}$$

where the Maxwellian M_w is such that $M_w(\xi) \mid \xi \cdot n \mid$ integrates to unity. This is the kernel corresponding to the Maxwell model, according to which a fraction $(1 - \alpha)$ of molecules undergoes a specular reflection, while the remaining fraction α is diffused with the Maxwellian distribution of the wall M_w. This is the only model for the scattering kernel that appeared in the literature before the late 1960*s*. Since this model was felt to be somehow inadequate to represent the gas-surface interaction, Nocilla[24] proposed assuming that the molecules are reemitted according to a drifting Maxwellian with a temperature that is, in general, different from the temperature of the wall. While this model is useful as a tool to represent experimental data and has been used in actual calculations, expecially in free-molecular flow[14], when interpreted in the light of later developments, it does not appear to be tenable, unless its flexibility is severely reduced[16,21,22]. While the idea of a model like Nocilla's can be traced back to Knudsen[17], the full development of these ideas led to the so-called Cercignani-Lampis (CL) model[11]. From the point of view taken in this section, this model can be easily obtained by taking the eigenfunctions ψ_k to be products of Hermite polynomials in the tangential components of ξ times Laguerre polynomials in the square of the normal component $\xi_n = \xi \cdot n$ of the same vector. The reason for the different treatment of the components will be clear if one thinks that the range of ξ_n is $[0, \infty)$ rather than $(-\infty, \infty)$ and the weight factor in the natural scalar product $(.,.)_+$ is not $M_w(\xi)$ but rather $\xi_n M_w(\xi)$. The eigenvalues

λ_k are taken to be subsequent powers of two parameters. Then the series in Eq. (4.1) can be summed up by means of the so-called Miller-Lebedev formula[3,8] to yield:

$$R(\xi' \to \xi) = \frac{[\alpha_n \alpha_t (2-\alpha)_t]}{\pi/2} \beta_w^2 \xi_n$$
$$\times \exp\left[-\beta_w \frac{\xi_n^2 + (1-\alpha_n)\xi_n'^2}{\alpha_n} - \beta_w \frac{\mid \xi_t - (1-\alpha_t)\xi_t' \mid^2}{\alpha_t(2-\alpha_t)}\right] \tag{4.4}$$
$$\times I_0[\beta_w \frac{2(1-\alpha_n)^{1/2}\xi_n \xi_n'}{\alpha_n}]$$

where, if T_w is the wall temperature, $\beta_w = (2RT_w)^{-1}$, while I_0 denotes the modified Bessel function of first kind and zeroth order defined by

$$I_0(y) = (2\pi)^{-1} \int_0^{2\pi} \exp(y\cos\phi) d\phi. \tag{4.5}$$

The two parameters α_t and α_n have a simple meaning; the first is the accommodation coefficient for the tangential components of momentum, the second is the accommodation coefficient for ξ_n^2, hence for the part of kinetic energy associated with the normal motion. This model became rather popular because it was found by other methods as well: through an analogy with Brownian motion by I. Kuščer et al.[22], under a special mathematical assumption by T. G. Cowling[12], and through an analogy with the scattering of electromagnetic waves from a surface by M. M. R. Williams[30,31]. Finally it followed from the solution of the steady Fokker-Planck equation describing a (somewhat artificial) physical model of the wall[8], as was mentioned in the previous section. Also, the comparison with the data from beam-scattering experiments was quite encouraging[3,6,9,11]. In spite of this, it must be clearly stated that there are no physical reasons why this model should be considered better than others. In particular we remark that any linear combination of scattering kernels with positive coefficients adding to unity is again a kernel that satisfies all the basic properties. Thus, from a kernel with two parameters, such as the CL model, we can construct a general model containing an arbitrary function of those parameters.

We want to mention another method to produce scattering kernels in a simple way. This method was described about fifteen years ago[3,9] but does not appear to have ever been used (except in Ref. 7). The starting point is any positive function $K(\xi, \xi')$, defined for $\xi_n \geq 0$ and $\xi_n' \geq 0$, symmetric in its arguments and such that

$$H(\xi') = \int_{\xi_n > 0} K(\xi, \xi') M_w(\xi) \xi_n d\xi \tag{4.6}$$

is not larger than unity for any ξ' ($\xi'_n \geq 0$). Then if we form the kernel

$$R(\xi' \to \xi) = \xi_n M_w(\xi) K(\xi, -\xi') + \xi_n M_w(\xi) \frac{(1 - H(-\xi'))(1 - H(\xi))}{\int (1 - H(w)) w_n M_w(w) dw} \tag{4.7}$$

the latter satisfies both reciprocity and normalization, and positivity follows from the additional assumption $H(\xi) \leq 1$.

This procedure was used in Ref. 7 to construct a model chosen to be as close as possible to Nocilla's model and still satisfy the requirements expressed by Eqs. (2.10) and (2.12). Then the model is completely evaluated. The kernel reads as follows:

$$R(\xi' \to \xi) = (2/\pi) {\beta_w}^2 b \xi_n \exp(-\beta_w (1 - a^2)^{-1} \mid \xi - a\xi'_R \mid^2) + \xi_n \exp(-\beta_w \mid \xi \mid^2) \frac{(1 - H(-\xi'))(1 - H(\xi))}{\int (1 - H(w)) w_n \exp(-\beta_w |w|^2) dw} \tag{4.8}$$

where a and b are the parameters of the kernel, ξ'_R is the reflected velocity $\xi' - 2n(n \cdot \xi')$, and $H(\xi)$ is given by

$$\begin{aligned} H(\xi') &= \int_{\xi_n > 0} K(\xi, \xi') M_w(\xi) \xi_n d\xi \\ &= b(1 - a^2)^2 [\exp(-a^2 \xi'^2_n \beta_w (1 - a^2)^{-1}) \\ &\quad + a\pi^{1/2} \xi'_n {\beta_w}^{1/2} (1 - a^2)^{-1/2} \operatorname{erfc}(-a\xi'_n \beta^{1/2}{}_w (1 - a^2)^{-1/2})]. \end{aligned} \tag{4.9}$$

Here erfc is the complementary error function defined by

$$\operatorname{erfc}(y) = 2\pi^{-1/2} \int_0^y \exp(-t^2) dt. \tag{4.10}$$

The integral in the denominator of Eq. (4.8) can be easily evaluated to give

$$\begin{aligned} \int (1 - H(w)) w_n M_w(w) dw = (\pi/2) \beta_w^{-2} \{ 1 - b[(1 - a^2)^3 \\ + a(1 - a^2)^{3/2} (\pi - \cos^{-1} a) + a^2 (1 - a^2)^2) \}. \end{aligned} \tag{4.11}$$

Unfortunately, at a second look this model does not satisfy the requirement of being positive, Eq. (2.8), for the case $a > 0$ (which seems to be the most interesting from a physical standpoint) because $H(\xi')$ turns out to grow indefinitely for large (positive) values of ξ'_n and thus the requirement $H(\xi') \leq 1$ is not satisfied everywhere. One can easily modify the kernel in such a way as to eliminate this defect, but this requires introducing a further parameter.

All the above models contain pure diffusion according to a nondrifting Maxwellian as a limiting case. The use of the latter model ($\alpha = 1$ in Eq. (4.3)) is justified for low-velocity flows over technical surfaces, but is inaccurate for flows with orbital velocity. In fact, the elaboration of the measurements of lift-to-drag ratio for the shuttle orbiter in the free-molecular regime[2] implies a significant departure from diffuse fully equilibrated reemission of molecules at the wall.

Even more remote from experimental facts is the model based on the assumption that the molecules are specularly reflected at the wall ($\alpha = 0$ in Eq. (4.3)). This model was considered and criticized by Maxwell as early as 1879[23]. Yet it is frequently used in pure mathematical papers because it is easy to deal with. In addition, it is one of the two deterministic models of gas-surface interaction (ξ' uniquely determines ξ), satisfying all the physical requirements laid down in Section 2. The other model with such a property is reverse reflection or the "bounce-back" boundary condition based on the kernel $\delta(\xi + \xi')$[26,28]. The latter condition appears to be physically odd, at least from a molecular viewpoint, but has the interesting property of implying a no-slip boundary condition. Because of this property this boundary condition has received a lot of attention in connection with the simulation of incompressible flows by means of discrete kinetic theory[4].

Problems

1. Prove that if in Eq. (4.1) there is only an eigenvalue different from unity, $\lambda = 1 - \alpha$, then Eq. (4.3) holds. (Hint: Take into account that the right-hand side of Eq. (4.1) with $\lambda_k = 1$ for any k converges to $\delta(\xi - \xi' + 2n(\xi' \cdot n))$; see Ref. 3).
2. Take the eigenfunctions ψ_k of Eq. (3.12) to be products of Hermite polynomials in the tangential components of ξ times Laguerre polynomials in the square of the normal component $\xi_n = \xi \cdot n$ of the same vector and the eigenvalues λ_k are taken to be subsequent powers of two parameters $(1-\alpha_t)$ and $(1-\alpha_n)$. Prove that the series in Eq. (4.1) can be summed up by means of the so-called Miller-Lebedev formula to yield Eq. (4.4) (see Refs. 3 and 8).
3. Check that the kernel in Eq. (4.7) satisfies reciprocity and normalization.
4. Check Eq. (4.11).

8.5 A Remarkable Inequality

It is remarkable that, for any scattering kernel satisfying the three properties of normalization, positivity, and preservation of equilibrium, a simple inequality, involving an arbitrary convex function C holds. To be precise we can prove the following.

(8.5.1) Theorem. *Let $g \to C(g)$ $(g \in [0,\infty))$ be a strictly convex function, f the distribution function at a wall at rest $\partial\Omega$, M_w the wall Maxwellian, and $g = f/M_w$. If a boundary condition of the kind specified by Eq. (2.5) holds, with a kernel $R(\xi' \to \xi)$ satisfying Eqs. (2.8), (2.10), and (2.14), and $\xi \cdot n M_w C(g)$ is integrable in ξ (for a. e. $x \in \partial\Omega$ and t), then the following inequality holds*

$$\int_{\Re^3} \xi \cdot n M_w C(g(\xi)) d\xi \leq 0. \tag{5.1}$$

The equality sign in Eq. (5.1) holds if and only if g=const. a. e., unless $R(\xi' \to \xi)$ is proportional to a delta function.

Proof. Let us define:

$$W(\xi') = \frac{R(\xi' \to \xi) \,|\, \xi' \cdot n \,|\, M_w(\xi')}{|\, \xi \cdot n \,|\, M_w(\xi)} \tag{5.2}$$

and note that $W(\xi')$ is an L^1-function such that

$$\int W(\xi') d\xi' = 1 \tag{5.3}$$

with the integral restricted to $\xi' \cdot n \leq 0$.

Let us remark that, thanks to Jensen's inequality[15,25,32], for any convex function C of the type indicated in the statement of the theorem for any non-negative function such that $Wg \in L^1$, we have

$$C\Big(\int W(\xi') g(\xi') d\xi'\Big) \leq \int W(\xi') C(g(\xi')) d\xi' \tag{5.4}$$

because of Eq. (2.14) (with the same *proviso* on the integral). We can rewrite Eq. (5.4) more explicitly as follows:

$$\begin{aligned} &C\Big(\int_{\xi' \cdot n < 0} \frac{R(\xi' \to \xi) |\xi' \cdot n| M_w(\xi')}{|\xi \cdot n| M_w(\xi)} g(\xi') d\xi'\Big) \\ &\leq \int_{\xi' \cdot n < 0} \frac{R(\xi' \to \xi) |\xi' \cdot n| M_w(\xi')}{|\xi \cdot n| M_w(\xi)} C(g(\xi')) d\xi'. \end{aligned} \tag{5.5}$$

The integral on the left-hand side, however, is, because of the boundary condition (2.5), nothing but $g(\xi)$ $(\xi \cdot n \geq 0)$. Hence multiplying Eq. (5.5) by $|\, \xi \cdot n \,|\, M_w(\xi)$ and integrating with respect to ξ for $\xi \cdot n > 0$, with due use of Eq. (2.10), gives

$$\int_{\xi \cdot n > 0} |\xi \cdot n| M_w(\xi) C(g(\xi)) d\xi \leq \int_{\xi' \cdot n < 0} |\xi' \cdot n| M_w(\xi') C(g(\xi')) d\xi', \tag{5.6}$$

which is Eq. (5.1). The equality sign applies iff it applies a. e. in Eq. (5.5); this can happen only if just one value of g appears on both sides; that is if either g=constant or R has support at just one point (for a. a. given ξ). In the second case positivity implies that R is a delta function. □

This theorem was stated by Darrozès and Guiraud[13] in a particular case; the proof given here is due to one of the authors[8]. The result mentioned by Darrozès and Guiraud is important when discussing the H-theorem in the presence of walls and can be obtained as a corollary of the theorem.

(8.5.2) Corollary. *If Eqs. (2.5), (2.8), (2.10), and (2.14) hold, then:*

$$\mathcal{J}_n = \int \xi \cdot nf \; logf d\xi \leq -\beta_w \int \xi \cdot n|\xi|^2 f d\xi \qquad (x \in \partial\Omega). \tag{5.7}$$

Equality holds if and only if f coincides with M_w(the wall Maxwellian) on $\partial\Omega$ (unless the kernel in Eq. (2.5) is a delta function).

Proof. In fact if we take

$$C(g) = g\text{log}g \qquad (g > 0), \qquad\qquad C(0) = 0, \tag{5.8}$$

Eq. (5.1) gives:

$$\int \xi \cdot ng\text{log}gM_w d\xi \leq 0. \tag{5.9}$$

In terms of $f = M_w g$ this inequality takes the form indicated in Eq. (5.7). □

We remark that if the gas does not slip upon the wall, the right-hand side of Eq. (5.7) equals $-q_n/(RT_w)$ where q_n is the heat flow along the normal, according to its definition given in Chapter 3. If the gas slips on the wall with velocity u, then one must add the power of the stresses $p_n \cdot u$ $(= p_{ij}n_iu_j)$ to q_n. In this case, however, the right-hand side of Eq. (5.7) still equals $-q_n^{(w)}/(RT_w)$, where $q_n^{(w)}$ is the heat flow in the solid at the wall. In fact $q_n + p_n.u = q_n^{(w)}$ because the normal energy flow must be continuous through the wall (unless there are energy sources concentrated at the surface) and stresses have vanishing power in the solid, since the latter is at rest.

There is another interesting corollary of the theorem proved in this section. In fact, if we let $c(g) = |g|^p$ in Eq. (5.1) [or, better, in its equivalent form (5.6)], we obtain the following.

(8.5.3) Corollary. *The operator A, defined in Eq. (3.6), in the L^p-space ($p >$ 1) of the functions of ξ ($\xi \cdot n > 0$) with respect to the measure $|\xi \cdot n|M_w(\xi)d\xi$*

has norm unity. In other words if $\|g\|_p$ *denotes the norm in this space we have:*

$$\|Ag\|_p \leq \|g\|_p. \tag{5.10}$$

In addition, the equality sign holds iff g *is a. e. constant (unless* $R(\xi' \to \xi)$ *is proportional to a delta function).*

Problem

1. Check that Eq. (5.10) holds.

References

1. K. Bärwinkel and S. Schippers, "Nonreciprocity in noble-gas metal-surface scattering," in *Rarefied Gas Dynamics: Space-Related Studies,* E. P. Muntz, D. P. Weaver and D. H. Campbell, eds., 487–501, AIAA, Washington (1989).
2. R. C. Blanchard, "Rarefied flow lift to drag measurement of the shuttle orbiter," Paper No. ICAS 86-2.10.1, 15th ICAS Congress, London (September 1986).
3. C. Cercignani, *The Boltzmann Equation and Its Applications*, Springer, New York (1987).
4. C. Cercignani, "Kinetic theory with 'bounce-back' Boundary Conditions," *Transp. Theory Stat. Phys.*, **18**, 125–131 (1989).
5. C. Cercignani, *Mathematical Methods in Kinetic Theory*, Plenum Press, New York (1969).
6. C. Cercignani, "Models for gas-surface interactions: Comparison between theory and experiment," in *Rarefied Gas Dynamics*, D. Dini et al., eds., vol. I, 75–96, Editrice Tecnico-Scientifica, Pisa (1974).
7. C. Cercignani "Scattering kernels for gas-surface interaction," in *Proceedings of the Workshop on Hypersonic Flows for Reentry Problems*, Vol. I, 9–29, INRIA, Antibes (1990).
8. C. Cercignani, "Scattering kernels for gas-surface interactions," *Transport Theory and Stat. Phys.* **2**, 27–53 (1972).
9. C. Cercignani, *Theory and Application of the Boltzmann Equation*, Scottish Academic Press, Edinburgh and Elsevier, New York (1975).
10. C. Cercignani, in *Transport Theory*, G. Birkhoff et al., eds., SIAM-AMS Proceedings, Vol. I, p. 249, Providence (1968).
11. C. Cercignani and M. Lampis, "Kinetic models for gas-surface interactions," *Transport Theory and Stat. Phys.* **1**, 101–114 (1971).
12. T. G. Cowling, "On the Cercignani-Lampis formula for gas-surface interactions," *J. Phys. D. Appl. Phys.* **7**, 781–785 (1974).
13. J. S. Darrozès and J. P. Guiraud, "Généralisation formelle du théorème *H* en présence de parois. Applications," *C. R. A. S.* (Paris) **A262**, 1368–1371 (1966).
14. F. Hurlbut and F. S. Sherman, "Application of the Nocilla wall reflection model to free-molecule kinetic theory," *Phys. Fluids* **11**, 486–496 (1968).
15. J. L. W. V. Jensen, "Sur les fonctions convexes et les inégalités entre les valeurs moyennes," *Acta Mathematica* **30**, 175–193 (1906).
16. M. N. Kogan, *Rarefied Gas Dynamics*, Plenum Press, New York (1969).
17. M. Knudsen, *The Kinetic Theory of Gases*, Methuen, London (1950).
18. A. Kundt and E. Warburg, *Pogg. Ann. Phys.* **155**, 337 (1875).
19. I. Kuščer, "Phenomenological aspects of gas-surface interaction," in *Fundamental Problems in Statistical Mechanics* **IV**, E. G. D. Cohen and W. Fiszdon, eds., 441–467, Ossolineum, Warsaw (1978).

20. I. Kuščer, "Reciprocity in scattering of gas molecules by surfaces," *Surface Science* **25**, 225 (1971).
21. I. Kuščer, in *Transport Theory Conference*, AEC Report ORO-3588-1, Blacksburgh, Va. (1969).
22. I. Kuščer, J. Možina, and F. Krizanic, "The Knudsen model of thermal accommodation," in *Rarefied Gas Dynamics*, Dini et al., eds., Vol. I, 97–108, Editrice Tecnico-Scientifica, Pisa (1974).
23. J. C. Maxwell, "On stresses in rarified gases arising from inequalities of temperature," *Phil. Trans. Royal Soc.* **170**, 231–256, Appendix (1879).
24. S. Nocilla, "On the interaction between stream and body in free-molecule flow," in *Rarefied Gas Dynamics*, L. Talbot, ed., 169, Academic Press, New York (1961)
25. W. Rudin, *Real and Complex Analysis*, McGraw-Hill, London (1970).
26. J. Schnute and M. Shinbrot, "Kinetic theory and boundary conditions for fluids," *Can. J. Math.* **XXV**, 1183–1215 (1973).
27. S. F. Shen and I. Kuščer, "Symmetry of accommodation coefficients and the parametric representation of gas-surface interaction," in *Rarefied Gas Dynamics*, Dini et al., eds. Vol. I, 109–122 (1974).
28. M. Shinbrot, "Entropy change and the no-slip condition," *Arch. Rational Mech. Anal.* **67**, 351–363 (1978).
29. M. Smoluchowski de Smolan, "On conduction of heat by rarefied gases," *Phil. Mag.* **46**, 192–206 (1898).
30. M. M. R. Williams, "A phenomenological study of gas-surface interactions," *J. Phys. D. Appl. Phys.* **4**, 1315–1319 (1971).
31. M. M. R. Williams, *J. Phys. D: Appl. Phys.* **4**, 1315 (1971).
32. A. Zygmund, *Trigonometrical Series*, Dover, New York (1955).

9

Existence Results for Initial-Boundary and Boundary Value Problems

9.1 Preliminary Remarks

The global existence of a weak solution for the Cauchy problem for the Boltzmann equation, first obtained by DiPerna and Lions[13], was presented in Chapter 5. The proof applies to non-negative data with finite energy and entropy. In this chapter, we shall first deal with the initial boundary value problem, which arises when we consider the time evolution of a rarefied gas in a vessel Ω whose boundaries are kept at constant temperature. We shall assume that Ω is a bounded open set of $\Re^3$ with a sufficiently smooth boundary $\partial\Omega$. On $\partial\Omega$ we impose a linear boundary condition of the form envisaged in the previous chapter, when the molecules are assumed to be reemitted by the surface with negligible delay. This boundary condition is given by Eq. (8.2.5) with a kernel enjoying the properties expressed by Eqs. (8.2.8), (8.2.10) and (8.2.14). For convenience, we rewrite these equations here in a form appropriate for this chapter; $n = n(x)$ will always denote the inner normal at $x \in \partial\Omega$.

$$\gamma_D^+ f(t,x,\xi) = \int_{\xi'\cdot n<0} K(\xi' \to \xi; x,t)\gamma_D^- f(t,x,\xi')d\xi' \equiv K\gamma_D^- f \tag{1.1}$$

$(x \in \partial\Omega, \quad \xi \cdot n > 0)$

$$K(\xi' \to \xi; x, t) \geq 0 \tag{1.2}$$

$$\int_{\xi \cdot n > 0} K(\xi' \to \xi; x, t) |\xi \cdot n| d\xi = |\xi' \cdot n| \tag{1.3}$$

$$M_w(\xi) = \int_{\xi' \cdot n < 0} K(\xi' \to \xi; x, t) M_w(\xi') d\xi' \tag{1.4}$$

where M_w is the wall Maxwellian and $\gamma_D^{\pm}$ are the so-called trace operators on $E^{\pm} = \{(t, x, \xi) \in \partial\Omega \times \Re^3 \times [0, T] \mid \pm \xi \cdot n(x) > 0\}$. These operators permit us to define "the values taken on the boundary" $\gamma_D^{\pm} f$ (a.e. in $\xi \in \Re^3$ and $x \in \partial\Omega$) by a function f for which this concept is not a priori defined [such as a function $f \in L^1(\Omega \times \Re^3)$]. Of course, one must show that these operators are well defined. Actually this is one of the main points in the extension of the proof of the DiPerna and Lions theorem to the case under consideration.

The first existence theorem that we shall deal with was proved by Hamdache[15] in the case when the trace $\gamma_D^{+} f$ is a linear combination (with weights summing up to unity) of the right-hand side of Eq. (1.1) and a given function. Here, we shall restrict ourselves to the case when the weight of the second term is zero because, as remarked by Hamdache[15], the real problems arise in this case.

Hamdache's result is important not only for its own sake, but also because it leads to discussing the long time behavior of the solution and its trend to a Maxwellian distribution when $t \to \infty$, a problem that has been hotly debated for many years, since Boltzmann first established his H-theorem. Here we shall discuss in detail the results of Desvillettes[11] and Cercignani[8], who have shown that the solution can be proved to tend asymptotically to a Maxwellian (in a weak sense). This result will be proved in Section 5.

More difficult cases refer to a boundary along which the temperature is not constant or moving boundaries. The case of a nonisothermal boundary has been treated by Arkeryd and Cercignani[2], but only with a cutoff for large velocities, and we shall not discuss it here. Another topic that should be treated in this chapter is the pure boundary value problem that arises when we look for the asymptotic state of general initial-boundary value problems. Results about these problems, however, are scanty. Thus we shall restrict ourselves to the particular case of solutions close to equilibrium, while the general case will be the subject of a few remarks at the end of the chapter.

One of the key points in the proofs will be Corollary 8.5.2, proved in the previous chapter, which we restate here in the form of a lemma appropriate to the use to be made of it in this chapter:

(9.1.1) Lemma. *If Eqs. (1.1), (1.2), (1.3), and (1.4) hold, then:*

$$\int \xi \cdot n\gamma_D f \log \gamma_D f d\xi \leq -\beta_w \int \xi \cdot n|\xi|^2 \gamma_D f d\xi \tag{1.5}$$

(a.e. in t and $x \in \partial\Omega$) where β_w is the inverse temperature at a point $x \in \partial\Omega$. Unless the kernel in Eq. (1.1) is a delta function, equality holds if and only if the trace $\gamma_D f$ of f on $\partial\Omega$ coincides with M_w (the wall Maxwellian).

The trace operator γ_D has been informally defined in the lines following (1.4). We will give a formal definition in Section 2 of this chapter.

The other results to be discussed in this chapter refer to solutions of initial-boundary value problems when the data are close to an absolute Maxwellian.

9.2 Results on the Traces

Before proving Hamdache's theorem (in a slightly more general form), we have to prove some trace results giving the L^1-regularity of the trace of f on the boundary and to study the semigroup generated by the free-streaming operator. These will be done in this and the next section, respectively.

Let us first review the general results of Ukai[21] on the traces of the solutions. To this end we define

$$\Lambda f = \frac{\partial f}{\partial t} + \xi \cdot \frac{\partial f}{\partial x} \tag{2.1}$$

and assume that $\partial\Omega$ is piecewise C^1.

We denote by $S^t(x,\xi)$ the pair $(x+\xi t, \xi)$ that gives the position and velocity of a molecule initially located at (x,ξ) as long as $x+\xi t$ stays in Ω. Denote the forward ($t>0$) stay time in Ω by $t^+(x,\xi)$ and the backward one by $t^-(x,\xi)$. Then $S^t(x,\xi) \in \Omega \times \Re^3$ for $-t^-(x,\xi) < t < t^+(x,\xi)$ and $S^t(x,\xi) \in \partial\Omega \times \Re^3$ for $t = t^\pm(x,\xi)$ if $t^\pm(x,\xi) < \infty$. Let us also define $\Sigma^\pm = \{(x,\xi) \in \partial\Omega \times \Re^3 \mid \pm\xi \cdot n(x) > 0\}$ and remark that $S^t(x,\xi)$ exists for $(x,\xi) \in \Sigma^\mp$ with $t^\pm = 0, t^\mp > 0$. In any case $S^t \in \overline{\Sigma^\mp}$ at $t = \pm t^\pm$. It is now convenient to write $r = (t,x,\xi)$ and for $T>0$ define

$$D = (0,T)\times\Omega\times\Re^3; \qquad V^\pm = \{T^\pm\}\times\Omega\times\Re^3 \qquad (\text{where } T^+ = 0, T^- = T)$$

$$E^\pm = (0,T)\times\Sigma^\pm; \qquad \partial D^\pm = E^\pm \cup V^\pm \qquad (\text{same sign throughout}) \tag{2.2}$$

The world line of a molecule (Fig. 28) passing through $r = (t,x,\xi) \in D \cup \partial D^+ \cup \partial D^-$ is given by

$$R^s(r) = (t+s, x+\xi s, \xi) \qquad -s^-(r) \leq s \leq s^+(r) \tag{2.3}$$

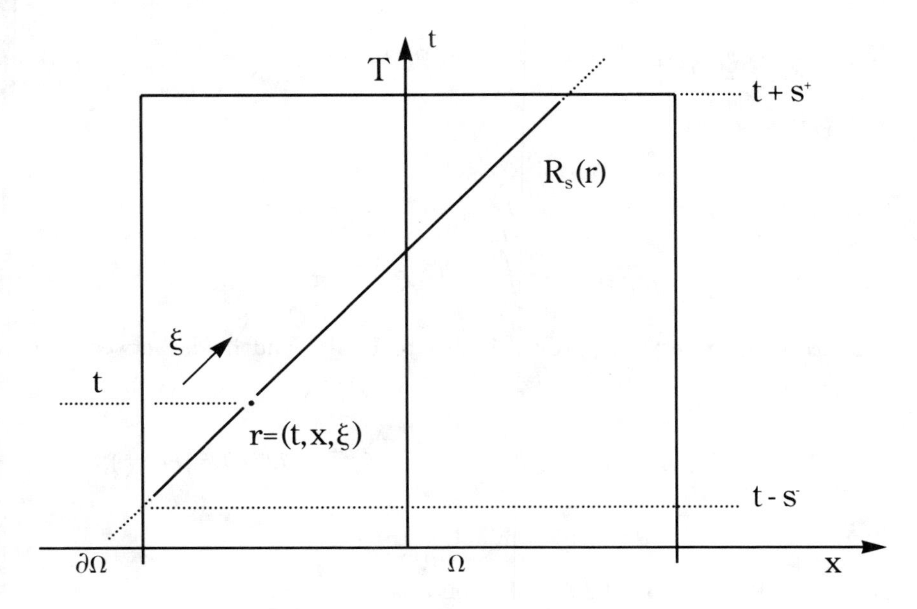

FIGURE 28.

where $s^{\pm}(r) = \min(T^{\mp} \mp t, t^{\pm}(x,\xi))$ and $T^{\mp}$ are defined as in Eq. (2.2). Obviously

$$
\text{(2.4)} \qquad \begin{aligned} &R^s(r) \in D \qquad (-s^-(r) < s < s^+(r)), \\ &R^s(r) \in \overline{\partial D^{\mp}} \qquad (s = s^{\pm}(r)). \end{aligned}
$$

Clearly if $f \in L^1(D)$, then $f(R^s(r))$ as a function of s is in $L^1(-s^-(r), s^+(r))$ for almost all $r \in \partial D^{\pm}$ and

$$
\text{(2.5)} \qquad \int_D f(r)dr = \int_{\partial D^{\pm}} \int_{s^-(r)}^{s^+(r)} f(R^s(r))dsd\sigma^{\pm}
$$

holds, where

$$
\text{(2.6)} \qquad \begin{aligned} dr &= dtdxd\xi; \\ d\sigma^{\pm} &= |n(x) \cdot \xi| dtd\sigma d\xi \quad (\text{on } E^{\pm}); \\ d\sigma^{\pm} &= dxd\xi \quad (\text{on } V^{\pm}) \end{aligned}
$$

$n(x)$ is the inward normal to $\partial\Omega$ and $d\sigma$ the usual measure on $\partial\Omega$. We set

$$
(f,g) = \int_D f\overline{g}dr
$$

$$< \phi, \psi >_{\pm} = < \phi, \psi >_{E^{\pm}} + < \phi, \psi >_{V^{\pm}} = \int_{\partial D^{\pm}} \phi \overline{\psi} d\sigma^{\pm} \tag{2.7}$$

where, of course:

$$\begin{aligned} < \phi, \psi >_{E^{\pm}} &= \int_{E^{\pm}} \phi \overline{\psi} |n(x) \cdot \xi| dt d\sigma d\xi, \\ < \phi, \psi >_{V^{\pm}} &= \int_{V} \phi \overline{\psi} dx d\xi \quad (\text{for } t = T^{\pm}). \end{aligned} \tag{2.8}$$

The trace theorem to be proved is due to Ukai[21] and holds between the spaces:

$$W^p = \{f \in L^p(D) | \Lambda f \in L^p(D)\}, \qquad L^{p,\pm}_{\theta} = L^p(\partial D^{\pm}; \theta d\sigma^{\pm}),$$

$$\theta = \theta(r) = \min(1, s^+(r) + s^-(r)) \tag{2.9}$$

where Λf is defined in the distribution sense.

The trace operators $\gamma^{\pm}_D$ are first defined on $C^1_0(\overline{D})$ by

$$\gamma^{\pm}_D f = f|_{\partial D^{\pm}} \qquad f \in C^1_0(\overline{D}). \tag{2.10}$$

Then, following Ukai[21], we can prove the following.

(9.2.1) Theorem. *Let $p \in [1, \infty]$. $\gamma^{\pm}_D$ have extensions in $\mathbf{B}(W^p, L^{p,\pm}_{\theta})$, the space of bounded linear operators from W^p to $L^{p,\pm}_{\theta}$. We denote these extensions again by $\gamma^{\pm}_D$. Thus it holds that*

$$\| \gamma^{\pm}_D f \|_{L^{p,\pm}_{\theta}} \leq C \| f \|_{W^p} = C(\| f \|_{L^p(D)} + \| \Lambda f \|_{L^p(D)}). \tag{2.11}$$

Proof. Let $f \in W^p$ and let $\hat{f}(s, r) = f(R^s(r))$. Since $\widehat{(\Lambda \psi)} = \partial \hat{\psi} / \partial s$ holds for $\psi \in C^1_0(\overline{D})$, we can deduce from Eqs. (2.5) and the distributional definition of Λ that for almost all $r \in \partial D^{\pm}$, $\hat{f}(s, r)$ is absolutely continuous in s and

$$\hat{f}(s, r) = \hat{f}(s', r) + \int_{s'}^{s} \widehat{(\Lambda f)}(\tau, r) d\tau \tag{2.12}$$

holds for any $s, s' \in [-s^-(r), s^+(r)]$. Let us now define $\gamma^{\pm}_D f = \hat{f}(s^{\pm}(r), r)$, for $r \in \partial D^{\pm}$; this coincides with Eq. (2.10) when $f \in C^1_0(\overline{D})$) (obviously $C^1_0(\overline{D}) \subset W^p$).

In order to prove Eq. (2.11), we first consider the case $p = \infty$ and let $s = s^{\pm}(r) = 0$ in Eq. (2.12), while s' is arbitrary in $(-s^-(r), s^+(r))$. Eq.

(2.11) (with $p = \infty$) easily follows. If p is finite, Eq. (2.12) gives (thanks to elementary inequalities)

(2.13)
$$\begin{aligned}|\gamma_D^{\pm} f(r)|^p \leq & 2^{p-1}(|f(s',r)|^p + |\int_{s'}^{s^{\pm}(r)} (\widehat{\Lambda f})(\tau,r)d\tau|^p) \\ \leq & \quad 2^{p-1}(|f(s',r)|^p + |s^{\pm}(r) - s'|^{p-1} \int_{-s^-(r))}^{s^+(r))} |(\widehat{\Lambda f})(\tau,r)|^p d\tau).\end{aligned}$$

If we now integrate both sides first with respect to s' on $(-s^-(r), s^+(r))$ and then with respect to r over $\partial D^{\pm}$, in view of Eq. (2.5) we are done. Remark that since $s^-(r) + s^+(r) \leq T$ by definition, θ and $s^-(r) + s^+(r)$ are equivalent as weight functions as long as $T < \infty$. □

We cannot remove the weight function θ if $p < \infty$ in Eq. (2.11). For this reason, some authors[5,27] have obtained just $L^{p,\pm}_{\text{loc}}$ -traces. In order to solve the initial-boundary value problem, however, the $L^{p,\pm}_{\theta}$-traces are not adequate. We need $L^{p,\pm}$ traces defined by

$$L^{p,\pm} = L^p(\partial D^{\pm}; d\sigma^{\pm}). \tag{2.14}$$

We remark that $L^{p,\pm} = L^{p,\pm}_{\theta}$ for $p = \infty$ but $L^{p,\pm} \subseteq L^{p,\pm}_{\theta}$ if $p < \infty$. Let us also define, for future use,

$$\hat{W}^p = \{f \in W^p(D)|\ \gamma_D^{\pm} f \in L^{p,\pm}\} \subset W^p. \tag{2.15}$$

If we impose suitable boundary conditions, then we can make some progress in the direction of proving that $f \in \hat{W}^p$. To this end it is expedient, again following Ukai[21], to prove the following.

(9.2.2) Theorem. *Let $f \in W^p$, $p \in [1, \infty)$. If $\gamma_D^{\pm} f \in L^{p,\pm}$ (only one sign throughout), then $\gamma_D^{\mp} f \in L^{p,\mp}$. In this case, the following relation holds:*

$$\| \gamma_D^- f(r) \|^p_{L^{p,-}} = \| \gamma_D^+ f(r) \|^p_{L^{p,+}} + p\text{Re} \int_D |f|^{p-2} \bar{f} \Lambda f dr \tag{2.16}$$

where "Re" denotes the real part.

Proof. If $f \in W^p$, then $|f|^p \in W^1$ and $\Lambda |f|^p = p\text{Re}|f|^{p-2}\bar{f}\Lambda f$. Then if we replace f by $|f|^p$ in Eq. (2.12) and set $s = -s^-(r), s' = s^+(r), r \in \partial D^+$, Eq. (2.16) follows by integration on ∂D^- and use of Eq. (2.5). □

This theorem immediately allows to deduce the existence of the traces when f is assigned on ∂D^+, as a function of $L^{p,+}$. The situation is more complicated if the boundary conditions are less trivial. We shall assume that boundary conditions of the form (1.1) are satisfied and prove the existence of the traces on the boundary under suitable assumptions.

There is an important comment to be made on the case $p = 1$. One might doubt, in fact, the results that have been just discussed, because a smooth sequence of functions with bounded $L^{1,-}$ norm might tend to a measure. It is easy, however, to check that the trace exists locally around any $r = (x, \xi, t) \in \partial\Omega \times \Re^3 \times \Re_+$ with $\xi \cdot n(x) \neq 0$. In fact if we apply Eq. (2.12) to ϕf, where $f \in C_0^\infty(\Omega \times \Re^3 \times \Re_+)$ with $\phi = 1$ in a neighborhood of r on the boundary and with sufficiently small support not to involve any outgoing point if r is ingoing and conversely, we obtain, e.g.:

$$\widehat{\phi f}(s_+(r), r) = \int_{s_- r}^{s_+(r)} \widehat{\Lambda(\phi f)}(\tau, r) d\tau. \tag{2.17}$$

This is sufficient to prove that $\gamma_D f$ is locally in L^1 outside of $\xi \cdot n(x) = 0$; the latter set is killed, however, in the $L^{1,\pm}$ norms.

In order to simplify the formal aspects of the treatment, we suppress the time derivative from Λ, and correspondingly the parts of the boundaries $\partial D^\pm$ corresponding to $t = 0$ and $t = T$ and integrations with respect to t; in fact these details do not play any role in what follows. Let us remark that when Eq. (1.1) holds, we easily obtain for $\phi \in W^\infty(D)$ and $f \in W^1(D)$:

$$\begin{aligned} < \gamma_D^- \phi(r), \gamma_D^- f(r) >_- = & < \gamma_D^+ \phi(r), K \gamma_D^- f(r) >_+ \\ & + \int_D \phi(\Lambda f) dr + \int_D f(\Lambda \phi) dr. \end{aligned} \tag{2.18}$$

If we take the value of ϕ on the boundary to be L^∞ and (distributionally) $\Lambda\phi = 0$, Eq. (2.12) shows that

$$\| \gamma_D^\pm \phi \|_{L^\infty, \pm} = \| \phi \|_{L^\infty(D)} \tag{2.19}$$

and in Eq. (2.18) the last term disappears.

We can now introduce the operator P, which reflects ξ, defined by $(Pf(\xi) = f(-\xi))$, and rewrite (2.18) as follows:

$$< \gamma_D^- \phi(r) - (PK)^\dagger P \gamma_D^+ \phi(r), \gamma_D^- f(r) >_- = \int_D \phi(\Lambda f) dr$$

$$(\Lambda \phi = 0) \tag{2.20}$$

where $(PK)^\dagger$ is the dual operator of PK with respect to $< \cdot, \cdot >_-$. Let us put:

$$\psi = \gamma_D^- \phi(r) - (PK)^\dagger P \gamma_D^+ \phi(r). \tag{2.21}$$

We remark that if we take $\gamma_D^- \phi = 1$, then by Eq. (2.12) we have $\gamma_D^+ \phi = 1$ and thanks to the fact that K is mass preserving:

$$(2.22) \qquad < \gamma_D^- \phi(r), \gamma_D^- f(r) >_- = < \gamma_D^+ \phi(r), K \gamma_D^- f(r) >_+$$

for any $\gamma_D^- f \epsilon L^{1,-}$, and hence

$$(2.23) \qquad \psi = \gamma_D^- \phi(r) - (PK)^\dagger P \gamma_D^+ \phi(r) = 0.$$

Let us denote by λ_D^+ the operator that carries $\gamma_D^- \phi(r)$ into $\gamma_D^+ \phi(r)$ (via $\Lambda \phi = 0$; i.e., by Eq. (2.12), λ_D^+ deposits the value $\gamma_D^- \phi(r)$ taken at a point x of the boundary as a value for $\gamma_D^+ \phi(r)$ on the next intersection with the boundary of the half straight line through x directed and oriented as $-\xi$). Eq. (2.23), which holds for $\gamma_D^- \phi = 1$ (and hence $\gamma_D^+ \phi = \lambda_D^+ \gamma_D^- \phi = 1$), then shows that the operator $I - (PK)^\dagger P \lambda_D^+$ does not have an inverse in $L^{\infty,-}$. Thus Eq. (2.20) cannot provide us with a good estimate for the norm of $\gamma_D^- f(r)$ in $L^{1,-}$. Let us then consider for any function ϕ of $L^{\infty,-}$ the decomposition into a constant part

$$(2.24) \qquad P_M \phi = < \phi, M_w^- >_- / < 1, M_w^- >_-$$

plus the remainder

$$(2.25) \qquad P_O \phi = \phi - P_M \phi.$$

All this makes sense if the boundary has a finite measure. We can now assume that the operator $I - (PK)^\dagger P \lambda_D^+$ has a bounded inverse in the subspace O of the functions having the form $P_O \phi$. Then, by allowing only functions of this kind and taking the supremum with respect to such functions we obtain the boundedness of the part of $\gamma_D^- f(r)$ that lies in any complete subspace of functions $f \in L^{1,-}$ for which $< P_M \phi, f >_- = 0$. One such subspace is obtained by decomposing any $f \in L^{1,-}$ into $cM_w^- + g$, where c is a factor defined by

$$(2.26) \qquad c = < 1, f >_- / < 1, M_w^- >_- .$$

Then

$$(2.27) \qquad < 1, g >_- = 0.$$

It is clear that in this way $L^{1,-}$ is decomposed into two subspaces M' and O'; we remark that $< P_M \phi, P_{O'} f >_- = 0$, and hence $P_{O'} \gamma_D^- f(r)$ is bounded in $L^{1,-}$. In this way we have proved that, although the traces of f may not exist, it makes sense to talk of $P_{O'} \gamma_D^- f(r)$, where $P_{O'}$ is the projector into O'. In order to talk about $\gamma_D^- f(r)$, we must have additional information on

f. To this end, it is enough to assume that not only $f \in W^1$, but also that $|\xi|^2 f \in L^1$ and $|\xi|^2 \Lambda f \in L^1(D)$. In fact if this holds, then by taking ϕ as a function that equals $n(x)$ on the boundary, we obtain through Eq. (2.18) (see also Hamdache[15], Lemma 4.3)

$$\| \, |\xi \cdot n| \gamma_D^- f(r) \, \|_{L^{1,-}} \leq h \, \| \, (1 + |\xi|^2) f \, \|_{W^1} \tag{2.28}$$

where h is a constant. Now let $k > 0$ be any fixed constant; we define

$$\eta_k(\xi \cdot n) = \begin{cases} |\xi \cdot n|/k & \text{if } |\xi \cdot n| \leq k \\ 1 & \text{if } |\xi \cdot n| \geq k. \end{cases} \tag{2.29}$$

Then

$$\| \, \eta_k(\xi \cdot n) \gamma_D^- f(r) \, \|_{L^{1,-}} \leq C \, \| \, (1 + |\xi|^2) f \, \|_{W^1} \qquad (C = h/k). \tag{2.30}$$

Since, however, $\gamma_D^- f = P_{O'} \gamma_D^- f + P_{M'} \gamma_D^- f$, we can infer

$$\| \, \eta_k(\xi \cdot n) P_{M'} \gamma_D^- f \, \|_{L^{1,-}} \leq \| \, \eta_k(\xi \cdot n) P_{O'} \gamma_D^- f(r) \, \|_{L^{1,-}} + C \, \| \, (1 + |\xi|^2) f \, \|_{W^1}$$

$$\leq \| \, P_{O'} \gamma_D^- f(r) \, \|_{L^{1,-}} + C \, \| \, (1 + |\xi|^2) f \, \|_{W^1} \, . \tag{2.31}$$

But

$$\begin{aligned} &\| \, \eta_k(\xi \cdot n) P_{M'} \gamma_D^- f \, \|_{L^{1,-}} \\ &= \| \, P_{M'} \gamma_D^- f \, \|_{L^{1,-}} < \eta_k, M_w^- >_- / < 1, M_w^- >_- \\ &= H \, \| \, P_{M'} \gamma_D^- f \, \|_{L^{1,-}} \end{aligned} \tag{2.32}$$

where $H = H(k, \beta_w)$ (β_w is the inverse temperature in M_w) is a constant. Eq. (2.31) hence becomes

$$\begin{aligned} \| \, P_{M'} \gamma_D^- f \, \|_{L^{1,-}} \leq & (1/H) \, \| \, P_{O'} \gamma_D^- f(r) \, \|_{L^{1,-}} \\ & + (C/H) \, \| \, (1 + |\xi|^2) f \, \|_{W^1} \end{aligned} \tag{2.33}$$

and $P_{O'} \gamma_D^- f \in L^{1,-}$ implies $P_{M'} \gamma_D^- f \in L^{1,-}$ and $\gamma_D^- f = P_{O'} \gamma_D^- f + P_{M'} \gamma_D^- f \in L^{1,-}$. We have thus proved the following.

(9.2.3) Theorem. *(See Ref. 9.) Let* $f \in W^1, |\xi|^2 f \in L^1, |\xi|^2 \Lambda f \in L^1(D)$. *If the boundary condition (1.1) applies and* $I - (PK)^\dagger P \lambda_D^+$ *has a bounded inverse in the subspace* O *of* $L^{\infty,-}$, *then* $\gamma_D^\mp f \in L^{1,\mp}$.

Theorem 9.2.3 is the result that is needed in order to deal with sufficiently general boundary operators K; Hamdache's[15] results refer, apart from the deterministic conditions of specular and reverse reflection, only to operators with kernels having compact support in $\Re^3 \times \Re^3$ for almost any $\{x,t\} \in \partial\Omega \times [0,T]$, which excludes practically all the typical cases.

One must, of course, prove that the criterion in Theorem 9.2.3 is actually satisfied by any reasonable boundary condition for sufficiently smooth boundaries. So far an explicit proof has been given[7] for the important case of a boundary diffusing the particles according to a Maxwellian distribution. The proof is based on the fact that, in this case, the inverse operator to be constructed is the identity plus an operator whose range is in the subspace of the functions independent of ξ. Then everything is reduced to finding this last part of the operator; this leads to a linear integral equation for a function of $x \in \partial\Omega$. This equation has appeared before in the literature and can be solved in an L^2-framework by means of Fredholm theorems provided the singularity of the kernel $b(x,x')$ at $x = x'$ is sufficiently weak. The most general discussion of this integral equation is due to Maslova[19], who treated the case in which the boundary is a Lyapunov surface (i.e., essentially, the angle between two neighboring points x and x' of the surface is less, in absolute value, than $\Lambda|x-x'|^\lambda$, with $\Lambda > 0$ and $0 < \lambda \leq 1$ given constants). We can thus prove that Theorem 9.2.3 applies to the case of diffuse reflection on a Lyapunov boundary. For further details we refer to the paper by Cannone and Cercignani[7].

As a final comment, we point out that there is an alternative way to deal with the trace problem, namely, the strategy we followed in Chapter 4 in connection with the derivation of the BBGKY hierarchy. The idea there was to use the special flow representation in order to show that traces had to exist. The same strategy could be used for the more general stochastic boundary conditions considered in this chapter, but some theory of stochastic processes would have to be introduced to do so. We preferred the purely functional analytic strategy presented here.

Problems

1. Obtain an explicit expression of the costant $H = H(k, \beta_w)$ in Eq.(2.32).
2. Prove by an explicit calculation that Theorem 9.2.3 applies to the case of a slab with a purely diffusing boundary. (Assume dependence on just one space coordinate, say x.) (Hint: Evaluate the operator $I - (PK)^\dagger P\lambda_D^+$ explicitly and prove that it has an inverse in the subspace O.)
3. Fill in the details of the proof that Theorem 9.2.3 applies to the case of diffuse reflection on a Lyapunov boundary (see Cannone and Cercignani[7]).

9.3 Properties of the Free-Streaming Operator

As in the previous sections, we shall assume that f is assigned at $t = 0$ and satisfies the boundary condition (1.1). We first study the problem

$$(\Lambda + \lambda)f = 0 \qquad \text{in } D \qquad (\lambda \in \Re) \tag{3.1}$$

$$\gamma_D^+ f(x, \xi, t) = K\gamma_D^- f \qquad \text{on } \partial D \tag{3.2}$$

$$f(x, \xi, 0) = f_0(x, \xi). \tag{3.3}$$

The parameter λ is introduced for the sake of more flexibility when obtaining the estimates; in fact if f solves Eq. (3.1-3), then $\hat{f} = e^{\lambda t} f$ satisfies Eqs.(3.1–3) with $\lambda = 0$. If the norm of K is (in some space) less than unity, then we can use iteration methods to solve the problem; since, however, we assume that (1.3) is true, then the right assumption is $\| K \|= 1$. The boundary will be assumed to be piecewise C^1. We use the notation of Section 2 and set $Y^{p,\pm} = L^p(\Sigma^\pm |\ |n(x) \cdot \xi| d\sigma d\xi)$. In addition we assume $\| K \| \leq 1$ in $B(Y^{p,-}, Y^{p,+})$ because in this way we can obtain intermediate results, which are useful in the case $\| K \|= 1$. Denote the dual of K by $K^\dagger$. Then for $p \in [1, \infty)$ we have, automatically, $\| K^\dagger \| \leq 1$ in $B(Y^{q,-}, Y^{q,+})$ with $p^{-1} + q^{-1} = 1$. For $p = \infty$ this is an extra assumption (always true in the physically interesting cases). We shall also assume that K does not act on t; hence we may replace $Y^{p,\pm}$ by $L^{p,\pm} = L^p(E^\pm |\ |n(x) \cdot \xi| dt d\sigma d\xi)$.

The weak solution is defined, as usual, through a sort of Green's formula, which can be established[2] in the same way as (2.18):

$$\begin{aligned} &< \gamma_D^- f(r), \gamma_D^- \phi(r) >_- - < \gamma_D^+ f(r), \gamma_D^+ \phi(r) >_+ \\ &\qquad\qquad = (f, (\Lambda - \lambda)\phi) + ((\Lambda + \lambda)f, \phi) \end{aligned} \tag{3.4}$$

where $< \phi, \psi >_\pm$ are defined as in Eq. (2.7). We take as space of test functions

$$W_p^\dagger = \{\phi \in \hat{W}^q |\ \gamma_D^- \phi = K^\dagger \gamma_D^+ \phi,\ \phi(., T) = 0\} \tag{3.5}$$

where $\hat{W}^q$ was defined by Eq. (2.15). If f satisfies (3.1–3) in a strong sense, then Eq. (3.4) gives

$$(f, (\Lambda - \lambda)\phi) = - < f_0, \phi >_{V^+} . \tag{3.6}$$

We take this as a definition of a weak solution.

(9.3.1) Definition. *Let* $f_0 \in L^p(\Omega \times \Re^3)$. $f \in L^p(D)$ *is called a weak solution of (3.1–3) if Eq. (3.6) holds for any* $\phi \in W_p^\dagger$.

(9.3.1) Theorem. *If* $p \in [1, \infty]$ *and* $f_0 \in L^p(\Omega \times \Re^3)$, *a weak (actually mild) solution* $f \in L^p(D)$ *exists for* $\lambda > 0$ *if* $\| K \| < 1$. *If* K *carries non-negative functions into functions of the same kind and* f_0 *is non-negative, then* f *is also non-negative.*

Proof. This theorem can be proved in many ways. Ukai[21] gives a proof that is valid only if $p \in (1, \infty]$. Here we follow a different strategy. We first consider the case when Eq. (3.2) is replaced by

$$\gamma_D^+ f(x, \xi, t) = f^+ \qquad \text{on } \partial D^+ \tag{3.7}$$

where $f^+ \in L^{p,+}$ is a given function. Then the solution can be written in an explicit way by means of Eq. (2.12):

$$\hat{f}(s, r) = \hat{g}(r) e^{-\lambda(s + s^-(r))} \tag{3.8}$$

where $\hat{g}(r)$ equals f_0 or f^+, according to whether $s = -s^-(r)$ corresponds to $t = 0$ or a point of the boundary of the space domain. It is clear, thanks to Eq. (2.5), that the solution constructed in this way is in L^p. If we now go back to the original boundary condition (3.2), we find a solution for that problem provided there is a function $\hat{g}$ such that

$$\hat{g}(r) = \mathcal{K}[\hat{g}(r^*) e^{-\lambda(s^+(r_*))}] + g_0 \qquad \text{for } r \in \partial D^+ \tag{3.9}$$

where $\mathcal{K}$ is 0 on $t = 0$ and K on the boundary of the space domain, while g_0 is f_0 for $t = 0$ and 0 on the boundary of the space domain. r^* is the other point where the relevant world line through r intersects ∂D. Since $\| K \| < 1$ (and hence $\| \mathcal{K} \| < 1$), Eq. (3.9) can be solved explicitly by means of a perturbation series. The part on non-negativity is obvious by glancing at the details of this constructive proof. □

We can now characterize the weak solutions by the following.

(9.3.2) Theorem. *(See Ref. 9.) Any weak solution* $f \in L^p(D)$ *satisfies*

(i) $f \in W^p$, $\quad (\Lambda + \lambda) f = 0$,

(ii) $f(.,0) = f_0 \in L^p(\Omega \times \Re^3)$.

(iii) If we let $f_\epsilon = \chi_\epsilon f$ *(where* χ_ϵ *is the characteristic function of the set* $s(r) \equiv s_+(r) + s_-(r) > \epsilon$*), then* $\gamma_D^+ f_\epsilon - K \gamma_D^- f_\epsilon \to 0$ $(\epsilon \to \infty)$ *weakly if* $p \in (1, \infty)$ *or weak-* if* $p = \infty$, *in* $L^{p,+}$.

Proof. To prove (i) take $\phi \in C_0^1(D)$ in (3.6) and recall the distributional definition of $\Lambda\phi$. To prove (ii) we need two steps. We first let $\phi_\delta = \psi_\delta * \chi_\epsilon \phi$ where $\psi_\delta *$ is a mollifier, $\phi \in C_0^1(D)$ and χ_ϵ is as in statement (iii). Then,

$\phi_\delta \in C_0^\infty(D)$ converges to $\chi_\epsilon \phi$ in $L^p(D)$, and using (2.5) and the remark in the first lines of the proof of Theorem 9.2.1, $\Lambda\phi_\delta$ converges to $\chi_\epsilon \Lambda\phi$ in L^p. This is applied to the definition of Λf in a distributional sense to see that

$$\Lambda(f_\epsilon) = \chi_\epsilon \Lambda f. \tag{3.10}$$

Hence $f_\epsilon \in \hat{W}_p$. We can use Green's formula (3.4) with f_ϵ in place of f. Take $\phi \in C_0^1(D \cup V^-)$ [see (2.2)], which is in $W_p^\dagger(D)$. Then, (3.10), part (i) and (3.6) prove (ii). We use (3.10) for ϕ, too, and note that $\chi_\epsilon \phi \in W_p^\dagger$. For the proof of (iii), note that Green's formula (3.4), again with a general $\phi \in W_p^\dagger$, taking account of (i) and (ii), gives (iii). □

We can now provide some estimates for the weak solutions. This can be done with the following.

(9.3.3) Theorem. *When* $\| K \| < 1$, *the mild solution is unique with the estimates*

$$\begin{aligned} \lambda p \| f \|_{L^p(D)}^p + (1 - \| K \|^p) \| \gamma^- f \|_{L^p,+}^p + \| f(T) \|_{L^p(\Omega \times \Re^3)}^p \\ \leq \| f_0 \|_{L^p(\Omega \times \Re^3)}^p \end{aligned} \tag{3.11}$$

for $p \in [1, \infty)$ *and*

$$\| f \|_{L^\infty(D)}, \quad \| \gamma^- f \|_{L^\infty,+}, \quad \| f(T) \|_{L^\infty(\Omega \times \Re^3)} \leq \| f_0 \|_{L^\infty(\Omega \times \Re^3)} \tag{3.12}$$

for $p = \infty$.

Proof. Let us consider Eq. (2.16) for functions of, say, C_0^1, not necessarily solutions of $\Lambda f = -\lambda f$; if $\gamma_D^+ f = K \gamma_D^- f$, with $\| K \| < 1$, then:

$$\begin{aligned} (1 - \| K \|^p) \| \gamma^- f \|_{L^p,+}^p + \| f(.,T) \|_{L^p(\Omega \times \Re^3)}^p \\ \leq \| f_0 \|_{L^p(\Omega \times \Re^3)}^p + p \operatorname{Re} \int_D |f|^{p-2} \bar{f} (\Lambda f) dr. \end{aligned} \tag{3.13}$$

If we now take a sequence $\{f_n\}$ of C_0^1 functions that approach a solution f in W^p, the limit as $n \to \infty$ of Eq. (3.13) (with f_n in place of f) is Eq. (3.11). This relation also proves that $f \in \hat{W}^p$. The case of $f \in W^\infty$ follows from the previous theorem because the weight θ is not essential in the case $p = \infty$ and $\hat{W}^\infty = W^\infty$. The uniqueness of the solution now follows from linearity and estimates (3.11) and (3.12). □

We can now consider the case $\| K \| = 1$ and prove the following.

(9.3.4) Theorem. *When* $\| K \| = 1$, *if* K *carries non-negative functions into functions of the same kind and* f_0 *is non-negative, then (3.1–3) have a mild non-negative solution* $f \in L^p(D)$, *with the estimate* ($\lambda > 0$):

$$\| f(.,T) \|_{L^p(\Omega\times\Re^3)} \leq \| f_0 \|_{L^p(\Omega\times\Re^3)} . \tag{3.14}$$

Remark. The problem of uniqueness for $\| K \| = 1$ has been solved[5] only with additional conditions on K.

Proof. Let us replace K by μK with $\mu \in (0,1)$ in (3.2); then, by the previous theorem, we have a unique strong solution f^μ satisfying (3.11) ($p \in [1,\infty)$) or (3.12) ($p = \infty$), which give uniform estimates for f^μ in $L^p(D)$ and $f^\mu(.,T)$ in $L^p(\Omega \times \Re^3)$. Hence, taking a nondecreasing sequence of values of μ converging to unity, we obtain a nondecreasing sequence of functions $f^\mu \to f$ in $L^p(D)$ and $f^\mu(.,T) \to h$ in $L^p(\Omega \times \Re^3)$, pointwise a. e. and strongly ($p \in [1,\infty]$). This f is clearly a weak solution with $f(.,T) = h$, and going to the limit in (3.11) ($p \in [1,\infty)$) or (3.12) ($p = \infty$), we obtain (3.14). □

So far we have assumed $\lambda > 0$. We have already remarked, however, that the constant λ can be removed and thus all the results remain true with some changes in the estimates. In particular, the following corollary holds.

(9.3.5) Corollary. *Theorem 9.3.4 is also true for* $\lambda = 0$.

Since $f(T) \in L^p(\Omega \times \Re^3)$ by (3.14) and since $T > 0$ may be arbitrary, we can introduce the solution operator $U(t)$ ($t \in \Re^+$), which carries $f_0 = f(.,0)$ into $f(.,t)$:

$$U(t) f_0 = f(.,t). \tag{3.15}$$

Then it is not hard to prove the following.

(9.3.6) Theorem. *If* $p \in [1,\infty)$, $U(t)$ *is a* C_0*-semigroup on* $L^p(\Omega \times \Re^3)$.

Remark. The continuity property is lacking for $p = \infty$.

In the sequel we shall need a generalization of these results to the case when the parameter λ in Eq. (3.1) is replaced by a non-negative function $l(t,x,\xi) \in L^1((0,T)\times\Omega\times{\Re^3}_{\text{loc}})$. Then the above treatment carries through. The main difference arises in the definition of the spaces W^p and $W_p^\dagger$ and in the proof of the analogue of Theorem 9.3.1. In fact, W^p is now replaced by W_l^p, such that $f \in L^p$ and $(\Lambda + l) f \in L^p$ and $W^{p\dagger}$ is replaced by $W_l^{p\dagger}$, such that $f \in L^p$ and $(\Lambda - l) f \in L^p$ while Eq. (3.8) must be replaced by

$$\hat{f}(s,r) = \hat{g}(r) e^{-\int_{-s^-(r)}^{s} l(R^{s'}(r))ds'} \tag{3.16}$$

and Eq. (3.9) is replaced by

$$\hat{g}(r) = \mathcal{K}[\hat{g}(r^*)e^{-\int_{s^-(r_*)}^{s^+(r_*)} l(R^{s'}(r_*))ds'}] + g_0 \qquad \text{for } r \in \partial D. \tag{3.17}$$

Thus we may conclude with this theorem.

(9.3.7) Theorem. *When* $\| K \| = 1$, *if* K *carries non-negative functions into functions of the same kind and* f_0 *is non-negative, then the problem*

$$(\Lambda + l)f = 0 \qquad \textit{in } D \tag{3.18}$$

(where $0 \leq l = l(t, x, \xi) \in L^1((0,T) \times \Omega \times \Re^3_{loc})$*) with the boundary and initial conditions (3.2–3) has a mild non-negative solution* $f \in L^p(D)$ *with the estimate:*

$$\| f(.,T) \|_{L^p(\Omega\times\Re^3)} \leq \| f_0 \|_{L^p(\Omega\times\Re^3)} \, . \tag{3.19}$$

The solution can be written as $U_l(t)f_0$*, where, if* $p \in [1, \infty)$*,* $U_l(t)$ *is a* C_0*-semigroup on* $L^p(\Omega \times \Re^3)$*.*

We shall also have to deal with sequences of non-negative functions $l_k \in L^1((0,T) \times \Omega \times \Re^3_{loc})$. In this case, if $\{l_k\}$ converges to l in $L^1((0,T) \times \Omega \times \Re^3_{loc})$, then $\{F_k\}$, where

$$F_k = \int_{-s^-(r)}^{0} l_k(s, x - \xi(t+s), \xi)ds \tag{3.20}$$

is a bounded sequence in $C([0,T], L^1(\Omega \times \Re^3{}_{loc}))$ and converges for any $t \in R_+$, a. e. in (x, ξ) to

$$F = \int_{-s^-(r)}^{0} l(s, x - \xi(t+s), \xi)ds. \tag{3.21}$$

Associated with the sequence $\{l_k\}$ we now have the sequence of solutions $\{U_{l_k}(t)f_0\}$ (for the sake of simplicity we restrict our attention to the case $p = 1$), which is pointwise dominated by $U(t)f_0$. Thus $\{U_{l_k}(t)f_0\}$ converges to $U_l(t)f_0$ because of the dominated convergence theorem, and thanks to the fact that all the relations that we need apply, such as (3.16) and (3.17), we can pass to the limit when we replace l by l_k and let n go to ∞ [in $[0,T]$ for a. e. (x, ξ)].

We remark that we can also solve

$$(\Lambda + l)f = g \qquad \text{in } D \qquad (\lambda \in \Re) \tag{3.22}$$

with initial and boundary conditions (3.2) and (3.3), when $g \in L^1((0,T) \times \Omega \times \Re^3_{loc})$. The solution is

$$f = U_l(t)f_0 + \int_0^t U_l(t-s)g ds. \tag{3.23}$$

We also remark that the traces do exist and satisfy Eq. (1.1) almost everywhere in $[0,T] \times \partial\Omega \times \Re^3$, because this is true of any function of the form $U_l(\tau)g$, $\tau > 0$.

We finally notice that $\{U_l g^\nu\}$ is an increasing sequence when $\{g^\nu\}$ is such a sequence.

Problems

1. Prove Theorem 9.3.6.
2. Why does (3.19) hold the way it is without a constant C_{0T} in front of the norm of f_0?

9.4 Existence in a Vessel with Isothermal Boundary

In order to deal with the existence theorem in a vessel at rest, with constant temperature along the boundary, it is convenient to remark that there is an absolute Maxwellian naturally associated with the problem, i.e. the wall Maxwellian M_w; an exception is offered by specular reflecting boundaries. In the latter case, M_w will mean the absolute Maxwellian with zero bulk velocity and total mass and energy equal to the total mass and energy of the gas at time t=0. Eq. (1.5) gives:

$$\int \xi \cdot n\gamma_D f \log \gamma_D f d\xi + \beta_w \int \xi \cdot n|\xi|^2 \gamma_D f d\xi \leq 0 \tag{4.1}$$

(a.e. in t and $x \in \partial\Omega$). Then the modified H-functional:

$$H = \int f \log f d\xi dx + \beta_w \int |\xi|^2 f d\xi dx \tag{4.2}$$

will decrease in time as a consequence of the Boltzmann equation and inequality (4.1). Thus H is bounded if it is bounded initially.

Let us divide the subset of $\Omega \times \Re^3$ where $f < 1$ into two subsets $\Delta^\pm = \{(x,\xi) : \pm \log f < \mp\beta_w \xi^2/2)\}$. Then (since $-f\log f$ is a growing function in $(0, e^{-1})$ and less than f for $f > e^{-1}$)

$$-\int_{\Delta^+} f \log f d\xi dx \leq \int f d\xi dx + \beta_w \int \xi^2 \exp[-\beta_w \xi^2/2] d\xi dx \leq C \tag{4.3}$$

and in Δ^-

$$-\int_{\Delta^-} f \log f d\xi dx \leq [\beta_w/2] \int \xi^2 f d\xi dx. \tag{4.4}$$

Then Eq. (4.2) implies that both $\int f|\log f|d\xi dx$ and $\int |\xi|^2 f d\xi dx$ are separately bounded in terms of the initial data. It is then easy to prove that the mass and entropy relations take on the following form:

$$\int f(\cdot,t)d\xi dx = \int f(.,0)d\xi dx \tag{4.5}$$

$$\begin{aligned} &\int f\log f(.,t)d\xi dx + \beta_w \int |\xi|^2 f(.,t)d\xi dx \\ &\quad + \int_0^t \int e(f)(\cdot,s)d\xi dx ds \\ &\quad \le \int f\log f(\cdot,0)d\xi dx + \beta_w \int |\xi|^2 f(.,0)d\xi dx \\ &\quad + m^3 \parallel \frac{\partial\beta}{\partial x} \parallel_{L^\infty} \int f_0 d\xi dx \end{aligned} \tag{4.6}$$

where

$$e(f)(x,\xi,t) = \frac{1}{4}\int_{\Re^3}\int_{S+}(f'f'_* - ff_*)\log(f'f'_*/ff_*)B(v,V\cdot n)d\xi_* dn. \tag{4.7}$$

The theorem that we will prove now was first discussed by Hamdache[15], but here we shall follow the previously quoted paper by Arkeryd and Cercignani[2], which, although devoted to the case of nonisothermal boundaries, contains a slightly different proof of Hamdache's theorem, with an extension to more general boundary conditions, to a more detailed study of the boundary behavior, and for the full class of collision operators of the Di Perna and Lions [13] existence context. As hinted at in Section 1, we will use the equivalent concepts of exponential, mild, and renormalized solutions as defined by DiPerna and Lions[13], and such solutions will be found as limits of functions solving truncated equations.

The existence theorem to be proved reads as follows.

(9.4.1) Theorem. *Let $f^0 \in L^1(\Omega\times\Re^3)$ be such that*

$$\int f^0(1+|\xi|^2)d\xi dx < \infty; \qquad \int f^0|\log f^0|d\xi dx < \infty. \tag{4.8}$$

Then there is a solution $f \in C(\Re_+, L^1(\Omega\times\Re^3))$ of the Boltzmann equation such that $f(.,0) = f^0$, which also satisfies relations (4.5) and (4.6).

Proof. We shall only sketch a proof of the theorem, since the argument is rather similar to the one presented by Di Perna and Lions[13] for the case of $\Re^3$; the main differences are the necessity to have trace estimates and the fact that we do not have separate energy and entropy estimates; both aspects have been dealt with. We shall mention any important modification in the course of the proof. We first introduce a smooth truncation of the Boltzmann equation and prove an existence theorem for the truncated equation. We choose as truncated collision term

$$Q^k(f,f) = (1 + k^{-1} \int f d\xi)^{-1} \int_{\Re^3} \int_{S^2} d\xi_* dn (f' f'_* - f f*) B_k \tag{4.9}$$

where

$$B_k = B \wedge k \quad \text{for} \quad \xi^2 + \xi_*^2 \le k^2 \,, \quad B_k = 0 \quad \text{otherwise} \tag{4.10}$$

We also set

$$l_k = \int d\xi_* dn B_k f_{k*} (1 + k^{-1} \int f_k d\xi)^{-1} \tag{4.11}$$

$$\tilde{U}_k^{\pm}(r,s) = \exp(\pm \int_{-s^-(r)}^{s} l_k(R^\tau(r)) d\tau). \tag{4.12}$$

Further, $l = l_\infty$ and $\tilde{U}^\pm = \tilde{U}^\pm_\infty$ are analogously defined with $f_\infty = f = \lim f_k$. Subsequences $\{k_j\}$ of $\{k\}$, sometimes necessary from one step to the next, will still be denoted by $\{k\}$.

Then for $f, g \in L^1(\Omega \times \Re^3)$,

$$\| Q^k(f,f) - Q^k(g,g) \|_{L^1(\Omega\times\Re^3)} \le C_k \| f - g \|_{L^1(\Omega\times\Re^3)} \,. \tag{4.13}$$

Hence, for $f_0 \in L^1(\Omega \times \Re^3)$, the mild Boltzmann equation

$$f_k = U(t) f_0 + \int_0^t U(t-s) Q^k(f_k, f_k) ds \tag{4.14}$$

with the desired boundary behavior can be solved by a contraction mapping argument.

If (4.8) holds for f_0, then by Green's formula (3.4) f_k satisfies (4.5). Via suitable smooth approximations it can also be shown that, thanks to (4.3) and (4.4), f_k also satisfies (4.6), even with $f_k(t,.) \log f_k(t,.)$ replaced by $|f_k(t,.) \log f_k(t,.)|$, if we add a suitable k-dependent constant on the right-hand side. In particular

$$\forall T > 0 \qquad \sup_{t\in[0,T]} \sup_k \int f_k(\cdot, t) d\xi dx < \infty \tag{4.15}$$

$$\forall T > 0 \qquad \sup_{t\in[0,T]} \sup_k \int \xi^2 f_k(\cdot,t)d\xi dx < \infty \tag{4.16}$$

$$\forall T > 0 \qquad \sup_{t\in[0,T]} \sup_k \int f_k |\log f_k(\cdot,t)| d\xi dx < \infty \tag{4.17}$$

$$\sup_k \int_0^\infty \int e_k(f_k)(\cdot,s)d\xi dx ds < \infty \tag{4.18}$$

$$\begin{aligned} e_k(f_k)(x,\xi,t) =& \frac{1}{4}(1+k^{-1}\int f_k d\xi)^{-1} \int_{\Re^3}\int_{S_+} (f'_k f'_{k*} - \\ & f_k f_{k*}) \log(f'_k f'_{k*}/f_k f_{k*}) B_k(V, V\cdot n) d\xi_* dn. \end{aligned} \tag{4.19}$$

By (3.4), with a test function that is a suitable extension to the interior of the function $n(x)$ as defined on the boundary, it then follows that

$$\int_{E^\pm} |\xi\cdot n(x)|^2 \gamma_D^\pm f_k d\sigma d\xi ds \le C_{0T}, \tag{4.20}$$

where C_{0T} only depends on T and the right hand-side of (4.8) (but not on k). Under the conditions of, e.g., Theorem 9.2.3, these traces also belong to $L^{1\pm}$, but that is not known to be true in general.

We can now apply the Dunford-Pettis criterion to our sequence to conclude from (4.15–17) that $\{f_k\}$ has a subsequence that converges weakly to some function f.

It is then easy to show that the sequences $\{\frac{Q_+^k(f_k,f_k)}{1+f_k}\}$, $\{\frac{Q_-^k(f_k,f_k)}{1+f_k}\}$ are in a weakly compact set of $L^1((0,T)\times\Omega\times\mathbf{B}_R)$, where $\mathbf{B}_R$ is the ball of radius R in velocity space. This is proved in exactly the same way as in the case[13] of $\Re^3$.

We denote by $l = Lf$ the function multiplying f in the loss term of the Boltzmann equation and by $U_l(t)$ the semigroup associated with it according to Theorem 3.6. When we solve the truncated equation, we deal with l_k and $U_{l_k}(t)$. According to Eq. (3.23) we have

$$f_k = U_{l_k}(t)f_k^0 + \int_0^t U_{l_k}(t-s)Q_+^k(f_k,f_k)ds \tag{4.21}$$

Using the velocity-averaging lemma we shall prove that f is a solution of the Boltzmann equation that retains a fairly weak control of the traces.

As in the case of $\Re^3$, one exploits the fact that all the terms are non-negative to go to the limit in Eq. (4.21). Let us denote by $\alpha^\nu(s)$ the minimum $\nu\wedge s = \min(\nu,s)$ and put $f_k^\nu = \alpha^\nu(f_k)$. We may assume that f_k^ν tends (weakly) to some f^ν in $L^1((0,T)\times\Re^3\times\Re^3)$ and as a consequence f^ν (different from $\alpha^\nu(f)$ in general) converges to f in a monotonous way.

Given ν, we can apply the averaging lemma to study the convergence of f_k^ν, which is also bounded by ν; then $Q_+^k(f_k^\nu, f_k)$ converges weakly to $Q_+(f^\nu, f)$ in $L^1((0,T)\times\Omega\times\mathbf{B}_R)$ for any $R > 0$. On the other hand, Eq. (4.21) implies

$$f_k \geq U_{l_k}(t)f_k^0 + \int_0^t U_{l_k}(t-s)Q_+^k(f_k^\nu, f_k)ds. \tag{4.22}$$

For a. e. $r \in D$ either the x-component of $R^s(r)$ is nontangential to $\partial\Omega$ at $-s = s^-(r) > 0$ and belongs to an open C^1-component of $\partial\Omega$, or the x-component belongs to Ω at $-s = s^-(r) = 0$. (The analogous situation for $s = s^+(r)$ will not be discussed.) For such an r, there is a neighborhood $\mathcal{N}$ of $R^{-s^-(r)}(r)$ in ∂D^+, so that the world lines emanating from that neighborhood have the same properties as $R^s(r)$ and have an x-component staying in Ω through a neighborhood of r. Let $\psi_\mathcal{N}$, or, for convenience, simply ψ, be the characteristic function of $\mathcal{N}$ prolonged with value unity along the corresponding world lines $R^s(.)$, and with $\psi = 0$ otherwise. We have to abandon the semigroup U_{l_k} because it is not clear that it will survive when we take the limit as k goes to ∞. Thus we shall prove that the (weak) limit of f_k is a mild (and hence renormalized) solution of the Boltzmann equation and defer to Theorem 9.4.2 the discussion of its tie with the boundary conditions. To this end, for $r' \in \mathcal{N}$ and $0 < s < s^-(r)$, we rewrite Eq. (4.21) for $t = s$ and $t = \delta$, solve the second of these relations for f_k^0, and insert the resulting expression into the first of them; then we argue on the new relation as we did before to pass from (4.21) to (4.22), to obtain:

$$\begin{aligned} f_k(R^s(r')) \geq & \tilde{U}_k^-(r',s)\tilde{U}_k^+(r',\delta)f_k(R^\delta(r')) \\ & + \int_\delta^s \tilde{U}_k^-(r',s)\tilde{U}_k^+(r',\tau)Q_+^k(f_k^\nu, f_k)(R^\tau(r'))d\tau. \end{aligned} \tag{4.23}$$

For $R^{-s^-(r)}(r) = (t,x,\xi) \in E^+$ pick a product neighborhood of (t,x,ξ): $I_t \times \mathcal{N}_x \times \mathcal{N}_\xi \subset \mathcal{N}$. For $0 \leq s \leq s^-(r)$ let $\mathcal{N}_s$ be the projection in $[0,T] \times \Omega$ of $R^s(I_t \times \mathcal{N}_x \times \{\xi\})$. Take a smaller product neighborhood of (t,x,ξ);

$$\mathcal{N}' = I_t' \times \mathcal{N}_x' \times \mathcal{N}_\xi' \subset I_t \times \mathcal{N}_x \times \mathcal{N}_\xi \tag{4.24}$$

so that

$$\{R^s(r')|r' \in \mathcal{N}', 0 \leq s \leq S^-(r)\} \subset \bigcup_s \mathcal{N}_s \times \mathcal{N}_\xi'. \tag{4.25}$$

Denote by $\mathcal{N}_{\delta s}$ the subset
(4.26)
$$\mathcal{N}_{\delta s} = \{R^{s'}(r')|r' \in \mathcal{N}', 0 \leq s' \leq s^-(r), R^{s'}(r') \in \bigcup_{\delta\leq s''\leq s} \mathcal{N}_{s''} \times \mathcal{N}_\xi'\}.$$

Set $\psi_{\mathcal{N}_{\delta s}} = \psi$. It is a consequence of the averaging lemma and the estimates of f_k that for the integral over $\mathcal{N}_{\delta s}$,

$$\lim_{\nu\to\infty}\lim_{k\to\infty}\int|\int(f_k^\nu\psi - f\psi)d\xi|dxdt = 0. \tag{4.27}$$

It follows that for a. e. $0 < s'' < s^-(r)$,

$$\lim_{\nu\to\infty}\lim_{k\to\infty}\int_{\mathcal{N}_{s''}}|\int_{\mathcal{N}'_\xi}(f_k^\nu\psi - f\psi)d\xi|dxdt = 0. \tag{4.28}$$

We are now going to use a technique that was introduced in a paper by Arkeryd and Cercignani[3], devoted to the Enskog equation, and adapted to the present case by the same authors[2]. Multiplying (4.23) by ψ and integrating, we get, by averaging and using (4.28) for a. e. δ and s with the same double limit and after letting the support of ψ shrink to a Lebesgue world line, that

$$\begin{aligned} f(R^s(r')) \geq &\tilde{U}^-(r',s)\tilde{U}^+(r',\delta)f(R^\delta(r')) \\ &+ \int_\delta^s \tilde{U}^-(r',s)\tilde{U}^+(r',\tau)Q_+(f,f)(R^\tau(r'))d\tau \end{aligned} \tag{4.29}$$

for a. e. $r' \in E^+$ and a. e. $0 < \delta < s < s^+(r')$. An analogous reasoning gives (4.29) in the case $R^{-s^-(r)}(r) \in V^+$.

Having obtained the last inequality, we now prove that the opposite inequality also holds, in order to be able to conclude that the equality sign applies in (4.29). To this end, let us now denote by $f_k^\nu = \nu\log(1 + f_k/\nu)$ so that

$$\begin{aligned} f_k^\nu = &U_{l_k}(t)f_k^{0\nu} + \int_0^t U_{l_k}(t-s)\frac{Q_+^k(f_k,f_k)}{1+f_k/\nu}ds \\ &+ \int_0^t U_{l_k}(t-s)[l_k(f_k^\nu - \frac{f_k}{1+f_k/\nu})]ds. \end{aligned} \tag{4.30}$$

Rewriting (4.30) similarly to (4.23) and arguing as for (4.29) from here we get for a. e. $r' \in E^+$ and a. e. $0 < \delta < s < s^+(r')$

$$\begin{aligned} f(R^s(r')) \leq &\tilde{U}^-(r',s)\tilde{U}^+(r',\delta)f(R^\delta(r')) \\ &+ \int_\delta^s \tilde{U}^-(r',s)\tilde{U}^+(r',\tau)Q_+(f,f)(R^\tau(r'))d\tau, \end{aligned} \tag{4.31}$$

which, together with (4.29), implies that the equality sign holds in (4.29). Thus when $\delta \to 0$

(4.32)

$$f(r) = \tilde{U}^-(r,0)f(R^{-s^-(r)}(r)) + \int_{-s^-(r)}^0 \tilde{U}^-(r,0)\tilde{U}^+(r,s)Q_+(f,f)(R^s(r))ds$$

for any r on a. e. world line in D. Eq. (4.32) allows us to conclude that f is a solution of the Boltzmann equation.

Finally the entropy inequality can also be proved, by starting from the truncated equation and arguing as in the case of $\Re^3$. □

We are now in a position to study the boundary condition satisfied by these solutions. These boundary conditions have been left unspecified so far. In fact, Eq. (4.32) only says that f is a mild solution (and hence a renormalized solution) of the Boltzmann equation but gives no indication whether this solution satisfies the boundary conditions. This would be automatic if f were an ordinary weak solution. As a matter of fact, nobody has proved so far that the traces of f satisfy (1.1). The only result that has been proved, by Hamdache[15] and, in a more explicit way, by Arkeryd and Cercignani [2], is that Eq. (1.1) is replaced by an inequality. Following these papers we prove the following.

(9.4.2) Theorem. *There is a solution, as in Theorem 4.1, that satisfies*

$$\gamma_D^+(f) \geq K(\gamma_D^- f) \quad \textit{a. e. on } E^+. \tag{4.33}$$

Proof. It follows from (4.20) that

$$\int_{E^+} |\xi \cdot n|^2 \gamma_D^+ f_k^\nu \, d\sigma d\xi' ds \leq C_{0T} \tag{4.34}$$

and

$$\int_{E^-} |\xi \cdot n|^2 \gamma_D^- f_k^\nu \, d\sigma d\xi' ds \leq C_{0T}. \tag{4.35}$$

Given $\epsilon > 0$, consider the subset $E_\epsilon^\pm \subset E^\pm$ where the $\partial\Omega$-projection is in the open C^1 part of $\partial\Omega$ and $s_+ + s_- > \epsilon$. ψ_ϵ or, for convenience, simply ψ, will hereafter denote the characteristic function of a bounded Borel set in $E_\epsilon^\pm$ prolonged with value unity along the word lines and equal to zero otherwise.

The particular Borel sets actually considered will also be required to be contained in the intersection of $E_\epsilon^\pm$ with a product neighborhood $I_t \times \mathcal{N}_x \times \mathcal{N}_\xi$, where $\mathcal{N}_x$ is contained in a C^1 piece of $\partial\Omega$ and $\xi' \cdot n(x') \neq 0$ for $x' \in \overline{\mathcal{N}}_x$, $\xi' \in \overline{\mathcal{N}}_\xi$. We shall finally require that $f(R(\cdot)) \to f(\cdot)$ uniformly when $\delta \to 0$ on the Borel sets considered.

To prove the trace statement, it is enough to prove for such ψs that

$$< \gamma_D^+ f, \gamma_D^+ \psi >_{E_+} \geq < K\gamma_D^- f, \gamma_D^+ \psi >_{E_+} . \tag{4.36}$$

With f_k^ν as in (4.30) and $f^\nu = \text{w-}\lim_{k\to\infty} f_k^\nu$ in $L^1(D)$, we have $f^\nu \uparrow f$ pointwise a. e. and strongly in $L^1(D)$ when $\nu \to \infty$. So for a. e. (small) $\delta > 0$, $\psi f^\nu \uparrow \psi f$ strongly in $L^1(\mathcal{N}_\delta \times \mathcal{N}_\xi)$. By averaging, similarly to (4.27–28),

$$\text{w-}\lim_{k\to\infty} \psi f_k^\nu = \psi f_\nu \quad \text{in } L^1_{\mathcal{N}_\delta}, \tag{4.37}$$

and outside a set of arbitrarily small measure

$$\lim_{k\to\infty} \psi(r) \int_0^s l_k(R^\tau(r))d\tau = \psi(r) \int_0^s l(R^\tau(r))d\tau \tag{4.38}$$

uniformly over $r \in E_+ \cap \text{supp}\psi$, and with the right-hand side uniformly bounded. Let $\tilde{\psi} = \psi$ for the remaining set of word lines in supp ψ and $\tilde{\psi} = 0$ otherwise.

For $\delta > \delta' > \delta'' > 0$ it follows from (4.30) that

$$\begin{aligned}
0 &\leq \int_{\delta'}^{\delta''} \tilde{U}_k^+(r,\tau)[\tilde{\psi} l_k(f_k^\nu - \frac{f_k}{1+f_k/\nu})(R^\tau(r))]d\tau \\
&\leq \tilde{U}_k^+(r,\delta')[\tilde{\psi}(f_k^\nu(R^{\delta'}(r))] - \tilde{U}_k^+(r,\delta'')[\tilde{\psi}(f_k^\nu(R^{\delta''}(r))] \\
&\leq \tilde{U}_k^+(r,\delta)[\tilde{\psi}(f_k(R^{\delta}(r))] - \tilde{U}_k^+(r,\delta'')[\tilde{\psi}(f_k(R^{\delta''}(r))] \\
&\quad + \int_{\delta''}^{\delta'} \tilde{U}_k^+(r,\tau)[l_k(f_k(R^\tau(r))]d\tau \\
&\leq (j+1)\{\tilde{U}_k^+(r,\delta)[\tilde{\psi}(f_k(R^{\delta}(r))] - \tilde{U}_k^+(r,\delta'')[\tilde{\psi}(f_k(R^{\delta''}(r))]\} \\
&\quad + \frac{2}{\log j} \int_0^\delta e(f_k)(R^\tau)d\tau.
\end{aligned} \tag{4.39}$$

In the limit $k \to \infty$, the first two inequalities give

$$0 \leq \tilde{U}^+(r,\delta')[\tilde{\psi}(f^\nu(R^{\delta'}(r))] - \tilde{U}^+(r,\delta'')[\tilde{\psi}(f^\nu(R^{\delta''}(r))]. \tag{4.40}$$

The first two terms in the last member of (4.39) give, in the limit when $k \to \infty$,

$$(j+1)\{\tilde{U}^+(r,\delta)[\tilde{\psi}(f(R^{\delta}(r))] - \tilde{U}^+(r,\delta'')[\tilde{\psi}(f(R^{\delta''}(r))]\},$$

which is bounded by

$$(j+1)\{\tilde{U}^+(r,\delta)[\tilde{\psi}(f(R^{\delta}(r))] - \tilde{\psi} f(r)\}.$$

Recalling that f_k satisfies (4.18), and the uniform convergence $f \circ R^\delta \to f$, we may conclude that

$$\lim_{\delta\to 0} \int |\tilde{U}^+[\tilde{\psi} f^\nu_{|\mathcal{N}_\delta}] - \tilde{\psi} f^\nu_{|E_+}| dt d\sigma d\xi = 0 \tag{4.41}$$

uniformly with respect to ν. This, together with

$$\text{s-}\lim_{\nu\to\infty} \tilde{\psi} f^\nu = \tilde{\psi} f \text{ in } L^1(\mathcal{N}_\delta \times \mathcal{N}_\xi), \tag{4.42}$$

implies that

$$\text{s-}\lim_{\nu\to\infty} \gamma_D^+[\tilde{\psi} f^\nu] = \gamma_D^+[\tilde{\psi} f] \text{ in } L^{1+}. \tag{4.43}$$

An analogous result holds for L^{1-}.

By concavity and Jensen's inequality, (1.1) implies

$$\gamma_D^+ f_k^\nu \geq K(\gamma_D^- f_k^\nu). \tag{4.44}$$

Now in $L^{1\pm}$

$$\text{w-}\lim_{k\to\infty} \gamma_D^\pm[\tilde{\psi} f_k^\nu] = \gamma_D^\pm[\tilde{\psi} f^\nu] \tag{4.45}$$

and so

$$< \gamma_D^+ f^\nu, \gamma_D^+ \tilde{\psi} >_{E_+} \geq < K\gamma_D^- f^\nu, \gamma_D^+ \tilde{\psi} >_{E_+} . \tag{4.46}$$

By (4.43) this gives

$$< \gamma_D^+ f, \gamma_D^+ \tilde{\psi} >_{E_+} \geq < K\gamma_D^- f, \gamma_D^+ \tilde{\psi} >_{E_+} \tag{4.47}$$

and from here finally

$$< \gamma_D^+ f, \gamma_D^+ \psi >_{E_+} \geq < K\gamma_D^- f, \gamma_D^+ \psi >_{E_+} . \tag{4.48}$$

□

Remark. If the traces of the solutions are in $L^{1\pm}$ (as in the case of Maxwellian diffusion at the boundary), and if $Q(f,f)$ belongs to $L^1(D)$ then there is equality in (4.33).

Problem

1. Prove that if Eq. (1.1) is replaced by an inequality (with $\geq$ in place of $=$) and the total mass is conserved, then the equality sign must apply and the original equality is recovered a. e. (Hint: Integrate the Boltzmann equation with respect to x, ξ, t.)

9.5 Rigorous Proof of the Approach to Equilibrium

Discussions of equilibrium states in kinetic theory are as old as the theory itself; actually these states were discussed even before the basic evolution equation of the theory, i.e., the Boltzmann equation, was formulated. The recent work on the mathematical aspects of kinetic theory has also led to new results on this problem.

The aim of this section is to discuss the trend to equilibrium, following the approach of Desvillettes[11] and Cercignani[8], which is based on a remark by DiPerna and Lions[12]. The main result is the following.

(9.5.1) Theorem. *Let $f(x,\xi,t)$ be the solution of the Boltzmann equation, with initial data $f_0(x,\xi)$ such that*

$$(5.1)\qquad f_0 \geq 0; \qquad \int_\Omega \int_{\Re^3} f_0(x,\xi)(1+|\xi|^2+|\log f_0(x,\xi)|)dxd\xi < +\infty.$$

Let f also satisfy the boundary condition (1.1),where the kernel is such that Eqs. (1.2–4) hold (M_w is a constant and uniform Maxwellian). Then, for every sequence t_n going to infinity, there exist a subsequence t_{n_k} and a local Maxwellian $M(x,\xi,t)$ such that $f_{n_k}(x,\xi,t) = f(x,\xi,t_{n_k}+t)$ converges weakly in $L^1(\Omega \times \Re^3 \times [0,T])$ to $M(x,\xi,t)$ for any $T > 0$. Moreover, M satisfies the free transport equation

$$(5.2)\qquad \frac{\partial M}{\partial t} + \xi \cdot \frac{\partial M}{\partial x} = 0$$

and the boundary condition (1.1).

Proof. According to the proof presented in the previous section

$$(5.3)\qquad \begin{aligned} &\int_0^T \int_{\Re^3}\int_\Omega\int_{S^2}\int_{\Re^3} [f(x,\xi',t)f(x,\xi'_*,t) - f(x,\xi,t)f(x,\xi_*,t)] \\ &\quad \times \{\log[f(x,\xi',t)f(x,\xi'_*,t)] - \log[f(x,\xi,t)f(x,\xi_*,t)]\} \\ &\quad \times B(V, V\cdot n)d\xi_* dndxd\xi dt \\ &+ \sup_t \int_\Omega\int_{\Re^3} f(x,\xi,t)(1+|\xi|^2+|\log f(x,\xi,t)|)dxd\xi < +\infty. \end{aligned}$$

Thus $f_n(x,\xi,t) = f(x,\xi,t+t_n)$ is weakly compact in $L^1(\Omega\times\Re^3\times[0,T])$ for any sequence t_n of non-negative numbers and any $T > 0$. If $t_n \to \infty$, then there exist a subsequence t_{n_k} and a function $M(x,\xi,t)$ in $L^1(\Omega\times\Re^3\times[0,T])$ such that f_{n_k} converges weakly to M in $L^1(\Omega\times\Re^3\times[0,T])$ for any $T > 0$. In order to prove that M is a Maxwellian, we remark that, since the first integral in Eq. (5.3) is finite, then

$$\begin{aligned} &\int_{t_{n_k}}^{T+t_{n_k}} \int_{\Re^3}\int_\Omega\int_{S^2}\int_{\Re^3} [f(x,\xi',t)f(x,\xi'_*,t) - f(x,\xi,t)f(x,\xi_*,t)] \\ &\{\log\frac{f(x,\xi',t)f(x,\xi'_*,t)}{f(x,\xi,t)f(x,\xi_*,t)}\}B(V,V\cdot n)d\xi_* dndxd\xi dt \to 0 \end{aligned}$$

$(k \to \infty)$, and thus

$$(5.4)\qquad \begin{aligned} &\int_0^T \int_{\Re^3}\int_\Omega\int_{S^2}\int_{\Re^3} [f_{n_k}(x,\xi',t)f_{n_k}(x,\xi'_*,t) \\ &\quad - f_{n_k}(x,\xi,t)f_{n_k}(x,\xi_*,t)]\{\log[f_{n_k}(x,\xi',t)f_{n_k}(x,\xi'*,t)] \\ &\quad - \log[f_{n_k}(x,\xi,t)f_{n_k}(x,\xi_*,t)]\}B(V,V\cdot n)d\xi_* dndxd\xi dt \to 0 \end{aligned}$$

$(k \to \infty)$. But for all smooth non-negative functions ϕ, ψ with compact support:

(5.5)
$$\begin{aligned}&\int_{\Re 3}\int_{S^2}\int_{\Re^3} f_{n_k}(x,\xi',t)f_{n_k}(x,\xi'_*,t)\phi(\xi)\psi(\xi_*)B(V,V\cdot n)d\xi_* dn d\xi\\ &\to \int_{\Re^3}\int_{S^2}\int_{\Re^3} M(x,\xi',t)M(x,\xi'_*,t)\phi(\xi)\psi(\xi_*)B(V,V\cdot n)d\xi_* dn d\xi\end{aligned}$$

(a. e. in $\Omega \times [0,T]$ when $k \to \infty$) and

(5.6)
$$\begin{aligned}&\int_{\Re^3}\int_{S^2}\int_{\Re^3} f_{n_k}(x,\xi,t)f_{n_k}(x,\xi_*,t)\phi(\xi)\psi(\xi_*)B(V,V\cdot n)d\xi_* dn d\xi\\ &\to \int_{\Re^3}\int_{S^2}\int_{\Re^3} M(x,\xi,t)M(x,\xi_*,t)\phi(\xi)\psi(\xi_*)B(V,V\cdot n)d\xi_* dn d\xi\end{aligned}$$

(a. e. in $\Omega \times [0,T]$ when $k \to \infty$). This was proved by DiPerna and Lions[12] with the following kind of argument (we just consider (5.6) because (5.5) is analogous). First, remark that for any Borel set $A \in \mathcal{B}(\mathbf{B}_R \times \mathbf{B}_R \times S^2)$ with respect to the measure $d\mu = B(V, V\cdot n)d\xi_* dn d\xi$, where $\mathbf{B}_R$ is the ball of radius R in velocity space

(5.7)
$$\begin{aligned}&\int_A f_{n_k}(x,\xi,t)f_{n_k}(x,\xi_*,t)B(V,V\cdot n)d\xi_* dn d\xi\\ \le& M^2\mu(A) + \int_{E_R} f_{n_k}(x,\xi,t)f_{n_k}(x,\xi_*,t)[\chi(f_{n_k}(x,\xi,t)\ge M)\\ &+\chi(f_{n_k}(x,\xi_*,t)\ge M)]B(V,V\cdot n)d\xi_* dn d\xi\\ \le& M^2\mu(A) + [\psi(M)]^{-1}\overline{K}(t,x)\overline{N}(t,x) \qquad (\forall M>0)\end{aligned}$$

where χ denotes the characteristic function of a set and $\psi(t) \in C([0,\infty))$ is an increasing function such that $\psi \to \infty$ as $t \to \infty$, $\psi(t)(\log t)^{-1} \to 0$ as $t \to \infty$, while

(5.8)
$$\begin{aligned}\overline{K} =& \sup_k \int_{E_R} f_{n_k}(x,\xi,t)f_{n_k}(x,\xi_*,t)[\psi(f_{n_k}(x,\xi,t)) + \psi(f_{n_k}(x,\xi_*,t))]\\ &B(V,V\cdot n)d\xi_* dn d\xi \times (1+\int_{\Re^3} f_{n_k} d\xi)^{-1}\\ \overline{N} =& \sup_k (1+\int_{\Re^3} f_{n_k} d\xi).\end{aligned}$$

Here $\overline{K}$ and $\overline{N}$ are functions independent of f_{n_k}. This proves that the product $f_{n_k}(x,\xi,t)f_{n_k}(x,\xi_*,t)$ is weakly compact in $L^1(E_R, d\mu)$ for a. a. $(x,t) \in \Omega \times (0,T)$. We already know that

$$f_{n_k}(x,\xi,t)f_{n_k}(x,\xi_*,t)[1+\int_{\Re^3} f_{n_k} d\xi]^{-1}$$

converges to the corresponding function,

$$M(x,\xi,t)M(x,\xi_*,t)[1+\int_{\Re^3} M d\xi]^{-1},$$

weakly in $L^1(E_R, d\mu)$ for a. a. $(x,t) \in \Omega \times (0,T)$. But the denominator converges a. e. to $1+\int_{\Re^3} M d\xi$; hence $f_{n_k}(x,\xi,t)f_{n_k}(x,\xi_*,t)$, which has been shown to be weakly compact, converges weakly to $M(x,\xi,t)M(x,\xi_*,t)$ in $L^1(E_R, d\mu)$ for a. a. $(x,t) \in \Omega \times (0,T)$. It is then possible to extract a subsequence (which we still denote by f_{n_k}), such that

$$\begin{aligned}(5.9)\qquad &\int_{\Re^3}\int_{S^2}\int_{\Re^3} [f_{n_k}(x,\xi',t)f_{n_k}(x,\xi'_*,t) - f_{n_k}(x,\xi,t)f_{n_k}(x,\xi_*,t)] \\ &\{\log[f_{n_k}(x,\xi',t)f_{n_k}(x,\xi'_*,t)] - \log[f_{n_k}(x,\xi,t)f_{n_k}(x,\xi_*,t)]\} \\ &B(V, V\cdot n)d\xi_* dn d\xi \to 0\end{aligned}$$

(a. e. in $\Omega \times [0,T]$ when $k \to \infty$) for a dense denumerable set in $C(\Re^3)$ of non-negative smooth functions ϕ and ψ. But then the convexity of the function $C(f,g) = (f-g)(\log f - \log g)$ $(\Re_+ \times \Re_+ \to \Re_+)$ implies that we can pass to the limit and obtain

$$\begin{aligned}(5.10)\qquad &[M(x,\xi',t)M(x,\xi'_*,t) - M(x,\xi,t)M(x,\xi_*,t)] \\ &\log\frac{M(x,\xi',t)M(x,\xi'*,t)]}{[M(x,\xi,t)M(x,\xi_*,t)}B(V,V\cdot n) = 0\end{aligned}$$

(a. e. in ξ_*, n, x, ξ_*, t). Then, since $C(f,g)$ is non-negative and $B(V, V\cdot n)$ strictly positive:

$$(5.11)\qquad M(x,\xi',t)M(x,\xi'_*,t) = M(x,\xi,t)M(x,\xi_*,t)$$

(a. e. in ξ, n, x, ξ_*, t)). Then, as proved in Chapter 3, M is a Maxwellian and is, thanks to the property of weak stability, a renormalized solution of the Boltzmann equation satisfying the boundary condition (1.1). Accordingly $Q(M,M) = 0$ and Eq. (5.2) holds. □

Theorem 5.1 tells us that the solutions of the Boltzmann equation with the boundary conditions (1.1) behave (in the case of a boundary at constant temperature) as Maxwellians satisfying the free transport equation, Eq. (5.2). These Maxwellians are well known since Boltzmann[6], and they were already discussed in Chapter 3. They have the following form:

$$(5.12)\quad M = \exp[a_0 + b_0\cdot\xi + c_0|\xi|^2 + d_0|x-\xi t|^2 + e_0\cdot(x-\xi t) + f_0\cdot(x\wedge\xi)]$$

where $a_0, c_0, d_0 \in \Re$ and $b_0, e_0, f_0 \in \Re^3$ are constants. Now if we impose the condition that $M(x,.,t)$ is an L^1 function for any $t \geq 0$, we see that c_0 must be negative and d_0 nonpositive. We exclude now from our considerations the cases in which the kernel K is a delta function; in fact, the only significant situations in which a Dirac delta occurs are exceptional and can easily be treated in detail as shown by Desvillettes[11]. In the other cases, there is only one Maxwellian that is compatible with the boundary conditions, i.e., a Maxwellian with no drift and constant temperature; this immediately implies that b_0, d_0, e_0, f_0 are zero. Hence M is a uniform Maxwellian, which coincides with M_w. Thus we have the following result, embodying the results of Desvillettes[11] and Cercignani[8].

(9.5.2) Theorem. *The Maxwellian M in Theorem 5.1 is uniform, with the exception of specularly reflecting boundaries having rotational symmetry about an axis ℓ. In the latter case the Maxwellian M might describe a solid body rotation of the gas about ℓ.*

Recently L. Arkeryd[1] has proved that f actually tends to a Maxwellian in a strong sense for a periodic box, but his argument also works in other cases; his proof uses techniques of nonstandard analysis and, as such, is outside the scope of this book. Subsequently P.-L. Lions[17] obtained the same result without resorting to nonstandard analysis.

Problem

1. Consider the case of a specularly reflecting boundary and prove that a solid body rotation of the gas about an axis is compatible with the boundary conditions if the boundary has rotational symmetry with respect to that axis (as mentioned in Theorem 9.5.2).

9.6 Perturbations of Equilibria

Let us consider again the initial-boundary value problem for the Boltzmann equation

$$\Lambda f = Q(f,f) \qquad \text{in } D \tag{6.1}$$

$$\gamma_D^+ f(x,\xi,t) = K\gamma_D^- f \qquad \text{on } \partial D^+ \tag{6.2}$$

$$f(x,\xi,0) = f_0(x,\xi). \tag{6.3}$$

Let $U(t)$ be as in (5.15) ($\lambda = 0$). Then the initial-boundary value problem reduces to solving the following integral equation

$$f(t) = U(t)f_0 + \int_0^t U(t-s)Q(f(s),f(s))ds. \tag{6.4}$$

It is easy to prove local existence of this equation (Problem 2) in several spaces, typically:

$$X_{\alpha,\beta} = \{f|\ (1+|\xi|^2)^{\alpha/2}\exp(-\beta|\xi|^2)f \in L^\infty(\Omega\times\Re^3)\} \tag{6.5}$$

with norm

$$\| f \|_{\alpha,\beta} = \| (1+|\xi|^2)^{\alpha/2} \exp(-\beta|\xi|^2) f \|_{L^\infty(\Omega\times\Re^3)} \tag{6.6}$$

A case in which one can say a lot more about the solutions of initial-boundary value problems is the case when the data are compatible with a solution close to a uniform Maxwellian distribution M. Then techniques akin to those in Chapter 7 can be used. To this end, let us introduce the perturbation h such that

$$f = M + M^{1/2}h \tag{6.7}$$

and first assume that M coincides with the wall Maxwellian, so that Eq. (1.4) is satisfied by M. Eqs. (6.1–3) can then be rewritten in the following way:

$$\Lambda f = Lh + \Gamma(h,h) \qquad \text{in } D \tag{6.8}$$

$$\gamma_D^+ h(x,\xi,t) = \hat{K}\gamma_D^- h \qquad \text{on } \partial D \tag{6.9}$$

$$h(x,\xi,0) = h_0(x,\xi) \tag{6.10}$$

where L and Γ are the same as in Chapter 7 (see Eqs. (7.1.4) and (7.1.5)), while

$$\hat{K} = M^{-1/2}KM^{1/2}. \tag{6.11}$$

Now the global solution can be found by the same technique used in Chapter 7, provided the linearized operator

$$B = -\xi\cdot\partial/\partial x + L \tag{6.12}$$

with the boundary conditions (6.9) generates a semigroup $T(t)$ with a nice decay.

In the previous chapter, we have met two different types of decay of the semigroup $T(t)$. One of them occurs in the pure initial value problem in $\Re^3$ and the other in the same problem for $\mathbf{T}^3$. We may expect that these two types of result also apply in the presence of boundary conditions; the first of them with a decay like $t^{-\alpha}(\alpha > 0)$ should apply to unbounded domains, while the second, with an exponential decay, should apply to bounded domains.

There are several papers dealing with the proofs of the behaviors conjectured. The case of a bounded domain has been considered by Guiraud[14] in the case of diffuse reflection and by Shizuta and Asano[20] in the case of specular reflection, both assuming that Ω is convex. The case of unbounded domains exterior to a bounded convex obstacle was treated by several Japanese authors[21,24]. Here it will be dealt with in Section 8 after studying the corresponding steady problem.

Problems

1. Prove an estimate for $Q(f,f)$ in $X_{\alpha,\beta}$ (see Ref. 2).
2. Use the estimate of Problem 1 to prove a local existence theorem for (6.4) (see Ref. 2).

9.7 A Steady Problem

We shall now deal with the steady flow of a gas past an obstacle $\mathcal{O}$ whose exterior will be denoted by Ω. At space infinity the distribution function will be a drifting Maxwellian M_v. Then the distribution function f satisfies:

$$\xi \cdot \partial f/\partial x = Q(f,f) \qquad \text{in } V \tag{7.1}$$

$$\gamma_D^+ f(x,\xi) = K\gamma_D^- f \qquad \text{on } \partial V \tag{7.2}$$

$$f(x,\xi) \to M_v(\xi) = \rho(2\pi/\beta)^{-3/2}\exp(-\beta|\xi - v|^2) \qquad (|x| \to \infty). \tag{7.3}$$

We remark that M_v $(v \neq 0)$ is not, in general, a solution of (7.1–3), because it violates the boundary condition (7.2). Since, however, it is a solution of the problem when $v = 0$, we can expect the solution to be close to M_v when v is small.

This conjecture has been exploited and proved in a paper by Ukai and Asano[25], which we shall follow in the sequel. We must warn the reader that here and in the next section, where we shall discuss the stability of the solution under consideration, we shall not reproduce the entire proofs; we shall rather try to give the line of thought and a few indications, that may give the reader the flavor of the proofs.

Since, in analogy with what was discussed in Chapter 7 for the pure initial value problem, the basic technique is to show that the solution is a small perturbation of M_v, we let

$$f = M_v + M_w^{1/2} h \tag{7.4}$$

where M_w is the wall Maxwellian. We also assume that the temperatures β and β_w of M_v and M_w are close in the sense that $|\beta - \beta_w| \leq \eta|v|$ with some $\eta \geq 0$.

In terms of h, Eqs. (7.1–3) read as follows

(7.5) $$\xi \cdot \partial h/\partial x = L_v h + \Gamma(h, h) \qquad \text{in } V$$

(7.6) $$\gamma_D^+ h(x, \xi) = \hat{K}\gamma_D^- h + s_v \qquad \text{on } \partial V$$

(7.7) $$h(x, \xi) \to 0 \qquad (|x| \to \infty)$$

where

(7.8) $$L_v h = 2M_w^{-1/2} Q(M_v, M^{1/2} h)$$

and Γ and $\hat{K}$ are as before, while

(7.9) $$s_v = M_w^{-1/2}(K\gamma_D^- M_v - \gamma_D^+ M).$$

We remark that L_v is not self-adjoint in $L^2(\Re_\xi^3)$ when $v \neq 0$.

Let B_v be the linearized Boltzmann operator

(7.10) $$B_v = -\xi \cdot \partial/\partial x + L_v \qquad \text{in } V$$

with the boundary conditions (7.7) and

(7.11) $$\gamma_D^+ h(x, \xi) = \hat{K}\gamma_D^- h \qquad \text{on } \partial V$$

and assume that it has an inverse B_v^{-1}. Then Eqs. (7.5–7) are equivalent to

(7.12) $$u + B_v^{-1}\Gamma(h, h) = \phi_v$$

where ϕ_v is a solution of the linear steady problem

(7.13) $$\xi \cdot \partial \phi/\partial x = L_v \phi \qquad \text{in } V$$

(7.14) $$\gamma_D^+ \phi(x, \xi) = \hat{K}\gamma_D^- \phi + s_v \qquad \text{on } \partial V$$

(7.15) $$\phi(x, \xi) \to 0 \qquad (|x| \to \infty).$$

Once B_v^{-1} and ϕ_v have been shown to exist, we can solve Eq. (7.12) by the implicit function theorem (see below). A delicate problem is posed by the existence of B_v^{-1}. As we shall see, $0 \in \sigma(B_v)$ and thus B_v^{-1} does not exist in $L^2(V)$. By means of the so-called principle of limiting absorption, familiar in scattering theory, it is possible, however, to find B_v^{-1} (in another function space) as a limit of $(B_v - \lambda I)^{-1}$ as $\lambda \to 0$.

To illustrate the method, let us first consider B_v in the special case $\Omega = \Re^3$. Then, as in Chapter 7, it is enough to study the Fourier-transformed operator $B_v(k) = -ik \cdot \xi + L_v$. L_v has the same properties as L ($= L_v$ for $v = 0$) except that it is not self-adjoint. In particular,

$$L_v = -\nu_v(\xi) + K_v \tag{7.16}$$

where $\nu_v = \nu(\xi - v)$ and K_v is an integral operator to which Theorem 7.2.4 applies continuously in v. Let L^2, L_r^∞ be defined in Section 2 of Chapter 7 and $k_0, \sigma_0, \mu_j(\kappa), S_1[\kappa]$ as in Theorem 7.2.6, and set

$$\Sigma(a,\sigma) = \{\lambda : \mathrm{Re}(\lambda) \geq -\sigma; -\mathrm{Re}(\lambda) \leq a|\mathrm{Im}(\lambda)|^2\}. \tag{7.17}$$

It is then possible to prove the following.

(9.7.1) Theorem. *Define B_v in $L^2(\Re_x^3 \times \Re_\xi^3)$ with the maximal domain. Then, for any $v_0 \geq 0$, there is an $a_0 > 0$ such that for all $v \in S_1[v_0]$, the following holds:*

$$\rho(B_v) \supset \Sigma(a_0, \sigma_0) \backslash \{0\}, \qquad 0 \in \sigma(B_v). \tag{7.18}$$

$$(\lambda I - B_v)^{-1} = \sum_{j=0}^{5} U_j(\lambda, v), \tag{7.19}$$

for all $\lambda \in \Sigma(a_0, \sigma_0) \backslash \{0\}$, where for $0 \leq j \leq n+1$,

$$U_j(\lambda, v) = \mathcal{F}_x^{-1} \chi(k)(\lambda - \lambda_j(k,v))^{-1} P_j(k,v) \mathcal{F}_x,$$

$$\chi(k) = 1 \quad (k \in S_1[\kappa_0]), \qquad \chi(k) = 0 \quad (k \notin S_1[\kappa_0]),$$

$$\lambda_j(k,v) = \mu_j(|\kappa|) + ik \cdot v$$

$$P_j(k,v) \in B^0(S_1[\kappa_0] \times S_1[v_0]; \mathbf{B}(L^2, L_r^\infty)), \qquad r \geq 0 \tag{7.20}$$

while for $j = 5$

$$U_5(\lambda, v) \in B^0(\Sigma(a_0, \sigma_0) \times S_1[v_0]; \mathbf{B}(L^2(\Re_x^3 \times \Re_\xi^3)). \tag{7.21}$$

Further, the operators U_j are mutually orthogonal and the P_js are mutually orthogonal projectors of L^2 with $P_j(k,0) = P_j(k), \sum_j P_j(0,v) = P_v = P_0$. Here $U \in B^0(P;V)$ means that U varies in V and is uniformly bounded with respect to parameters varying in set P.

For the proof, we refer to the original paper by Ukai and Asano[25].

According to (7.21) $U_5(0,v)$ is a bounded operator, whereas, since $[\lambda_j(k,v)]^{-1}$ has a singularity at $k=0$ as seen from the asymptotic expansion of $\mu_j(\kappa)$ given in Theorem 7.2.6, $U_j(0,v), 0 \leq j \leq 4$, are unbounded in $L^2(\Re_x^3 \times \Re_\xi^3)$. However, since this singularity is integrable, $U_j(\lambda,v)$ can be made continuous at $\lambda = 0$ (and, hence, $U_j(0,v)$ bounded), if the domain and range spaces are chosen appropriately. This is the principle of limiting absorption. In order to arrive at a more precise statement, we set

$$
\begin{aligned}
L_r^{p,s} &= L_r^{p,s}(\Re_x^3 \times \Re_\xi^3) \\
&=\{h = h(x,\xi) : (1+|\xi|^2)^{r/2} h \in L^s(\Re_\xi^3; L^p(\Re_x^3))\}.
\end{aligned} \tag{7.22}
$$

Then one can prove the following.

(9.7.2) Theorem. *Let* $1 \leq q \leq 2 \leq p \leq \infty$, $\theta \in [0,1)$, $m = 0,1$ *with*

$$
q^{-1} - p^{-1} > (2-m)/(3+\theta). \tag{7.23}
$$

Then for $0 \leq j \leq n+1$,

$$
|v|^\theta U_j(\lambda,v)(I - P_j(0,v))^m \in B^0(\Sigma(a_0,\sigma_0) \times S_1[v_0]; \mathbf{B}(L_0^{q,2}, L_r^{p,\infty})). \tag{7.24}
$$

Proof. It is enough to discuss the case $m=0$. By the interpolation inequality for the Fourier transform, and proceeding as in Theorem 7.4.1, we obtain:

$$
\| U_j(\lambda,v)h \|_{L_r^{p',\infty}} \leq C \| \mathcal{F}_x U_j(\lambda,v) \|_{L_r^{p',\infty}} \leq C\eta_j(v) \| h \|_L^{q,2} \tag{7.25}
$$

$(1/p + 1/p' = 1)$, where

$$
\eta_j(v) = \left\{ \int_{S_1(\kappa_0)} |\lambda - \lambda_j(k,v)|^{-1/\gamma} dk \right\}^\gamma \qquad (\gamma = 1/q - 1/p). \tag{7.26}
$$

After a lengthy calculation based on the asymptotic expansion of $\mu_j(\kappa)$, one can show that $\eta_j(v) \leq C_j |v|^{-\theta}$, so that $|v|^\theta U_j(\lambda,v)$ is uniformly bounded in $\mathbf{B}(L_0^{q,2}, L_r^{p,\infty})$ for $(\lambda,v) \in \Sigma(a_0,\sigma_0) \times S_1[v_0]$. The continuity in v and λ can be proved in a similar way. □

In order to pass from the case of $\Re_x^3$ to the complement Ω of an obstacle $\mathcal{O}$ in $\Re_x^3$, it is required to solve

$$
\lambda h + \xi \cdot \partial h/\partial x + \nu_v h = 0 \qquad \text{in } V \tag{7.27}
$$

$$
\gamma_D^+ h(x,\xi) = s \qquad \text{on } \partial V \tag{7.28}
$$

(7.29) $$h(x,\xi) \to 0 \qquad (|x| \to \infty)$$

for a given $s \in Y^{p,+} = L^p(\Sigma^+|\,|n(x)\cdot\xi|d\sigma d\xi)$. Assume

(7.30) $\mathcal{O}$ is a bounded convex domain and $\partial\mathcal{O} = \partial\Omega$ is piecewise C^2.

Then (7.27–29) can be easily (and explicitly) solved; denote the solution by $h = R_v(\lambda)s$, where $R_v(\lambda)$ is the solution operator. Let $\mathcal{E}$ be the operator extending an operator A from V to $\Re^3_x \times \Re^3_\xi$ by letting $\mathcal{E}A$ to be zero in V^c and $\mathcal{R}$ the restriction operator from $\Re^3_x \times \Re^3_\xi$ to V. Henceforth, also let B_v^∞ denote the Boltzmann operator in $\Re^3_x \times \Re^3_\xi$, in order to distinguish it from the Boltzmann operator B_v in V. Further, set

(7.31) $$\overline{K} = \gamma_D^+ - \hat{K}\gamma_D^-$$

and

(7.32) $$T_v(\lambda) = \overline{K}\mathcal{R}(\lambda I - B_v^\infty)^{-1}\mathcal{E}K_v R_v(\lambda).$$

After some manipulation we can write the resolvent $(\lambda I - B_v)^{-1}$ in an explicit fashion[25]:

(7.33)
$$\begin{aligned}(\lambda I - B_v)^{-1} &= \mathcal{R}(\lambda I - B_v^\infty)^{-1}\mathcal{E}\\ &\quad + S_v(\lambda)(I - T_v(\lambda))^{-1}\overline{K}(\lambda I - B_v^\infty)^{-1}\mathcal{E}\\ S_v(\lambda) &= R_v(\lambda) + \mathcal{R}(\lambda I - B_v^\infty)^{-1}\mathcal{E}K_v R_v(\lambda)\\ &= (\gamma_D^+\mathcal{R}(\lambda I - B_v^{\infty\dagger})^{-1}\mathcal{E})^\dagger.\end{aligned}$$

Originally, this representation formula is obtained in $L^2(V)$ for $\lambda \in \rho(B_v^\infty) \cap \rho(B_v)$ and $1 \in \rho(T_v(\lambda))$, but it can be used to define $(\lambda I - B_v)^{-1}$ in other spaces as long as the right-hand side makes sense. A crucial point is the existence of $(I - T_v(\lambda))^{-1}$, which is guaranteed by the following.

(9.7.3) Lemma. *Let $p \in [2,\infty], r > 3(p-2)/(2p)$. Then there are positive constants a_1, v_1, σ_1 such that*

(7.34) $$(I - T_v(\lambda))^{-1} \in B^0(\Sigma(a_1,\sigma_1) \times S_1[v_1]; \mathbf{B}(Y_r^{p,+})),$$

where $Y_r^{p,+} = \{h : (1+|\xi|^2)^{r/2}u \in Y^{p,+}\}$.

For the proof see Ukai and Asano[25].

We remark that in order to prove this lemma and the two subsequent theorems, one needs a space of at least three dimensions; no analogous result is known for two-dimensional flows. This is related to the so-called Stokes-paradox, which was discovered by Stokes for the Navier-Stokes equations and extended to the Boltzmann equation by Cercignani[10].

This lemma and Theorem 9.7.2 permit the evaluation of the right-hand side of Eq. (7.33). In addition, we need some estimates for $R_v(\lambda)$ and must use the same arguments as in Theorems 7.4.3 and 7.4.4. Let us define $L^{p,s}_r(V)$ in the same way as in (7.22) with $\Re^3_x \times \Re^3_\xi$ replaced by $V = \Omega \times \Re^3_\xi$, and set

$$X^p_r = L^{p,\infty}_{r-1/p} \cap L^{\infty,\infty}_r, \qquad Z^q = L^{2,2} \cap L^{q,2}. \tag{7.35}$$

One can prove the following.

(9.7.4) Theorem. *Let* $1 \le q \le 2 \le p \le \infty$, $r > 3/2$, $\theta \in [0,1)$, $m = 0, 1$ *with*

$$q^{-1} - p^{-1} > (2-m)/(3+\theta) \quad , \quad p^{-1} < 1 - 2/(3+\theta). \tag{7.36}$$

Also, let $\alpha \in [0,1]$ *and* $\gamma = 1 + p^{-1} - q^{-1}$. *Also with* a_1, v_1, σ_1 *of Lemma 9.7.3, set* $\overline{\Sigma} = \Sigma(a_1,\sigma_1) \times S_1[v_1]$. *Then*

(i) there is a constant $C \ge 0$ *such that for any* $(\lambda, v) \in \overline{\Sigma}$,

$$\begin{aligned} &|v|^{\theta\gamma} \parallel (\lambda I - B_v)^{-1}(I-P_v)^m(\nu_v I)^\alpha h \parallel_{L^{p,\infty}_{r-1/p}} \\ &\qquad \le C(\parallel h \parallel_{X^p_r} + \parallel (\nu_v I)^\alpha h \parallel_{Z^q}). \end{aligned} \tag{7.37}$$

(ii) Let $\epsilon > 0$ *and* $\delta > \theta\gamma$. *Let* h_v *be such that*

$$\begin{aligned} &h_v \in L^\infty(S_1[v_1]; X^p_r) \cap B^0(S_1[c_1]; X^p_{r-\epsilon}), \\ &(\nu_v I)^\alpha h_v \in B^0(S_1[c_1]; Z^q); \end{aligned} \tag{7.38}$$

then

$$|v|^{\theta\gamma}(\lambda I - B_v)^{-1}(I-P_v)^m(\nu_v I)^\alpha h_v \in B^0(\overline{\Sigma}; L^{p,\infty}_{r-\epsilon-1/p}). \tag{7.39}$$

If we compare this result with Theorem 9.7.2, the behavior of $(\lambda I - B_v)^{-1}$ near $v = 0$ is worse than that of $U_j(\lambda, v)$. Let $m = \alpha = 0$ and let $h \in X^p_r \cap Z^q$; then, for $v \in S_1[v_1]$ fixed, $(\lambda I - B_v)^{-1}h \in B^0(\Sigma(a_1,\sigma_1); L^{p,\infty}_{r-\epsilon-1/p})$ so that $B_v^{-1}h \in L^{p,\infty}_{r-\epsilon-1/p}$ exists as a limit of $-(\lambda I - B_v)^{-1}h$ as $\lambda \to 0$.

Using this inverse, we can solve Eqs. (7.12–14) in the form

$$\phi_v = R_v(0)h_v - B_v^{-1}K_v R_v(0)h_v, \tag{7.40}$$

and prove the following.

(9.7.5) Theorem. *Let $p \in [2, \infty]$, $\theta \in [0, 1)$ with*

$$p^{-1} < 1 - 2/(3 + \theta). \tag{7.41}$$

Let $r > 3$ and assume that

$$\begin{aligned} &h_v \in B^0(S_1[c_1]; Y_r^{\infty,-}), \\ &\| h_v \| = O(|v|) \qquad (v \to 0). \end{aligned} \tag{7.42}$$

Then ϕ_v solves (7.12–14) in L^p-sense and, with $\gamma = 2 - 1/p$,

$$\begin{aligned} &\phi_v \in B^0(S_1[c_1]; L_r^{p,\infty}), \\ &\| \phi_v \| = O(|v|^{1-\theta\gamma}) \qquad (v \to 0). \end{aligned} \tag{7.43}$$

For the proof, we refer to the paper by Ukai and Asano[21].

We remark that the second condition in (7.42) is satisfied if, as it was assumed before,

$$|\beta - \beta_w| \le \eta |v| \qquad (\eta \ge 0). \tag{7.44}$$

In order to solve the nonlinear problem (7.12), we must now estimate $B_v^{-1}\Gamma(g, h)$. This is done with the following.

(9.7.6) Lemma. *Let $\theta \in (0, 1)$ and $r > 5/2$. Assume that*

$$(3 + \theta)/(1 + \theta) < p < 3 + \theta \tag{7.45}$$

and let $\gamma = 1 + 2/p$. There is a constant $C \ge 0$ such that, for $v \in S_1[c_1]$,

$$\| B_v^{-1}\Gamma(g, h) \| \le C|v|^{-\theta\gamma} \| g \| \| h \| \qquad \text{in } X^p{}_r. \tag{7.46}$$

For the proof we refer again to the paper by Ukai and Asano[25].

This lemma and Theorem 9.7.5 enable us to apply the contraction mapping technique to solve the steady problem in the form (7.12). At a first glance, however, (7.46) does not seem to be good enough, because, if we choose $\theta \ne 0$, it diverges as $v \to 0$, while the choice $\theta = 0$ is excluded in Lemma 9.7.6. The nice behavior of ϕ_v indicated in (7.43), however, compensates for this defect of the estimate of $\| B_v^{-1}\Gamma(g, h) \|$. In fact, let $\theta \in [0, 2/7)$ and $p \ge 2$. Then, we can find α such that

$$\alpha_1 \equiv \theta(1 + 2p^{-1}) < \alpha < 1 - \theta(2 - p^{-1}) \equiv \alpha_2. \tag{7.47}$$

If we now put $h = |v|^\alpha g$ and rewrite (7.12) as

$$g = G_v(g) \equiv -|v|^\alpha B_v^{-1}\Gamma(g,g) + |v|^{-\alpha}\phi_v \quad (v \neq 0),$$

$$g = 0 \quad (v = 0) \tag{7.48}$$

then thanks to (7.43) and (7.46), the following estimates hold:

$$\| G_v(g) \| \leq C_1|v|^\sigma \| g \|^2 + C_2|v|^\tau \tag{7.49}$$

$(\tau = \alpha_2 - \alpha,\ \sigma = \alpha - \alpha_1)$

$$\| G_v(g) - G_v(h) \| \leq C_1|v|^\sigma(\| g \| + \| h \|) \| g - h \| \tag{7.50}$$

$(\sigma = \alpha - \alpha_1)$, where $\| \cdot \|$ is the norm in X_r^p, C_1 and C_2 are positive constants independent of v, g, and h. Since $\sigma, \tau > 0$, $G_v(g)$ is a contraction for small values of $|v|$, which proves the following.

(9.7.7) Theorem. *Let $\theta \in [0, 2/7)$, $r > 5/2$, and assume that (7.45) and (7.47) hold. Then there is a positive number v_0 ($\leq v_1$) such that for any $v \in S_1[v_0]$, Eq. (7.12) has a unique solution h_v in X_r^p satisfying*

$$\| h_v \|_{X_r^p} \leq C|v|^{\alpha+\tau} \qquad (\alpha + \tau = \alpha_2 = 1 - \theta(2 - 1/p)). \tag{7.51}$$

For the proof see Ukai and Asano[25].

Before ending this section, we make a few remarks. First, the continuity properties stated in Theorems 9.7.4 and 9.7.5 can be used to prove that

$$h_v \in B^0(S_1[v_0]; X_{r-\epsilon}^p) \qquad (\epsilon > 0). \tag{7.52}$$

Also, it can be shown that $h_v \in \hat{W}^p$ and satisfies Eq. (7.5) in L^p-sense.

We also explicitly remark that the exponent $\alpha_2 < 1$ in (7.52) is not quite satisfactory. In fact, heuristic considerations (see, e.g., Ref. 10) would suggest that the exponent should be unity (maybe with a different norm) so that the perturbation goes to zero linearly with v. As a matter of fact, a more refined result was proved by Ukai and Asano[26] by using a more sophisticated technique (Nash's implicit function theorem) supplemented by decay estimates of h_v for large values of x. According to this result, the solution in Theorem 9.7.7 satisfies (7.51) with $\alpha + \tau = 1$ ($\theta = 0$) in $L^{\infty,\infty}(\Omega \times \Re_\xi^3) \cap L_r^p(L^p(\Omega) \cap L^\infty(\Omega))$ $(1 < p < \infty,\ r > 4)$.

Problems

1. Prove Theorem 9.7.1 (see Ukai and Asano[25]).
2. Prove Lemma 9.7.3 (see Ukai and Asano[25]).
3. Prove Theorem 9.7.4 (see Ukai and Asano[25]).
4. Prove Theorem 9.7.5 (see Ukai and Asano[25]).
5. Prove Lemma 9.7.6 (see Ukai and Asano[25]).
6. Prove Theorem 9.7.7 (see Ukai and Asano[25]).

9.8 Stability of the Steady Flow Past an Obstacle

We shall now deal with the unsteady flow of a gas past an obstacle $\mathcal{O}$ whose exterior will be again denoted by Ω. As in the previous section, at space infinity the distribution function will be a drifting Maxwellian M_v. Then the distribution function f satisfies:

$$\partial f/\partial t + \xi \cdot \partial f/\partial x = Q(f,f) \qquad \text{in } D \tag{8.1}$$

$$\gamma_D^+ f(x,\xi,t) = K\gamma_D^- f \qquad \text{on } \partial D \tag{8.2}$$

$$f(x,\xi,t) \to M_v(\xi) = \rho(2\pi/\beta)^{-3/2}\exp(-\beta|\xi - v|^2)$$

$$(|x| \to \infty) \tag{8.3}$$

$$f(x,\xi,0) = f_0(x,\xi). \tag{8.4}$$

To solve this problem, following Ukai and Asano[24], we set

$$f = M_v + M_w^{1/2}(h_v + g) \tag{8.5}$$

where h_v is the steady solution, whose existence was discussed in the previous section. Eqs. (8.1–4) can be rewritten as follows

$$\partial g/\partial t + \xi \cdot \partial g/\partial x = L_v g + 2\Gamma(h_v, g) + \Gamma(g,g) \qquad \text{in } D \tag{8.6}$$

$$\gamma_D^+ g(x,\xi,t) = \hat{K}\gamma_D^- g \qquad \text{on } \partial D \tag{8.7}$$

$$g(x,\xi,t) \to 0 \qquad (|x| \to \infty) \tag{8.8}$$

$$f(x,\xi,0) = g_0(x,\xi). \tag{8.9}$$

By definition, the steady solution f_v is asymptotically stable if (8.6–8) have a global solution g that tends to zero as $t \to \infty$, whenever g_0 is sufficiently small. In order to prove the existence of such a solution, we transform (8.6–8) into the following integral equation:

$$g(t) = E_v(t)g_0 + \int_0^t E_v(t-s)[2\Gamma(h_v, g(s)) + \Gamma(g,g)(s)]ds \tag{8.10}$$

where $E_v(t)$ denotes the semigroup generated by B_v. In analogy with what we did in Chapter 7, we shall solve this equation by exploiting the decay properties of $E_v(t)$. We remark that one might have used the linear operator $B_v + 2\Gamma(h_v, .)$ as a generator of a semigroup, thus giving Eq. (8.10) a simpler aspect; it seems hard, however, to obtain the decay properties of that semigroup.

Taking the inverse Laplace transform of (7.34) gives an explicit formula for E_v:

$$E_v(t) = \mathcal{R}E_v^\infty(t)\mathcal{E} + (\gamma_D^+ \mathcal{R}E_v^{\infty\dagger}(t)\mathcal{E})^\dagger *_t D_v(t) *_t \overline{K}E_v^\infty\mathcal{E} \tag{8.11}$$

where $E_v^\infty(t)$ clearly means the semigroup generated by B_v^∞, $*_t$ is the convolution in t, and $D_v(t)$ the inverse Laplace transform of $(I - T_v(\lambda))^{-1}$, as defined in Chapter 7.

If we make use of Theorem 9.7.1 and proceed as in the proof of Theorem 7.4.3, we obtain the following.

(9.8.1) Theorem. *Let $1 \le q \le 2 \le p \le \infty$ and $m = 0, 1$. Then,*

$$\| E_v^\infty(t)(I - P_v)^m h \|_{L_r^{p,\infty}} \le C(1+t)^{-\gamma - m/2} \| h \|_{L_r^{p,\infty} \cap Z^q} \tag{8.12}$$

with $\gamma = 3(p-q)/(2pq)$ and $C \ge 0$ independent of v, t, h.

By means of this theorem one can obtain[24] the following.

(9.8.2) Lemma. *Let v_1 be the same as in Lemma 7.3. Then, for each $\theta \in [0,1)$, there is a constant $C \ge 0$ such that*

$$\| (D_v(t) - I)h \|_{Y_r^{\infty,+}} \le C|v|^{-\theta}(1+t)^{-\gamma} \| h \|_{Y_r^{\infty,+}} \tag{8.13}$$

holds for all $v \in S_1[v_1]$, with $\gamma = 1 + \theta/2$.

Theorem 9.8.1 and Lemma 9.8.2 are the estimates that we need to deal with Eq. (8.11).

We are now ready to deal with the existence theorem for Eq. (8.10), whose right-hand side we denote by $N(g)$. In order to evaluate the second (linear) term of $N(v)$, it is necessary to have $\gamma > 1$ in (8.12) $(m = 1)$ and

(8.13), while for the third (nonlinear) term, it suffices to have $\gamma > 1/2$. For the former, therefore, we should take $\theta > 0$ in (8.13) and a divergent factor $|v|^{-\theta}$ appears but this can be cancelled by means of (7.51). In any case, a careful choice of parameters is needed. Write p, θ of (7.45) as p_0, θ_0 and impose the additional condition $p_0 < 3$. Let α be as in (7.47) and let

$$\begin{aligned} &p \in [2,4] \cap (3, (1/2 - 1/p_0)^{-1}) \\ &q \in [1,2] \cap [1, (1/p + 1/3)^{-1}), \\ &\theta \in (0, \alpha), r > 5/2 \\ &\gamma = \min(3(p-q)/(2pq), (n/p_0 + 1)/2). \end{aligned} \tag{8.14}$$

Then $\gamma > 1/2$. Set

$$| \parallel g \parallel | = \sup_{t \geq 0} (1+t)^{\gamma} \parallel g \parallel_{X_r^p} . \tag{8.15}$$

We have

$$| \parallel N(g) \parallel | \leq C(\parallel g_0 \parallel_{L_r^{p,\infty} \cap Z^q} + (|v|^{-\theta} a + | \parallel g \parallel |) | \parallel g \parallel |) \tag{8.16}$$

$$| \parallel N(g) - N(h) \parallel | \leq C(|v|^{-\theta} a + | \parallel g \parallel | + | \parallel h \parallel |) | \parallel g - h \parallel |) \tag{8.17}$$

where $a = \parallel h_v \parallel$ in $X_r^p, p = p_0$. By (7.51), $|v|^{-\theta} a \to 0$ as $v \to 0$, so N is contractive if g_0 and v are sufficiently small. Thus we have proved the stability property stated in the following.

(9.8.3) Theorem. *Let us assume (8.14). Then there are positive constants a_0, a_1, v_0 such that for any $v \in S_1(v_0)$ and if $\parallel g_0 \parallel \leq a_0$ in $X_r^p \cap Z^q$, where Z^q was defined in (7.36), Eq. (8.10) has a global solution*

$$g = g(t) \in B^0([0,\infty); X_r^p), \qquad \parallel g(t) \parallel \leq a_1 (1+t)^{-\gamma}.$$

9.9 Concluding Remarks

In this chapter we have dealt with the existing results in the theory of existence and uniqueness for initial-boundary and pure boundary value problem for the Boltzmann equation. The most notable absence in this chapter concerns the existence of solutions far from equilibrium for pure boundary value problems and initial-boundary value problems for boundary data incompatible with a uniform Maxwellian solution. Concerning the first of these problems, there is a result in the case when the solution depends on just one space coordinate (say x) by Arkeryd, Cercignani, and Illner[4], which has, however, a rather serious restriction, i.e., a cutoff in the small values of the x-velocity component; in addition, the proof does not apply to hard-sphere molecules but only to particles interacting with soft potentials. As for the initial-boundary value problems with general boundary data, the difficulties lie with large velocities; this was already clear from an argument used by Kawashima[16] for a one-dimensional discrete velocity model and is fully confirmed by the recent paper by Arkeryd and Cercignani[2].

One should also mention the work done by Maslova in this field; for this we refer to her survey[18].

Problems

1. Prove Theorem 9.8.1 (see Ukai and Asano[24]).
2. Prove Lemma 9.8.2 (see Ukai and Asano[24]).

References

1. L. Arkeryd, "On the strong L^1 trend to equilibrium for the Boltzmann equation," *Studies in Appl. Math.* **87**, 283–288 (1992).
2. L. Arkeryd and C. Cercignani, "A global existence theorem for the initial boundary value problem for the Boltzmann equation when the boundaries are not isothermal," *Arch. Rat. Mech. Anal.* **125**, 271–288 (1993).
3. L. Arkeryd and C. Cercignani, "On the convergence of solutions of the Enskog equation to solutions of the Boltzmann equation," *Commun. in Partial Differential Equations* **14**, 1071–1089 (1989).
4. L. Arkeryd, C. Cercignani, and R. Illner, "Measure solutions of the steady Boltzmann equation in a slab," *Commun. Math. Phys.* **142**, 285–296 (1991).
5. R. Beals and V. Protopopescu, "Abstract time-dependent transport equations," *J. Math. Anal. Appl.* **121**, 370–405 (1987).
6. L. Boltzmann, "Über die Aufstellung und Integration der Gleichungen, welche die Molekularbewegungen in Gasen bestimmen," *Sitzungsberichte der Akademie der Wissenschaften Wien* **74**, 503–552 (1876).
7. M. Cannone and C. Cercignani, "A trace theorem in kinetic theory," *Appl. Math. Lett.* **4**, 63–67 (1991).
8. C. Cercignani, "Equilibrium states and trend to equilibrium in a gas according to the Boltzmann equation," *Rend. Mat. Appl.* **10**, 77–95 (1990).
9. C. Cercignani, "On the initial-boundary value problem for the Boltzmann equation," *Arch. Rational Mech. Anal.* **116**, 307–315 (1992).
10. C. Cercignani, "Stokes paradox in kinetic theory," *Phys. Fluids* **11**, 303 (1967).
11. L. Desvillettes, "Convergence to equilibrium in large time for Boltzmann and BGK equations," *Arch. Rat. Mech. Analysis* **110**, 73–91 (1990).
12. R. Di Perna and P. L. Lions, "Global solutions of Boltzmann's equation and the entropy inequality," *Arch. Rational Mech. Anal.* **114**, 47–55 (1991).
13. R. DiPerna, P. L. Lions, "On the Cauchy problem for Boltzmann equations," *Ann. of Math.* **130**, 321–366 (1989).
14. J. P. Guiraud, "An H-theorem for a gas of rigid spheres in a bounded domain," in *Théories cinétiques classiques et rélativistes*, G. Pichon, ed., 29–58, CNRS, Paris (1975).
15. K. Hamdache, "Initial boundary value problems for Boltzmann equation. Global existence of weak solutions," *Arch. Rat. Mech. Analysis* **119**, 309–353 (1992).

16. S. Kawashima, "Global solutions to the initial-boundary value problems for the discrete Boltzmann equation," *Nonlinear Analysis, Methods and Applications* **17**, 577–597 (1991).
17. P. L. Lions, "Compactness in Boltzmann's equation via Fourier integral operators and applications. I," Cahiers de Mathématiques de la décision no. 9301, CEREMADE (1993).
18. N. B. Maslova, "Existence and uniqueness theorems for the Boltzmann equation," Chapter 11 of *Dynamical Systems II*, Ya. G. Sinai, ed., Springer, Berlin (1982).
19. N. B. Maslova, "The solvability of stationary problems for Boltzmann's equation at large Knudsen numbers," *U.S.S.R. Comput. Maths. Math. Phys.*, **17**, 194–204 (1978).
20. Y. Shizuta and K. Asano, "Global solutions of the Boltzmann equation in a bounded convex domain," *Proc. Japan Acad.* **53A**, 3–5 (1977).
21. S. Ukai, "Solutions of the Boltzmann equation," in *Pattern and Waves-Qualitative Analysis of Nonlinear Differential Equations,* 37–96 (1986).
22. S. Ukai, *The transport equation* (in Japanese), Sangyo Toshio, Tokyo (1976).
23. S. Ukai and K. Asano, "On the existence and stability of stationary solutions of the Boltzmann equation for a gas flow past an obstacle," *Research Notes in Mathematics* **60**, 350–364 (1982).
24. S. Ukai and K. Asano, "On the initial boundary value problem of the linearized Boltzmann equation in an exterior domain," *Proc. Japan Acad.* **56**, 12–17 (1980).
25. S. Ukai and K. Asano, "Steady solutions of the Boltzmann equation for a gas flow past an obstacle. I. Existence," *Arch. Rational Mech. Anal.* **84**, 249–291 (1983).
26. S. Ukai and K. Asano, "Steady solutions of the Boltzmann equation for a gas flow past an obstacle. II. Stability," *Publ. RIMS, Kyoto Univ.* **22**, 1035–1062 (1986).
27. J. Voigt, "Functional analytic treatment of the initial boundary value problem for collisionless gases," Habilitationsschrift, Univ. München (1980).

10

Particle Simulation of the Boltzmann Equation

10.1 Rationale and Overview

Validity analysis, existence and uniqueness theorems, and qualitative results on the behavior of the solutions are certainly central to the understanding of rarefied gases. For real physical situations, however, like the flow pattern around an object that moves inside a rarefied gas, we need methods to actually calculate or approximate solutions of the Boltzmann equation. For most situations, it is hopeless to even look for explicit solutions of the Boltzmann equation. On the other hand, the five-dimensional integral in the collision operator makes numerical approximations a difficult topic. Specifically, recall that

$$Q(f,f) = \int_{\Re^3} \int_{S^2} |n \cdot (\xi - \xi_*)| \{ f' f'_* - f f_* \} dn d\xi_*.$$

Suppose we want to approximate the collision integral by a quadrature formula that requires the evaluation of the integrand at a number of points. Obviously, the integrand must decay fast enough at infinity to give us reasonable accuracy with a finite number of evaluation points.

For the sake of our argument, let us assume that the formula then requires twenty function evaluations to approximate a one-dimensional integral. Then $20^2 = 400$ evaluations would be needed to achieve the same accuracy for a comparable two-dimensional integral, 8,000 for a three-dimensional integral, 160,000 for a four-, and 3,200,000 for a five-dimensional integral as in $Q(f, f)$. Clearly, the numerical effort for such a procedure would be unreasonable, in particular because the evaluation would have to be repeated at each space point and each velocity on the grid, and, for time- dependent problems, for each time step.

Therefore, conventional numerical methods like difference or finite element approximations are only feasible when the dimension of the problem is a priori significantly reduced, like, say, for one-dimensional shock wave profiles. Even then, however, the particle methods described here offer attractive and probably physically reasonable alternatives.

Particle simulation methods avoid the cumbersome evaluation of the collision integral by replacing the density distribution f by a discrete measure of test particles, the "simulation gas." The number of test particles in the simulation can be anything between 100 and several million, depending on the capacity of the computer being used.

The key idea of particle simulation is simple and can be conveniently depicted in a diagram such as Fig. 29. Recall that the underlying physical reality is a gas with, say, 10^{25} particles. As it is impossible to keep a record of all of them, we introduced the Boltzmann equation as a way to keep track of the particle density, i.e., a function containing information about the average density, energy, temperature, etc. of the particle system. It is useful to recall at this time that the Boltzmann equation is only a mathematical approximation to the physical reality. The basic idea of particle simulation is really just to return to the particle level, but to restrict the number of particles to a tractable figure (which depends not only on the situation to be modeled, but also on the available computer). If, say, 10^5 particles are to be used, interaction rules must be given that will reflect the influence of the collisions on the behavior of the gas; it is clear that for any physically reasonable number of particles, it is hopeless to follow the time evolution of a system of, say, N hard spheres—not only would the analysis whether two particles will collide in the near future require large numerical effort, but the detailed calculation of the collision parameters and their implementation into the scheme must also be done.

Such an approach would not only be unreasonable from a mathematical point of view, it may also not be physically meaningful. After all, our target is the simulation of rarefied gas dynamics, not the solution of N- body problems with large N. Since the former is believed to evolve by averaging over the latter, some averaging should be part of the procedure.

Fig. 29 is a schematic representation of how particle simulations relate to the physical reality and the Boltzmann equation. There are two options to arrive at a simulation method. First, one can design procedures that are based on the fundamental properties of a rarefied gas alone, like free flow, the mean free path, and the collision frequency. Such schemes need not have an a priori relationship to the Boltzmann equation, but they will reflect many of the ideas and concepts employed in the derivation of the latter; in the best case, they will turn out to be consistent with and converge to solutions of the Boltzmann equation.

The Bird scheme, which has been successfully employed for the simula-

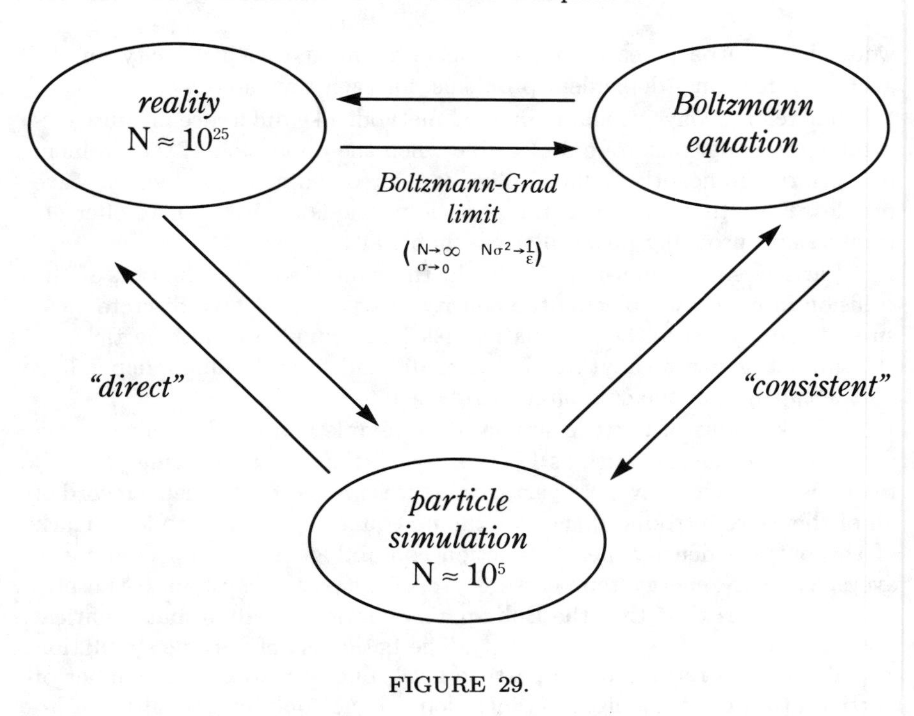

FIGURE 29.

tion of rarefied gases for decades, belongs to this category (see Bird [6]). The biggest advantage of this method is its practicability and success in applications such as reentry calculations for spacecraft. There are obvious relations to the derivation of the Boltzmann equation, but it has only recently been shown that the Bird simulation is actually convergent to solutions of the Boltzmann equation in the right limit (see W. Wagner [14]). We describe the essential ingredients of the Bird simulation scheme in Section 10.3.

The second option to arrive at simulation schemes is to actually start from the Boltzmann equation and derive simulation schemes that model the Boltzmann collision terms as accurately as possible. Consistency and convergence of such methods should be much easier to verify; the big question is whether the result will be practicable. Indeed, many authors have suggested particle simulations derived from the Boltzmann equation (see Ref. 9 for more references). Some of these procedures have been largely of theoretical interest, while others have been put to practical use. In this chapter, our main objective is to describe a type of simulation known as "low discrepancy method." These schemes were developed on the basis of a method suggested by Nanbu [11] and have been designed and tested by Babovsky [1] and Ploss [12].

The low discrepancy methods being used now are known to be consistent and convergent (Babovsky [1] and Babovsky and Illner [3]), and their practicability compares well with the Bird scheme. Recently, they have been modified for applications to steady rarefied gas flows (see Ref. 2.).

10.2 Low Discrepancy Methods

We derive these methods from the Boltzmann equation via a number of reduction steps. The first three of these steps are common to most of the simulation schemes being used, including the Bird simulation. The steps can be summarized in the following short list.

A. Time discretization.
B. Splitting (separation of free flow from interactions).
C. Local homogenization (over cells).
D. Weak formulation (necessary to allow point measures as solutions).
E. Measure formulation.
F. Collision simulation.

The most critical part of the procedure is step F, and this is where there are several ways to proceed. We describe steps A–F in detail.

A. Suppose that the rarefied gas is confined to a domain Λ with a deterministic boundary condition on $\partial\Lambda$. For $(x,\xi) \in \Lambda$, let

$$t \to \Phi_t(x,\xi)$$

denote the free flow operator, i.e.,

$$\Phi_{t_1}(x,\xi) = (x + t_1\xi, \xi)$$

before $x + t_1\xi$ reaches the boundary $\partial\Lambda$,

$$\Phi_{t_1}(x,\xi) = (x + t^*\xi + (t_1 - t^*)\xi', \xi')$$

after the first collision if $x + t^*\xi$ is the point where the trajectory reaches the boundary and ξ' is the post–collisional velocity, etc. Next, choose a time step Δt, and discretize the time interval during which the gas is to be simulated by $0, \Delta t, 2\Delta t, \ldots, j\Delta t$. If x, ξ and $t \in [0, \Delta t]$ are such that $x + t\xi \notin \partial\Lambda$, the time derivative

$$(\partial_t + \xi \cdot \partial_x)\, f(x + t\xi, \xi, j\Delta t + t) = \frac{d}{dt}[f(x + t\xi, \xi, j\Delta t + t)]$$

is to first order approximated by

$$\frac{1}{\Delta t}\big[f(\Phi_{\Delta t}(x,\xi), (j+1)\Delta t) - f(x,\xi,j\Delta t)\big].$$

If we substitute this difference approximation for the left-hand side of the Boltzmann equation (in mild form), we get

$$(2.1) \qquad \frac{1}{\Delta t}\left[f(\Phi_{\Delta t}(x,\xi),(j+1)\Delta t) - f(x,\xi,j\Delta t)\right] = Q(f,f)(x,\xi,j\Delta t).$$

We remark that this approximation is one of the crudest steps taken in the procedure—clearly, what we have done amounts to an Euler–type approximation to the derivative along characteristics.

Equation (2.1) can be rewritten in the equivalent form

$$(2.2) \quad f(y,\eta,(j+1)\Delta t) = f(\Phi_{-\Delta t}(y,\eta),j\Delta t) + \Delta t Q(f,f)(\Phi_{-\Delta t}(y,\eta),j\Delta t).$$

B. The splitting of the algorithm is nothing but the breakup of (2.2) into the following two steps: Let

$$(2.3) \qquad \tilde{f}(x,\xi,(j+1)\Delta t) = f(x,\xi,j\Delta t) + \Delta t Q(f,f)(x,\xi,j\Delta t)$$

(discretization of the spatially homogeneous Boltzmann equation), and then

$$(2.4) \qquad f(x,\xi,(j+1)\Delta t) = \tilde{f}(\Phi_{-\Delta t}(x,\xi),(j+1)\Delta t)$$

(free flow step).

Bogomolov [7] has proved the following result, which we mention because of its intrinsic practical interest. If, instead of computing $\tilde{f}$ from (2.3), one could solve the spatially homogeneous Boltzmann equation exactly on the time interval $[j\Delta t,(j+1)\Delta t)$, the truncation error per step of the whole procedure would be of order $O(t^3)$ instead of order $O(t^2)$. This is a surprising and useful result inasmuch as it shows that it is worthwhile to look for better approximations to the spatially homogeneous equation than the one given by (2.3). We will have further remarks on this at the end of the chapter.

The difficulties of the approximation are contained in (2.3), which includes the collision term. The free flow step (2.4) can be simulated numerically in an obvious way.

C. Local homogenization. This step is important for both practical and conceptual reasons. The practical reason is that in the collision simulation it is even for moderate particle numbers not feasible to keep track of the positions of particles in detail, for reasons we elaborated at the beginning of this chapter. The conceptual reason is that we already know that the Boltzmann equation only holds in a limit where the particle number goes to infinity. To make a collision count, Boltzmann in his classical derivation simply assumed that the spatial variation of the particle density over a cell with volume *dxdydz* could be neglected; it was the velocity dependence of the particles in such a cell that would make a difference.

This assumption of local homogeneity must also (implicitly) enter the derivation of the Boltzmann equation from hierarchy equations given in

Chapter 4 via regularity assumptions on the N- particle distribution function. We suggest that it would be a worthwhile exercise to analyze the regularity hypotheses made there from this point of view. Intuitively, the solution of the Boltzmann equation must be smooth enough on the scale of the particle diameter; in particle simulation methods this is replaced by homogeneity in a cell (whose side is much larger than the particle diameter but still much smaller than the mean free path, which is the smallest length over which we can describe spatial changes according to the Boltzmann equation).

For the numerical simulation, local homogenization is done as follows: We decompose the domain Λ into a number of cells, $\Lambda = \cup C_i$, where $C_i \cap C_j = \emptyset$ for $i \neq j$, except for possibly overlapping boundaries. The cells can be chosen to suit the particular geometry of the problem at hand. If $x \in C_i$, we approximate $f(x, \xi, j\Delta t)$ by its spatial average over C_i, i.e.,

$$\bar{f}(x, \xi, j\Delta t) := \frac{1}{\lambda^3(C_i)} \int_{C_i} f(y, \xi, j\Delta t) dy. \tag{2.5}$$

On C_i, $\bar{f}$ is constant with respect to x. We refer to $\bar{f}$ as the local homogenization of f with respect to the partition $\{C_i\}$. Clearly, if f is locally homogeneous, the simulation step (2.3) will result in a locally homogeneous function $\tilde{f}$. The free flow step (2.4) will destroy the local homogeneity. Therefore, for a simulation based on the assumption of local homogeneity, the free flow step (2.4) will have to be succeeded by a homogenization like (2.5) every time.

It is also desirable (but in practical situations very difficult, due to the limitations on computer memory) to have a large number of particles in each cell. In fact, convergence proofs for simulation methods only work if the number of particles per cell (initially) goes to infinity. In practice, often only twenty or so particles per cell are feasible, and the user compensates for this by repeating the calculation many times and taking averages over the runs; this eliminates some of the statistical fluctuations brought in through necessary random choices in the collision simulation (see step F). If the individual runs remain "close" to a solution of the Boltzmann equation in some sense, the average will do likewise. However, the averaging procedure does not bring us "closer" (it only removes noise); in order to achieve a truly better approximation of the equation, larger particle numbers per cell are necessary.

The steps we have described so far are not unique to the low discrepancy methods; in fact, they are universal to all particle simulations with which we are familiar (clearly, this is a somewhat unsatisfactory state—the Euler approximation of the flow derivative

$$(\partial_t + \xi \cdot \partial_x) f$$

done in step A is quite rough, and future research should be directed, if possible, to more accurate approximations of this derivative, which would be compatible with the splitting step and the final particle approximation).

The next step is necessary to allow point measures as solutions.

D. Weak formulation. We now consider only one cell C, arbitrary but fixed. Assume $f = f_C(\xi, j\Delta t)$ to be given and to be homogeneous in space in this cell. We simply write $f_j(\xi)$. Equation (2.3) becomes

$$\begin{aligned} f_{j+1}(\xi) = (1 - \Delta t \int\int B(n, \xi - \xi_*) f_j(\xi_*) dn d\xi_*) f_j(\xi) \\ + \Delta t \int\int B(n, \xi - \xi_*) f_j(\xi') f_j(\xi'_*) dn d\xi_*. \end{aligned} \tag{2.6}$$

Notice that we have now simply written f rather than $\tilde{f}$ on the left-hand side of (2.6). Recalling that f_j represents a particle density function, we certainly want to preserve non-negativity in the procedure. However, it does *not* automatically follow from (2.6) that $f_j \geq 0$ entails $f_{j+1} \geq 0$, because the collision kernel is unbounded even for the case of hard spheres (this difficulty is an artifact of our rough approximation to the flow derivative).

To guarantee that $f_{j+1} \geq 0$ for sufficiently small Δt, we have to truncate B such that

$$\textstyle\int B(n, \xi - \xi_*) dn \leq A \tag{2.7}$$

for some constant $A > 0$. For example, in the hard-sphere case we can simply multiply the collision kernel $|n \cdot (\xi - \xi_*)|$ by zero for large enough relative speed $|\xi - \xi_*|$ to achieve that, i.e., collisions between particles that move with large relative velocities will be disregarded, a simplification that seems acceptable for situations where there is only a small high-energy tail to the particle density function.

The next step after (2.7) is a renormalization of f_j. Assuming that we have total mass $\int_\Lambda \int f_j(x, \xi) d\xi dx = 1$ and spatial homogeneity in each cell C_i, it follows that

$$\sum_i \lambda^3(C_i) \int f_{j,i}(\xi) d\xi = 1,$$

where the indices j, i indicate that we are in the ith cell at time $j\Delta t$. By multiplying $f_{j,i}$ with $\frac{1}{\int f_{j,i}(\xi) d\xi}$, we renormalize in each cell such that

$$\int f_{j,i}(\xi) d\xi = 1$$

(this step is really not necessary, but it simplifies the following discussion.) With this renormalization, it is immediate that we only have to choose $\Delta t < \frac{1}{A}$ such that $f_j \geq 0$ will entail $f_{j+1} \geq 0$ (see Problem 1). In the sequel, we will suppress the cell index, because the remainder of the discussion only concerns the collision simulation in one cell.

To obtain the necessary measure version of (2.6), we multiply (2.6) by a continuous and bounded test function $\varphi \in C_b(\Re^3_\xi)$ and integrate:
(2.8)

$$\int \varphi(\xi) f_{j+1}(\xi) d\xi = \int \Big(1 - \Delta t \int \int B(n, \xi - \xi_*) f_j(\xi_*) d\xi_*\Big) f_j(\xi)\varphi(\xi) d\xi$$
$$+ \int \Delta t \int \int B(n, \xi - \xi_*) f_j(\xi') f_j(\xi'_*) dn d\xi_* \, \varphi(\xi) \, d\xi.$$

By using the invariance of the collision kernel under the collision transformation and the normalization $\int f_j(\xi) d\xi = 1$, (2.8) can be rewritten as
(2.9)

$$\int \varphi(\xi) f_{j+1}(\xi) \, d\xi = \int \int \varphi(\xi)\Big(1 - \Delta t \int B(n, \xi - \xi_*) \, dn\Big) f_j(\xi_*) f_j(\xi) \, d\xi_* \, d\xi$$
$$+ \int \int \Delta t \int B(n, \xi - \xi_*) \, \varphi(\xi') \, dn \, f_j(\xi_*) f_j(\xi) \, d\xi_* d\xi.$$

Let

$$K_{\xi,\xi_*}\varphi := \Big(1 - \Delta t \int B \, dn\Big)\varphi(\xi) + \Delta t \int B(\ldots)\varphi(\xi') \, dn,$$

then (2.9) can be written in compact form as

$$\int \varphi(\xi) f_{j+1}(\xi) d\xi = \int \int K_{\xi,\xi_*}\varphi \, f_j(\xi) f_j(\xi_*) \, d\xi \, d\xi_*. \tag{2.10}$$

E. Measure formulation. Let the probability measure μ_j be defined by $\mu_j(d\xi) = f_j(\xi) d\xi$. Then, μ_{j+1} is determined by μ_j by virtue of (2.10). It is clear that even though all the μ_js were so far absolutely continuous measures, this is not necessary any longer. The calculation of (approximations to) μ_{j+1} from (approximations to) μ_j is difficult because of the complicated relationship between φ and $K_{\xi,\xi_*}\varphi$. Therefore, we first need a good representation of $K_{\xi,\xi_*}\varphi$.

For $\xi, \xi_* \in \Re^3$ arbitrary but fixed, we define $S^2_+ := \{n \in S^2; n\cdot(\xi - \xi_*) > 0\}$ and

$$T_{\xi,\xi_*} : S^2_+ \longrightarrow \Re^3 \quad \text{by} \quad T_{\xi,\xi_*}(n) = \xi'.$$

Moreover, let $B^1 = \{y \in \Re^2; \|y\| < \frac{1}{\sqrt{\pi}}\}$ be the ball in $\Re^2$ centered at 0 and having area 1. We can then prove the following.

(10.2.1) Lemma. *Let $\Delta t < \frac{1}{A}$. Then, for all $\xi, \xi_* \in \Re^3$, there is a continuous function $\Phi_{\xi,\xi_*} : B^1 \longrightarrow S^2_+$ such that*

$$K_{\xi,\xi_*}\varphi = \int \int_{B^1} \varphi(T_{\xi,\xi_*} \circ \Phi_{\xi,\xi_*}(y)) \, dy.$$

Remark. Φ_{ξ,ξ_*} "makes a decision" whether particles with velocities ξ, ξ_* will collide at all during the time interval $[j\Delta t, (j+1)\Delta t]$, and if so, with what collision parameter n. Lemma 10.2.1 is due to Babovsky[1], who first recognized its usefulness for numerical simulation of the Boltzmann equation. The idea that it is useful to treat collisions and noncollisions between particles in a uniform way seems to appear for the first time in a paper by Koura [10].

Proof. We represent B^1 in polar coordinates as

$$B^1 = \{(r,\beta); 0 \le r \le \frac{1}{\sqrt{\pi}}, 0 \le \beta \le 2\pi\}$$

and define an $r_0 < \frac{1}{\sqrt{\pi}}$, depending on $\xi - \xi_*$ and $n \cdot (\xi - \xi_*)$, as

$$r_0^2 = \frac{1}{\pi} \Delta t \int_{S_+^2} B dn$$

(note that $\Delta t \int B dn < 1$ by assumption). As usual, n is represented by spherical coordinates, i.e., by a pair $(\gamma, \theta) \in [0, \frac{\pi}{2}] \times [0, 2\pi)$, where γ and θ denote the polar and azimuthal angles, respectively.

For $r \ge r_0$, let $\Phi_{\xi,\xi_*}(r,\beta) := (\frac{\pi}{2}, \beta)$, i.e., we set $\gamma = \frac{\pi}{2}$, $\theta = \beta$. These values correspond to a grazing collision, and therefore the velocities of the two particles remain unchanged. We have a "noncollision."

For $r \le r_0$, i.e., on an area $1 - \Delta t \int B\, dn$ of B^1, we define Φ such that the collision is nontrivial. Specifically, let $\theta = \theta(r, \beta) = \beta$. The polar angle $\gamma(r,\beta)$ is defined as a function of r only, such that $\gamma(r)$ is the inverse of an $r(\gamma)$ that satisfies $r(0) = 0$, $r(\frac{\pi}{2}) = r_0$, and

$$\begin{aligned} &\int_0^{2\pi} \int_0^{r_0} \varphi(T_{\xi,\xi_*}(\gamma(r), \beta)) r\, dr\, d\beta \\ =&\Delta t \int_0^{2\pi} \int_0^{\frac{\pi}{2}} \varphi(T_{\xi,\xi_*}(n)) B(|\xi - \xi_*|, \gamma) \sin\gamma\, d\gamma\, d\beta. \end{aligned} \tag{2.11}$$

Equation (2.11) will clearly hold if

$$r\, dr = \Delta t\, B(|\xi - \xi_*|, \gamma) \sin\gamma\, d\gamma$$

or

$$\frac{d}{d\gamma}\Big(\frac{1}{2} r^2(\gamma)\Big) = \Delta t\, B(|\xi - \xi_*|, \gamma)\, sin\gamma \tag{2.12}$$

with the initial condition $r^2(0) = 0$. It is readily verified that the solution of this equation satisfies $r(\frac{\pi}{2}) = r_0$. Summarizing, the mapping $\Phi_{\xi,\xi_*} : B^1 \longrightarrow [0, \frac{\pi}{2}] \times [0, 2\pi)$ is defined by

$$(\gamma, \theta) = \begin{cases} (\frac{\pi}{2}, \beta) \text{ if } r \ge r_0 \\ (\gamma(r), \beta) \text{ if } r < r_0 \end{cases}$$

where $r(\gamma)\frac{dr}{d\gamma} = \Delta t\, B(|\xi - \xi_*|, \gamma)\, \sin\gamma$. This completes the proof. □

Remark. The dependence of Φ on Δt and on $|\xi - \xi_*|$ is obvious. However, both of these quantities enter into Φ as scaling parameters. For Δt, this is immediate from (2.12). For $|\xi - \xi_*|$, it follows because for most B

$$B = |\xi - \xi_*|^p h(\gamma)$$

($p = 1$ for hard spheres). For Maxwell molecules, the dependence on $|\xi - \xi_*|$ vanishes entirely. In any case, the most important aspect of Φ is that Φ_{ξ,ξ_*} is essentially explicitly known and can be assumed to be computed prior to any gas simulation.

Lemma 10.2.1 enables us to rewrite (2.10) in a compact and practical form. Let $\Psi(\xi, \xi_*, y) = T_{\xi,\xi_*} \circ \Phi_{\xi,\xi_*}(y)$; then

$$\int_\xi \varphi(\xi) f_{j+1}(\xi)\, d\xi = \int_\xi \int_{\xi_*} \int_y \varphi(\Psi(\xi, \xi_*, y)) d^2y\, f_j(\xi) f_j(\xi_*)\, d\xi_*\, d\xi$$

or

$$\int_\xi \varphi(\xi)\, d\mu_{j+1}(\xi) = \int\int\int \varphi \circ \Psi(\xi, \xi_*, y)\, d^2y\, d\mu_j(\xi)\, d\mu_j(\xi_*). \tag{2.13}$$

By introducing a probability measure dM_j on $B^1 \times \Re^3 \times \Re^3$ as

$$dM_j := d^2y \times d\mu_j \times d\mu_j,$$

(2.13) becomes

$$\int \varphi d\mu_{j+1} = \int \varphi d(M_j \circ \Psi^{-1})$$

or

$$\mu_{j+1} = M_j \circ \Psi^{-1}. \tag{2.14}$$

We will refer to (2.14) as the time-discretized reduced Boltzmann equation. Equation (2.14) is well suited to introduce and analyze the next and final step in the procedure, the essential step.

F. Collision simulation. We have now reached the crucial step for the particle simulation. Fix $t = j\Delta t$, say $j = 0$, and focus on one arbitrary but fixed cell. Suppose that the probability measure μ_j describes the correct particle density in that cell at time $j\Delta t$, and assume further that we have a sequence of points $\xi_i^N(j)$, $i = 1, \ldots, N$, such that the sequence of discrete measures

$$\mu_j^N = \frac{1}{N} \sum_{i=1}^{N} \delta_{\xi_i^N(j)}$$

satisfies $\mu_j^N \to \mu_j$ *weak-$*$* in the sense of measures as $N \to \infty$. How do we then find a sequence of probability measures

$$\mu_{j+1}^N = \frac{1}{N}\sum_{i=1}^{N} \delta_{\xi_i^N(j+1)}^N,$$

with "post-collisional" velocities $\xi_i^N(j+1)$, such that *weak-**

$$\mu_{j+1}^N \to \mu_{j+1},$$

where μ_{j+1} is given by (2.14)?

Assume, for simplicity, that the function

$$\Psi(\xi,\eta,y) = T_{\xi,\eta} \circ \Phi_{\xi,\eta}(y)$$

is continuous as a function of ξ, η, and y (this requires at worst a special kind of truncation of the collision kernel B; as we have already truncated B to guarantee the conservation of positivity, the additional truncation is really no serious additional constraint. We also remark that continuity a.e. is sufficient, see Babovsky [1] and Billingsley [5]). Then clearly, if $\{M_j^N\}$ if a sequence of probability measures on $\Re_\xi^3 \times \Re_{\xi_*}^3 \times B^1$ such that $M_j^N \underset{w^*}{\longrightarrow} M_j$ as $N \to \infty$, it follows that $M_j^N \circ \Psi^{-1} \longrightarrow M_j \circ \Psi^{-1} = \mu_{j+1}$. Also, observe that if $(\xi_i^N, \xi_{*i}^N, y_i^N)_{i=1,\dots,N}$ is a sequence of triples such that

$$M_j^N := \frac{1}{N}\sum(\delta_{\xi_i} \times \delta_{\xi_{*i}} \times_{\delta_{y_i}}) \underset{w^*}{\longrightarrow} M_j, \tag{2.15}$$

the measures

$$\mu_{j+1}^N := M_j^N \circ \Psi^{-1} = \frac{1}{N}\sum \delta_{\Psi(\xi_i,\xi_{*i},y_i)}$$

are N-atomic and satisfy $\mu_{j+1}^N \underset{w^*}{\longrightarrow} \mu_{j+1}$.

Our work will therefore be completed if we can associate collision partners (with velocities ξ_{*i}^N) and generalized collision parameters y_i^N to the given velocities ξ_i^N such that (2.15) applies. This can actually be done in many ways and leads to interesting abstract research problems. We point out several approaches.

1. The traditional approach, which leads to a Monte Carlo-type simulation, is based on choosing the y_i^N and the *indices* of the collision partners randomly. Specifically, let $\{r_i\}_{i\in\mathbf{N}}$ be a sequence of independent, equidistributed random variables on $[0,1]$, and let $\{y_i\}_{i\in\mathbf{N}}$ be a corresponding sequence of independent, equidistributed random variables on B^1 (see Problem 2). If $|y_i| \le \sup_{|\xi-\xi_*|} r_0(\xi-\xi_*)$, (see the proof of Lemma 2.1) we are for sure in the situation where the particle with velocity ξ_i will not experience a collision, so we set

$$\xi_i^N(j+1) = \xi_i^N(j). \tag{2.16}$$

Otherwise, a collision is possible, and we need a collision partner. To choose it, let $c_i = [N \cdot r_i] + 1 \in \{1,\dots,N\}$ (as usual, $[z]$ is the largest integer less than z) and set

(2.17) $$\xi_{*i} = \xi_{c_i},$$

(2.18) $$\xi^N{}_i(j+1) = \Psi(\xi_i^N(j),\, y_i,\, \xi_{*i}^N(j)).$$

The formulas (2.16–18) summarize a numerical simulation procedure designed in the early and mid-1980s by Nanbu[11] and Babovsky [1].

This method, while simple, has some surprising (and some unpleasant) features, notably

- that the set of velocities for potential collision partners is identical with the given set $\{\xi_1^N, \ldots, \xi_N^N\}$. This means, effectively, that a "duplicate" of the gas is created for the collision simulation,
- moreover, the same velocity can be chosen repeatedly for collision partners,
- and therefore, while the procedure certainly preserves the mass (because the number of particles per cell is left invariant during the simulation), momentum and kinetic energy are not necessarily preserved. It can be proved (see Babovsky [1]) that they are preserved in the mean. However, as discovered by Greengard and Reyna [8], this conservation in the mean in really not good enough; they showed that the procedure described here entails a systematic decrease in temperature of order $1/N$ for a fixed time interval. Even though the procedure does converge in the limit $N \to \infty$ (the convergence is discussed in more detail later), this systematic error is a serious flaw for practical applications.

It should be intuitively convincing that this procedure guarantees the convergence $\mu_{j+1}^N \to \mu_{j+1}$ in some sense. Indeed, the central limit theorem can be invoked to this end. We state, without a detailed proof (see Babovsky[1] or Babovsky and Illner[3] for details), the following lemma.

(10.2.2) Lemma. *Let μ_j be absolutely continuous,*

$$dM_j = d\mu_j \times d\mu_j \times d^2y.$$

Then, as $N \to \infty$,

$$\frac{1}{N} \sum_{i=1}^{N} \delta_{\xi_i^N(j+1)} \xrightarrow[w^*]{} \mu_{j+1}$$

almost surely with respect to the random variables r_i, y_i.

Remarks

a) Note that it is important that N, the particle number per cell, be sent to infinity. Given this, the lemma follows from the preceding discussion, the central limit theorem, and the Borel–Cantelli lemma.

b) The most central part of the whole procedure is the requirement (2.15) that

$$\frac{1}{N}\sum_{i=1}^{N}(\delta_{\xi_i}\times\delta_{\xi_{*i}}\times\delta_{y_i})\underset{w^*}{\longrightarrow} d\mu_j\times d\mu_j\times d^2y$$

as $N\to\infty$. This is clearly a numerical version of "molecular chaos." It is natural that molecular chaos, which plays such a central part in the derivation of the Boltzmann equation, should also appear at some point in simulation procedures. It is one of the nicest features of low discrepancy methods that the appearance of molecular chaos is so transparent.

2. The creation of a duplicate of the gas, the possibility of repeated collisions, the resulting violation of the conservation laws, and the systematic decrease in temperature are undesirable features of the procedure described under 1, in particular because the particle number per cell is kept small in real calculations. The following simple modification, suggested by Babovsky[1], avoids some of these problems.

Suppose that (the particle number per cell) N is even. Then, choose randomly a permutation π of the set $\{1,\ldots,N\}$, group the velocities in pairs $(\xi_{\pi(1)},\xi_{\pi(2)})$, $(\xi_{\pi(3)},\xi_{\pi(4)}),\ldots$, choose independent and equidistributed random collision parameters $(y_i)_{i=1,\ldots,N/2}$ from B^1, and set

$$\begin{aligned}\xi_1' &= \Psi(\xi_{\pi(1)},y_1,\xi_{\pi(2)})\\ \xi_2' &= \Psi(\xi_{\pi(2)},y_1,\xi_{\pi(1)})\\ \xi_3' &= \Psi(\xi_{\pi(3)},y_2,\xi_{\pi(4)})\\ \xi_4' &= \Psi(\xi_{\pi(4)},y_2,\xi_{\pi(3)}).\end{aligned} \tag{2.19}$$

Clearly, there is no duplication of the velocity set in this method. The conservation laws are satisfied. If $|y_i|$ is large enough, there is no collision, and the calculation in (2.19) becomes trivial. If N is odd, one can simply delete one particle from the collision simulation.

3. Both methods described in 1 and 2 still need, to some degree, random variables. While it is unlikely that the use of random variables can be totally eliminated in the collision simulation, it is desirable to reduce it in order to suppress statistical fluctuations. We now present some ideas directed toward this target; they turn out to lead to independent interesting research questions.

Recall that we started the collision simulation with the question of how to find N-point approximations to $dM_j = d^2y\times d\mu_j\times d\mu_j$, given that $\mu_j^N = \frac{1}{N}\sum_{i=1}^{N}\delta_{\xi_i^N(j)}$ is an N-point approximation to μ_j. The dM_j^N we constructed so far were indeed based on μ_j^N. However, from (2.14) it is clear there is no real reason to use μ_j^N to find N-point approximations to

dM_j; we can actually forget μ_j^N and try to construct such approximations directly. We use an example suggested by Manfred Bäcker [4] to explain possible nonstatistical or semistatistical approaches to this problem.

Suppose you have a probability measure μ on the interval $[0,1]$, and your objective is to find an N-point approximation to $\mu \times \mu$ on $[0,1]^2$. If N happens to be a square integer ($N = k^2$), the following strategy avoids all random choices. Choose a partitioning of $[0,1]$, $0 = y_0 < y_1 < y_2 < \dots < y_{k-1} < y_k = 1$ such that $\int_{y_i}^{y_{i+1}} d\mu(x) = \frac{1}{k}$ for all i or, if an approximation $\mu^N = \frac{1}{N}\sum_{i=1}^N \delta_{x_i}$ of μ is already given, reorder the x_i such that $x_1 \leq x_2 \leq \dots \leq x_N$ and set $y_0 = 0,\ \ y_1 = x_k,\ y_2 = x_{2k}, \dots, y_{k-1} = x_{(k-1)k},\ y_k = 1$). Now let $I_j = [y_{j-1},\ y_j]$. Then the rectangles

$$R_{i,j} = I_i \times I_j$$

cover $[0,1]^2$, and there are clearly $k^2 = N$ of them. Let $(\bar{x}_i,\ \bar{x}_j)$ be the center of $R_{i,j}$; then we choose

$$\frac{1}{N}\sum_{i,j=1}^{k} \delta_{(\bar{x}_i,\bar{x}_j)}$$

as the approximation for $\mu \times \mu$ (see Fig. 30).

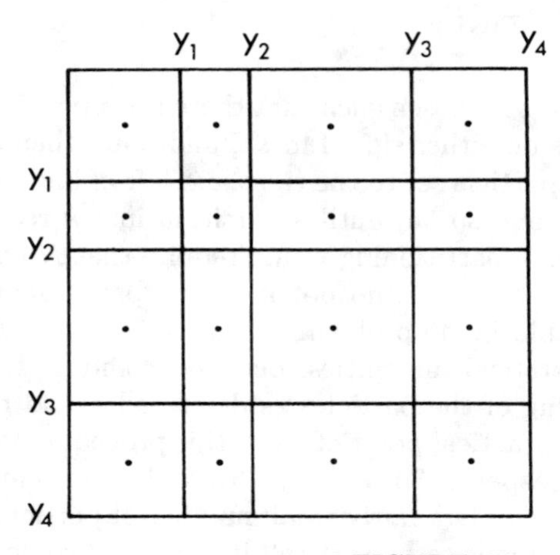

FIGURE 30.

All rectangles carry clearly equal contributions toward $\mu \times \mu$.

To generalize this idea to the setting at hand, we discuss the situation in two dimensions, where graphical explanations are easy.

Suppose an N-particle approximation μ_j^N to μ_j is given, where $N = k^2$ and

$$\mu_j^N = \frac{i}{N} \sum_{i=1}^{N} \delta_{\xi_i}, \qquad \xi_i \in \Re^2.$$

First, partition $\Re^2$ into k rectangles (some of them necessarily semi-infinite), where each rectangle contains (approximately) k particles. This can, of course, be done in many different ways; successive partitioning, as indicated in Fig. 31, is one possibility.

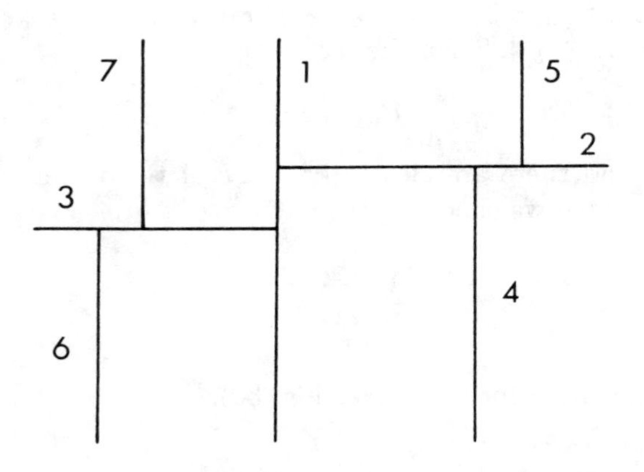

FIGURE 31.

The idea here is that line 1 is chosen such that there is (approximately) an equal number of particles on either side. Lines 2 and 3 are then chosen such that they partition the particle set to the right and left of line 1 in sets of approximately equal size; and so on, until a partition into k rectangles is achieved. Of course, a precise partitioning would require that N contains some power of 2 as a factor, but if we do not insist on partitioning into exactly equal sets, there should be no problem.

Actually these are theoretical difficulties, because it should be obvious that a precise partitioning of the particles as described would be prohibitively expensive. From a practical point of view, this procedure must be combined with Monte Carlo aspects. For example, to find a first partitioning line, one could choose randomly a fairly small number of particles from the total set available; if the line is to be vertical, its equation could be set as $x = X$, where X is the median of the x-coordinates of the velocities of the randomly chosen sample particles. Once X is given, it partitions the total set of particles, and the procedure can be repeatedly applied to the subsets. Many variations are conceivable.

In the end, one arrives at a partitioning of $\Re^3_\xi \times \Re^3_{\xi_*}$ into $k^2 = N$ (four-dimensional) rectangles of the type $R_i \times R_j$, where the R_i, R_j are the ones just constructed. As in the example, one places one collision pair into each

$R_i \times R_j$ (in the middle if $R_i \times R_j$ is finite, into a more or less randomly chosen spot otherwise) to obtain a point approximation to $\mu_j \times \mu_j$. The collision parameter, needless to say, remains to be chosen randomly, even though, at least in principle, the ideas can be extended to include the measure dy.

The fact that N has to be a square causes minor problems. If N is not a square, we can take advantage of the fact that many particles do not experience a collision anyway during the time interval under consideration. Suppose that $\overline{N}$ is the largest square integer less than or equal to N. Next choose Δt such that $r_0^2 \cdot \pi$, with the r_0 defined in the proof of Lemma 10.2.1, satisfies

$$r_0^2 \cdot \pi < \frac{\overline{N}}{N}.$$

Then choose $N - \overline{N}$ particles randomly from the N; these are the ones that will a priori not experience collisions. For the remaining $\overline{N} = k^2$ particles, we approximate $\mu_j \times \mu_j$ as described and then apply the collision transformation [as described by $\Psi(\xi, \xi_*, y) = T_{\xi,\xi_*} \circ \Phi_{\xi,\xi_*}(y)$] with a renormalized $\tilde{r}_0$, given by

$$\tilde{r}_0^2 = r_0^2 \cdot \frac{N}{\overline{N}}.$$

Before we summarize the end of the procedure, we have to voice a disclaimer and a strong word of caution. The ideas we described under 3 may be useful if the particle number per cell is very large (as it must be if the procedure is to converge to the Boltzmann equation). This is certainly no problem in the spatially homogeneous case, where all of $\Re_x^3$ can be treated as one cell and 100,000 to 500,000 test particles can be used. Indeed, in this case, it is to be expected that statistical fluctuations be reduced by a skillful implementation of the methods described under 3.

The reality is dramatically different in the spatially dependent case, where a careful partition of the physical space into cells (as described in step C) is inevitable. The total number of particles may still be $5 \cdot 10^5$, but if a partition into, say, 10,000 cells is deemed necessary, the number of particles per cell is a fortiori on the order of magnitude 50. The current practice is to use about twenty particles per cell, and limitations on memory are usually quoted as reasons why larger numbers are not feasible.

Needless to say, the procedures described under 3 are pointless in this case (they remain, of course, of theoretical interest), and the methods outlined under 1 or 2 are applied (or, frequently, the Bird method, described in the next section). It is evident that such small particle numbers per cell in combination with random choices should lead to large statistical fluctuations. To compensate for that, it is common practice to repeat a calculation many times and then take averages over the runs.

We return to the formal treatment of the low discrepancy methods. After following course 1, 2, or 3 and applying the collision transformation, all the post-collisional velocities $\xi_i(j+1)$ are known. All that remains to be done is to recall the positions and make the free flow step:

$$x_i(j+1) = x_i(j) + \Delta t \cdot \xi_i(j+1).$$

The resulting measure

$$\frac{1}{N} \sum_{i=1}^{N} \delta_{(x_i(j+1),\, \xi_i(j+1))}$$

(N here denotes the total particle number, not the number per cell) is the approximation to μ_{j+1}. At this point, one has various options. One can keep the old cells or construct a new partition. In any case, one can calculate such macroscopic quantities as the density per cell, the bulk velocity per cell, and the temperature per cell.

At the end of this section, we formulate (without proof) and explain the convergence theorem from Ref. 3. This convergence theorem uses a distance concept between measures known as "discrepancy," which we first define.

Definition. *Let μ, ν be probability measures on $\Re^N$. If $P \leq Q$ denotes the usual half-ordering on $\Re^N$ (if $P = (x_1, \ldots, x_N)$, $Q = (y_1, \ldots, y_N)$, then $P \leq Q$ iff $x_i \leq y_i$ for all $i = 1, \ldots, N$), then the discrepancy $D(\mu, \nu)$ between μ and ν is defined by*

$$D(\mu,\nu) = \sup_{Q} \Big| \int_{R(Q)} d\mu - \int_{R(Q)} d\nu \Big|$$

where $R(Q) = \{P \in \Re^N; \quad P \leq Q\}$.

If the limit measure is absolutely continuous, convergence in discrepancy is equivalent to weak-$*$ convergence (see Ref. 3).

The convergence theorem in Ref. 3 addresses the method described under 1. However, it certainly generalizes to the other two options we discussed. We need the following.

Assumptions

A1. For the initial value f_0, the Boltzmann equation has a global unique mild solution f on the time interval $[0, T]$, such that for some constants $\alpha_1 > 0$, $C_0 > 0$

$$\limsup_{x,\xi} |f(x,\xi,t)e^{\alpha_1 \xi^2}| \leq C_0 \qquad \text{for all } t \in [0,T]$$

(Maxwellian upper bounds).

A2. There is a constant $C_1 > 0$ such that $\int B(|\xi - \xi_*|, n)\, dn \leq C_1$ (truncation of the collision kernel).

A3. There is a $C_2 > 0$ and an $\alpha_2 > \alpha_1$ such that

$$\limsup_{t,x,\xi} |f(x+\Delta x,\xi,t) - f(x,\xi,t)| e^{\alpha_2 \xi^2} \leq C_2 \cdot \Delta x$$

for all Δx (assumption on spatial regularity, necessary because of the homogenization over cells).

Remark. A1 and A3 implicitly impose bounds and regularity properties on f_0.

(10.2.3) Theorem. *Suppose that A1–A3 hold, and assume that* $\{\mu_0^N = \frac{1}{N}\sum_{i=1}^N \delta_{(x_i,\xi_i)}\}$ *is a sequence of particle approximations to the initial value* f_0 *(i.e.,* $D(\mu_0^N,\ f_0\, dx\, d\xi) \to 0$ *as* $N \to \infty$*). Let* $(\Delta t)_n$ *and* $(\Delta x)_n$ *be sequences of time steps and cell sizes such that* $(\Delta t)_n \searrow 0$, $(\Delta x)_n \searrow 0$*. Then, there is a sequence* $N(n) \to \infty$ *such that*

$$D\big((\mu_{k\cdot(\Delta t)_n}^{N(n)},\quad f(k(\Delta t)_n,\, x,\, \xi)\big)\; dx\, d\xi \longrightarrow 0$$

as $n \to \infty$*, almost surely with respect to the* r_i *and* y_i*, for* $k \cdot (\Delta t)_n \in [0,T]$*.*

Remark. The result is weak inasmuch as it does not indicate how fast $N(n)$ has to grow relative to $(\Delta t)_n$ and $(\Delta x)_n$, but certainly the number of particles per cell has to go to infinity. There is no required relation between $(\Delta t)_n$ and $(\Delta x)_n$; this is typical for methods of characteristics, and, in fact, the free flow step is just a step of this type.

10.3 Bird's Scheme

Bird's scheme is much older than the low discrepancy schemes discussed in the previous section. It was designed in the 1960s[6] and has been applied consistently ever since, with good success. Recently, W. Wagner [14] has proved that the Bird simulation method does indeed converge to solutions of the Boltzmann equation in a suitable limit. The proof, based on a reinterpretation of the simulation scheme as a measure–valued stochastic process, is based on compactness arguments and therefore not constructive. For details, see Ref. 14. Still more recently, a constructive proof of the convergence based on the strategy used in the validation presented in chapter 4, i.e., by a direct control of the correlation functions, was given [13].

The derivation of the method is a priori independent of the Boltzmann equation (one could call the procedure "pre-Boltzmann"). Nevertheless, many of the steps closely resemble those taken by Boltzmann in his classical derivation of the equation.

Steps A–C from Section 10.2 are repeated identically. The difference is in the collision simulation in each cell.

Recall that the collision kernel $B(n, \xi - \xi_*)$ is really only a function of $|\xi - \xi_*|$ and $|n \cdot (\xi - \xi_*)|$. We define the differential cross section

$$\sigma(|\xi - \xi_*|, n)$$

by

$$\sigma(|\xi - \xi_*|, n) = \frac{B(n, \xi - \xi_*)}{|\xi - \xi_*|}. \tag{3.1}$$

To understand the formal derivation of the procedure, it is convenient to renormalize the Boltzmann density at time $j \cdot \Delta t$ such that

$$\iint f_i(x, \xi)\, d\xi\, dx = N,$$

where N is the total number of test particles being used. Then, if C is the cell in question,

$$N_C := \int_C \int f_j(x, \xi)\, d\xi\, dx$$

is the number of particles in C (please ignore the fact that $\int_C \int f_j(x, \xi)\, d\xi\, dx$ is in general no integer—one could define

$$N_C := \left[\int_C \int f_j(x, \xi)\, d\xi\, dx \right],$$

but we will prefer to work with real particle numbers; particle number, here, has more the meaning of a scaling parameter than of a natural number).

$$n_C := \frac{N_C}{\lambda^3(C)} \tag{3.2}$$

will be the particle number per unit volume in C, and, writing $f_{j,C}$ for the homogenization of f_j on $C \times \Re^3$,

$$\begin{aligned} \int \frac{1}{n_C} f_{j,C}(\xi)\, d\xi &= \frac{\lambda^3(C)}{N_C} \int f_{j,C}(\xi)\, d\xi \\ &= \frac{\lambda^3(C)}{N_C} \cdot \frac{1}{\lambda^3(C)} \iint f_j(x, \xi)\, d\xi\, dx \\ &= 1. \end{aligned}$$

Hence, $\frac{1}{n_C} f_{j,C}$ is a probability density (corresponding to our measure μ_j from Section 10.2).

We will write, with $V = \xi - \xi_*$,

$$\sigma_V(n) = \frac{B(|V|, |n \cdot V|)}{|V|}$$

for the differential cross section as a function of n alone, and define

$$\sigma_{\text{tot}}(V) = \int_{S^2} \sigma_V(n)\, dn \tag{3.3}$$

as the total cross section. The quantity

$$\sigma_{\text{tot}}(V) \cdot |V| = \int_{S^2} B\, dn$$

(cf. 2.7) is, in classical terminology, known as the volume of the collision cylinder for a time step $\Delta t = 1$. Hence $\sigma_{\text{tot}}(V)|V|\Delta t$ is the volume of the collision cylinder with base $\sigma_{\text{tot}}(V)$ and length $|V| \cdot \Delta t$.

We start a collision count. The quantity $f_{j,C}(\xi)\, d\xi$ is (by normalization) the number per unit volume of particles moving with a velocity from the cube with vertices $\xi = (\xi_1, \xi_2, \xi_3)$, $(\xi_1 + d\xi_1, \xi_2, \xi_3)$, $(\xi_1, \xi_2 + d\xi_2\ \xi_3)$, etc. Consequently,

$$\sigma_{\text{tot}}(|V|) \cdot |V| \cdot \Delta t \cdot f_{j,C}(\xi)\, d\xi$$

is the expected number per unit volume of collisions of particles moving with ξ_* with particles moving with ξ during Δt. Integrating over ξ, we find that

$$q(\xi_*) := \int_{\Re^3}\int_{S^2} B((|\xi - \xi_*|), |n \cdot (\xi - \xi_*)|)\, dn\, f_{j,C}(\xi)\, d\xi$$

is the number of collisions per unit volume and unit time of particles moving with velocity ξ_*, and

$$\frac{1}{2} \int q(\xi_*) f_{j,C}(\xi_*)\, d\xi_*$$

is then simply the number of collisions per unit volume and unit time (the factor $\frac{1}{2}$ is required because otherwise we would count collisions twice). The number of collisions in the cell C per unit time is then

$$N_{\text{coll}} = \frac{1}{2} \lambda^3(C) \int q(\xi_*) f_{j,C}(\xi_*)\, d\xi_*,$$

and it is reasonable to call

$$\Delta t_{\text{coll}} = \frac{1}{N_{\text{coll}}}$$

the mean free time between collisions in C.

Let

(3.4)
$$\overline{\sigma_{\text{tot}}(|\xi-\xi_*|)\cdot|\xi-\xi_*|} := \iint \sigma_{\text{tot}}(|\xi-\xi_*|)|\xi-\xi_*| \frac{f_{j,C}(\xi)}{n_C}\,\frac{f_{j,C}(\xi_*)}{n_C}\,d\xi\,d\xi_*,$$

be the expected value of $\sigma_{\text{tot}}(|\xi-\xi_*|)\cdot|\xi-\xi_*|$ with respect to the distribution $\frac{1}{n_C^2} f_{j,C}\otimes f_{j,C}$, then, by collecting terms, we see that

$$N_{\text{coll}} = \frac{1}{2}\lambda^3(C) n_C^2\, \overline{\sigma_{\text{tot}}(|\xi-\xi_*|)|\xi-\xi_*|}. \tag{3.5}$$

By using $N_C = \lambda^3(C)\cdot n_C$, it follows that

$$\Delta t_{\text{coll}} = \frac{2}{n_C\cdot N_C\, \overline{\sigma_{\text{tot}}(|\xi-\xi_*|)|\xi-\xi_*|}}. \tag{3.6}$$

The mean free time between collisions of one particular particle is then

$$N_C\cdot \Delta t_{\text{coll}} = \frac{2}{n_C\, \overline{\sigma_{\text{tot}}(|\xi-\xi_*|)|\xi-\xi_*|}}.$$

After these heuristic preparations, we are ready to introduce the Bird procedure. The probability measure $\frac{1}{n_C} f_{j,C}\,d\xi$ is, as in Section 10.2, approximated by the discrete measure

$$\frac{1}{N_C}\sum_{j=1}^{N_C}\delta_{\xi_j},$$

such that $\overline{\sigma_{\text{tot}}(|\xi-\xi_*|)|\xi-\xi_*|}$ must be approximated by

$$\frac{1}{N_C^2}\sum_{i,j=1}^{N_C}\sigma_{\text{tot}}(|\xi_i-\xi_j|)\cdot|\xi_i-\xi_j|, \tag{3.7}$$

and Δt_{coll} by

$$\frac{2}{(n_C/N_C)\sum_{i,j=1}^{N}\sigma_{\text{tot}}(|\xi_i-\xi_j|)|\xi_i-\xi_j|}. \tag{3.8}$$

There are $\binom{N_C}{2}$ potential collision pairs in the cell C. Three jobs are to be done in the Bird collision simulation:

1. We need to know how many collisions are expected during $[0,\Delta t]$.
2. If a collision happens, we need to choose the particles that collide.
3. We need to compute post-collisional velocities.

The considerations made earlier suggest simple formulas for these jobs. From (3.5) and (3.7), the number of expected collisions during $[0,\Delta t]$ is

$$\frac{1}{2}\Delta t \cdot n_C \cdot N_C \frac{1}{N_C^2} \sum_{i,j=1}^{N} \sigma_{\text{tot}}(|\xi_i - \xi_j|)|\xi_i - \xi_j|. \tag{3.9}$$

Unfortunately, this formula is not very useful in this form because N^2 operations are needed for the evaluation of (3.9).

A similar problem occurs in the second step. It is perfectly reasonable to choose the probability p_{ij} that particles i and j collide proportional to the volume of the collision cylinder, i.e.,

$$p_{ij} = k\sigma_{\text{tot}}(|\xi_i - \xi_j|) \cdot |\xi_i - \xi_j|. \tag{3.10}$$

Clearly, we want the normalization

$$\sum_{1 \le i < j \le N} p_{ij} = 1,$$

so it follows that

$$k = \frac{1}{\sum_{1 \le i < j \le N} \sigma_{\text{tot}}(|\xi_i - \xi_j|)|\xi_i - \xi_j|}. \tag{3.11}$$

Again, (3.11) involves N^2 calculations, which makes it unpractical.

Fortunately, not all the p_{ij} really need to be calculated for the selection of collision pairs. The costly calculations in (3.11) are avoided by a statistical trick known as the acceptance-rejection method, which, in the present context, applies as follows.

Choose a number g^* such that

$$g^* \ge \max\{\sigma_{\text{tot}}(|\xi_i - \xi_j|) \cdot |\xi_i - \xi_j|; \quad 1 \le i < j \le N_C\}$$

(such a number is easily found and does not require the calculation of all the $\sigma_{\text{tot}}(|\xi_i - \xi_j|) \cdot |\xi_i - \xi_j|$).

Then, choose $R \in [0,1]$ a random and equidistributed number, and choose a random index pair (l, m) such that $1 \le l < m \le N_C$. If $(\sigma_{\text{tot}}(|\xi_l - \xi_m|) \cdot |\xi_l - \xi_m|)/g^* \ge R$, we accept (ξ_l, ξ_m) as a collision pair. Otherwise, we reject it. This procedure is repeated until a pair is accepted (obviously, it helps if g^* is not chosen excessively large). Upon acceptance, a collision parameter is chosen randomly, and the post-collisional velocities ξ_l', ξ_m' are computed. It is clear that this procedure will choose collision pairs consistent with the unknown probabilities p_{ij}.

It remains to decide how many collisions should occur during $[0, \Delta t]$; as indicated, formula (3.9) is not feasible because of the N^2-effort. Instead, one uses (3.6) as a guideline to advance a "clock" if a collision happens. Specifically, let

$$\tau_k := \frac{2}{n_C N_C \sigma_{\text{tot}}(|\xi_r - \xi_s|) \cdot |\xi_r - \xi_s|}$$

if (r, s) is the kth accepted pair, and let $\tilde{N}_{\text{coll}}$ be defined by

$$\sum_{i=1}^{\tilde{N}_{\text{coll}}} \tau_i < \Delta t, \qquad \sum_{i=1}^{\tilde{N}_{\text{coll}}+1} \tau_i \geq \Delta t.$$

The $\tau_1, \ldots, \tau_{\tilde{N}_{\text{coll}}}$ are actually samples of random variables $T_1, \ldots, T_{\tilde{N}_{\text{coll}}}$, which are distributed according to

$$\text{Prob}\left\{ T_k = \frac{2}{n_C N_C \sigma_{\text{tot}}(|\xi_i - \xi_j|) \cdot |\xi_i - \xi_j|} \right\} = p_{ij}^{(k-1)},$$

where

$$1 \leq i < j < N_C$$
$$1 \leq k \leq \tilde{N}_{coll}.$$

The probabilities $p_{ij}^{(k-1)}$ are those from (3.10), except, as indicated by the index $(k-1)$, the velocity pairs that have undergone a collision are replaced by the post-collisional pairs.

It is easily checked from (3.6) that

$$E[T_1] = \Delta t_{\text{coll}}.$$

If the time step Δt is such that $\Delta t \ll \Delta t_{\text{coll}} \cdot N_C$ (the right-hand side is the mean free time between collisions of one particle), then the number of collisions will be small compared with N_C, and we expect that

$$E[T_k] \approx E[T_1] = \Delta t_{\text{coll}}$$

for all $1 \leq k \leq \tilde{N}_{\text{coll}}$. Therefore, we have $\Delta t = N_{\text{coll}} \cdot \Delta t \cdot \Delta t_{\text{coll}}$ (by definition) but also $\Delta t \approx \tilde{N}_{\text{coll}} \frac{1}{\tilde{N}_{\text{coll}}} \sum_{i=1}^{\tilde{N}_{\text{coll}}} \tau_i$. This shows that $\tilde{N}_{\text{coll}} \approx N_{\text{coll}} \cdot \Delta t$ if

$$\frac{1}{\tilde{N}_{\text{coll}}} \sum_{i=1}^{\tilde{N}_{\text{coll}}} \tau_i \approx \Delta t_{\text{coll}}.$$

The description of the collision simulation in Bird's method is complete.

We finish this section by giving a synopsis of the procedure, in four steps:

1. To choose collision pairs, first generate an upper bound

   $$g^* \geq \max\{\sigma_{\text{tot}}(|\xi_i - \xi_j|) \cdot |\xi_i - \xi_j|, \qquad 1 \leq i < j \leq N\}.$$

   Choose a random number $R \in [0, 1]$. For a randomly chosen index pair l, m, the collision pair (ξ_l, ξ_m) is accepted if

   $$\sigma_{\text{tot}}(|\xi_l - \xi_m|) \cdot |\xi_l - \xi_m| \geq R \cdot g^*.$$

   This is repeated *until* a pair is accepted.

2. If (l, m) is accepted, a "clock" is advanced by

$$\tau_1 = \frac{2}{n_C N_C \sigma_{\text{tot}}(|\xi_l - \xi_m|) \cdot |\xi_l - \xi_m|}.$$

3. With a random choice of the collision parameter n, convert the accepted velocities ξ_l, ξ_m into post-collisional velocities ξ_l', ξ_m'.
4. Repeat steps 1–3 until $\tau_1 + \ldots + \tau_{\tilde{N}_{\text{coll}}+1} \geq \Delta t$.

We mention that there are modifications of both the original Bird scheme, described here, and the low discrepancy schemes outlined in Section 10.2, which "interpolate" between the two in the sense that features of one method are transferred to the other. For example, there is a "no–time–counter" version of the Bird simulation, which avoids the advancement of the "clock" described in step 2. The idea is simply to allow "fake" collisions between particles, as allowed in the low discrepancy methods by virtue of the definition of Φ_{ξ,ξ_*} (see Lemma 2.1). In this version, $\frac{N_C}{2}$ randomly chosen velocity pairs are sequentially admitted for collision. If they are found to have really collided, they are replaced by their post-collisional velocity pair and returned into the original sample. The particles with these velocities are then readmitted for collision.

A modification of the low discrepancy scheme that would return post-collisional velocities immediately to the sample and admit them again for collision will lead to essentially the same procedure. In terms of an urn model, we originally drew random pairs once, and after testing them for collision, removed them from the sample. In the modification, they are returned to the sample. The difference seems small enough, but the readmission may not only be meaningful from a physical point of view (real particles may well collide twice during $[0, \Delta t]$), it may also lead to higher accuracy in the numerical approximation of the spatially homogeneous equation. This possibility is indicated by Bogomolov's result [7], outlined in step B of Section 10.2.

The useful idea of "fake" collisions was apparently first introduced by K. Koura in Ref. 10.

In spite of this similarity between the Bird scheme and the low discrepancy methods, there is a profound conceptual difference. In the Nanbu-Babovsky method, we aimed for a direct approximation of the Boltzmann equation. Here, the particles are just computational elements. On the other hand, if we ignore the fact that the particles have no position in a given cell, the Bird method is a genuine particle method; the correlations between particles generated by the dynamics are accounted for, and the state of the system is described by a probability measure that does not factorize. In other words, the simulation works on a level prior to the propagation of chaos.

So which is "the best" method? We do not dare to give a conclusive answer to this question. All the methods discussed are used with good success, for example, for the simulation of the space shuttle reentry. Their efficiency and accuracy depend on many more aspects than the ones we discussed here, e.g., how best to create and use the spatial grid, how to produce the random collision pairs, how to handle boundaries, and so on. For realistic cases, it is not reasonable to work with the Boltzmann equation for monoatomic gases. Inner degrees of freedom for the molecules have to be accounted for, gas mixtures must be considered, and chemical reactions are a reality. A detailed description of the necessary modifications and generalizations of the Boltzmann equation is beyond the scope of this book. The interested reader is referred to the vast literature survey offered by the publications in the Proceedings of the various Symposia on Rarefied Gas Dynamics (the most recent one was held in the summer of 1992 in Vancouver).

The theoretical analysis of particle simulation methods outlined in this chapter had two significant objectives (and, indeed, results). It clarified the relationships between the Boltzmann equation and particle simulation, and it led to improvements of the simulation schemes. There is certainly more room to move on both of these fronts, and work to this end continues. A particular challenge is to address situations where the mean free path between collisions becomes so small that we are in the realm of a fluid dynamical limit; the analytical theory for this situation will be discussed in the next chapter.

Problems

1. If $\int_\Lambda \int f_j(x,\xi)d\xi dx$ is not normalized to 1, how large can Δt be chosen in (2.6) such that $f_{j+1} \geq 0$ is guaranteed?
2. If $y \in B^1$ is described by polar coordinates (r,β), how do r and β, as random variables, have to be distributed such that y will be equidistributed on B^1? Find their distribution densities.

References

1. H. Babovsky, "A convergence proof for Nanbu's Boltzmann simulation scheme," *European J. Mech. B: Fluids* **8(1)**, 41–55 (1989).
2. H. Babovsky, "Time averages of simulation schemes as approximations to stationary kinetic equations," *Eur. J. Mech. B: Fluids* **11**, 199-212 (1992).
3. H. Babovsky and R. Illner, "A convergence proof for Nanbu's simulation method for the full Boltzmann equation," *SIAM J. Num. Anal.* **26(1)**, 45–65 (1989).
4. M. Bäcker, personal communication (1990).
5. P. Billingsley, *Convergence of probability measures,* Wiley, N.Y. (1968).
6. G. A. Bird, *Molecular gas dynamics,* Clarendon Press, Oxford (1976).
7. S. V. Bogomolov, "Convergence of the total-approximation method for the Boltzmann equation," *U.S.S.R. Comput. Math. Phys.* **28 (1)**, 79–84 (1988).
8. C. Greengard and L. Reyna, "Conservation of expected momentum and energy in Monte Carlo particle simulation," preprint, IBM Research Center, Yorktown Heights, NY (1992).

9. R. Illner and H. Neunzert, "On simulation methods for the Boltzmann equation," *Transport Theory Stat. Phys.* **16 (2&3)**, 141–154 (1987).
10. K. Koura, "Null-collision technique in the direct-simulation Monte Carlo method," *Phys. Fluids* **29 (11)**, 3509–3511 (1986).
11. K. Nanbu, "Interrelations between various direct simulation methods for solving the Boltzmann equation," *J. Phys. Soc. Japan* **52 (10)**, 3382–3388 (1983).
12. H. Ploss, "On simulation methods for solving the Boltzmann equation," *Computing* **38**, 101–115 (1987).
13. M. Pulvirenti, W. Wagner, and M. B. Zavelani, "Convergence of particle schemes for the Boltzmann equation," preprint (1993).
14. W. Wagner, "A convergence proof for Bird's direct simulation Monte Carlo method for the Boltzmann equation," *J. Stat. Phys.* **66 (3&4)**, 1011–1044 (1992).

11
Hydrodynamical Limits

11.1 A Formal Discussion

In Chapter 3 Section 8 we discussed the hydrodynamical limit for the Boltzmann equation in general terms and showed how a pure space-time scaling leads to the asymptotic limit $\epsilon \to 0$ of solutions of the Boltzmann equations

$$\partial_t f^\epsilon + \xi \cdot \nabla_x f^\epsilon = \frac{1}{\epsilon} Q(f^\epsilon, f^\epsilon). \tag{1.1}$$

We will use the abbreviation $D_t f := \partial_t f + \xi \cdot \nabla_x f$. From a formal point of view, we expect that

$$\epsilon D_t f^\epsilon \to 0 \ as \ \epsilon \to 0, \tag{1.2}$$

and if

$$f^\epsilon \to f^0, \tag{1.3}$$

the limit f^0 must satisfy

$$Q(f^0, f^0) = 0. \tag{1.4}$$

This implies, as we know from Section 3.2, that f^0 is a local Maxwellian:

$$f^0(x,\xi,t) \equiv M(x,\xi,t) = \frac{\rho(x,t)}{(2\pi T(x,t))^{3/2}} \exp(-\frac{|\xi - v(x,t)|^2}{2T(x,t)}). \tag{1.5}$$

The fields (ρ, v, T), which characterize the behavior of the local Maxwellian M in space and time, are expected to evolve according to fluid dynamical equations, which we are going to derive (repeating some of the arguments from Chapter 3). First, let us again emphasize that these fields are varying *slowly* on the space-time scales that are typical for the gas described in terms of the Boltzmann equation.

From the conservation laws (3.3.16)

$$\int \psi_\alpha Q(f,f)d\xi = 0, \ \alpha = 0, \dots, 4, \tag{1.6}$$

we readily obtain, as in Section 3.3,

$$\int \psi_\alpha(\partial_t f + \xi \cdot \nabla_x f)d\xi = 0. \tag{1.7}$$

This is a system of equations for the moments of f that is in general not closed. However, if we assume $f = M$ and use the identities (for M they are identities; for general f they are definitions given in (3.3.1–7), where $e = \frac{3}{2}T$)

$$\rho = \int M d\xi \tag{1.8}$$

$$\rho v = \int M\xi\, d\xi \tag{1.9}$$

$$w \equiv \frac{3}{2}\rho T + \frac{1}{2}\rho v^2 = \frac{1}{2}\int \xi^2 M d\xi, \tag{1.10}$$

we readily obtain from (1.7) that

$$\partial_t \rho + \operatorname{div}(\rho v) = 0 \tag{1.11}$$

$$\partial_t(\rho v_i) + \operatorname{div}(\int M\xi\xi_i) = 0 \tag{1.12}$$

$$\partial_t(\frac{3}{2}\rho T + \frac{1}{2}\rho v^2) + \frac{1}{2}\operatorname{div}(\int M\xi\xi^2) = 0. \tag{1.13}$$

These equations are nothing but Eqs. (3.3.18–20), specialized to the case of a Maxwellian density. This is of crucial importance if we want to write Eqs. (1.12) and (1.13) in closed form. To do so we have to express $\int M\xi\xi_i$ and $\int M\xi\xi^2$ in terms of the field (ρ, v, T). To this end, we use the elementary identities

$$\int M(\xi - v)_j(\xi - v)_i\, d\xi = \delta_{ij}\rho T \tag{1.14}$$

$$\int M(\xi - v)(\xi - v)^2 \, d\xi = 0, \tag{1.15}$$

which transform Eq. (1.12) into

$$\partial_t(\rho v_i) + \operatorname{div}(\rho v v_i) = -\partial_{x_i} p \tag{1.16}$$

with

$$p = \rho T. \tag{1.17}$$

Equation (1.17) is the perfect gas law with $R = 1$. Obviously, the p defined by Eq. (1.17) has the meaning of a pressure.

Recalling that the internal energy e is defined by

$$e = \frac{3}{2} T \tag{1.18}$$

we derive from this and Eq. (1.15)

$$\partial_t \big(\rho(e + \frac{1}{2} v^2)\big) + \operatorname{div}\big(\rho v(e + \frac{1}{2} v^2)\big) = -\operatorname{div}(pv). \tag{1.19}$$

The set of Eqs. (1.11), (1.16), and (1.19) express conservation of mass, momentum, and energy, respectively. For the convenience of the reader, we rewrite them in the compact form of a conservation law:

$$\partial_t \begin{pmatrix} \rho \\ \rho v_i \\ \rho(e + \frac{1}{2} v^2) \end{pmatrix} + \operatorname{div} \begin{pmatrix} \rho v \\ \rho v v_i + \frac{2}{3} \rho e d_i \\ \rho v(\frac{1}{2} v^2 + \frac{5}{3} e) \end{pmatrix} = 0. \tag{1.20}$$

Here, d_i denotes the unit vector in the ith direction. Complemented by the state equation for a perfect gas, Eq. (1.17), the system (1.20) is the Euler equations for a compressible gas.

For smooth functions, an equivalent way to write the Euler equations in terms of the field (ρ, v, T) is

$$\begin{cases} \partial_t \rho + \operatorname{div}(\rho v) = 0 \\ \partial_t v + (v \cdot \nabla) v + \dfrac{1}{\rho} \nabla p = 0 \\ \partial_t T + (v \cdot \nabla) T + \dfrac{2}{3} T \nabla_x \cdot v = 0. \end{cases} \tag{1.21}$$

However, in this form we lose the general structure of a conservation law as given in (1.20), i.e., the time derivative of a field equals the negative divergence of a current that is a nonlinear function of this field.

The arguments we have so far given are largely formal. Our main objective in this chapter is to make them as rigorous as possible, from both a conceptual and a mathematical point of view. Before doing this, we have to introduce two other formal tools, namely, the Hilbert expansion and the entropy method.

At this juncture, some comments on our limits are in order, because the reader may suspect an inconsistency in the passage from a rarefied to a dense gas. Recall that we derived the Boltzmann equation in a low density approximation ($N\sigma^2 = O(1)$). In the hydrodynamic limit, we have to take $N\sigma^2 = \frac{1}{\epsilon} \to \infty$. This, at first glance, seems contradictory, but there is really no problem. The Boltzmann equation holds for a perfect gas, i.e., for a gas such that the density parameter $\delta = N\sigma^3/V$, where V is the volume containing N molecules, tends to zero. The parameter

$$\frac{1}{Kn} = \frac{N\sigma^2}{V^{\frac{2}{3}}} = N^{\frac{1}{3}}\delta^{\frac{2}{3}}$$

may tend to zero, to ∞, or remain finite in this limit. These are the three cases that occur if we scale N as δ^{-m} ($m \geq 0$), for $m < 2$, $m > 2$, and $m = 2$, respectively. In the first case the gas is in free molecular flow and we can simply neglect the collision term (Knudsen gas); in the second we are in the fluid dynamic regime we are treating here, and we cannot simply "omit" the "small" term, i.e., the left-hand side of the Boltzmann equation, because the limit is singular. In the third case the two sides of the Boltzmann equation are equally important (Boltzmann gas); this is the case dealt with in the other chapters of this book.

We stress that the hydrodynamic equations are identical to the original "microscopic" dynamics (in the present case the Boltzmann evolution) seen on appropriate scales of space and time. It follows that compressible fluid dynamics is not a unique theory; there are many hydrodynamical regimes, associated with different microscopic dynamics and different scalings. This will become transparent in this chapter.

We briefly digress to discuss the question of whether Eqs. (1.20) apply to real fluids. In fact, they describe reasonably well flows in the high atmosphere, or all cases of moderate density for which the intermolecular forces are not relevant for the characterization of local equilibria. The state equation associated with (1.20) is indeed that of a perfect gas; for more general situations, one should investigate the hydrodynamical limit for particle systems interacting via Newton's laws.

The derivation of the Euler equations in this situation is hard even from a heuristic point of view. The key concept is again that of a local equilibrium, which, however, need not be Maxwellian any longer. In fact, the hydrodynamical regime of a particle system must be discussed in the context of a thermal equilibrium defined via general equilibrium statistical mechanics. We refrain here from a formal derivation, because this would take us beyond the purposes of this book, but we write down the expected Euler equations:

$$\partial \begin{pmatrix} \rho \\ \rho v_i \\ \rho(e + \frac{1}{2}v^2) \end{pmatrix} + \operatorname{div} \begin{pmatrix} \rho v \\ \rho v v_i + p d_i \\ v(\rho e + \frac{1}{2}\rho v^2 + p) \end{pmatrix} = 0. \tag{1.22}$$

Here, the pressure is a function of the density and the internal energy e, by means of a state equation $p = p(\rho, e)$, which must be computed from equilibrium statistical mechanics. Eqs. (1.22) reduce to Eqs. (1.20) when the state equation is that of a perfect gas.

As we will see in the sequel, it is possible to prove rigorous results about the hydrodynamical limit of the Boltzmann equation. In contrast, very little is known about the hydrodynamical limit of Hamiltonian systems, mainly because it is very difficult to prove that the local equilibrium structure is preserved in time. We refer the reader to Refs. 10, 20, and 22 for a conceptual discussion of this problem and the partial results available at the time of this writing.

11.2 The Hilbert Expansion

We return to the hydrodynamical limit for the Boltzmann equation.

In spite of the fact that we face a singular perturbation problem, Hilbert [13] proposed the following approach.

Try to find a solution of the initial value problem for the Boltzmann equation in the form

$$f = \sum_{n=0}^{\infty} \epsilon^n f_n. \tag{2.1}$$

By inserting this formal series into Eq. (1.1) and matching the various orders in ϵ, we obtain equations one can hope to solve recursively:

$$Q(f_0, f_0) = 0 \tag{2.2$_0$}$$

$$2Q(f_1, f_0) = D_t f_0 \equiv S_0 \tag{2.2$_1$}$$

$$\cdots$$

$$2Q(f_j, f_0) = D_t f_{j-1} - \sum_{1 \le i,j \le j-1 : i+k=j}^{j-1} Q(f_i, f_k) \equiv S_{j-1}, \tag{2.2$_j$}$$

$$\cdots$$

where $Q(f,g)$ denotes the symmetrized collision operator. The first equation, namely, Eq. $(2.2)_0$, gives

$$f_0 = M \tag{2.3}$$

with the five parameters (ρ, v, T) still unknown.

Eq. $(2.2)_1$ can be written as

$$\mathcal{L}_M f_1 = S_0 \tag{2.4}$$

where $\mathcal{L}_M$ is the linearized Boltzmann operator around M, i.e., $\mathcal{L}_M = 2Q(M, \cdot)$. Note that $\mathcal{L}_M = M^{\frac{1}{2}} L M^{-\frac{1}{2}}$, where L is the operator given in Eq. (7.1.4). By the Fredholm alternative, this equation has a solution if $S_0 = D_t f_0$ is orthogonal to $ker \mathcal{L}_M^* = ker \mathcal{L}_M$ in $L^2(\Re^3, M^{-1} d\xi)$. Indeed, $\mathcal{L}_M$ is symmetric in $L^2(\Re^3, M^{-1} d\xi)$, as follows from the symmetry of $\mathcal{L}$ in $L^2(\Re^3, d\xi)$ (see 7.1.9, or use direct inspection). Therefore, the solvability condition is

$$(M\psi_\alpha, D_t f_0)_{L^2(\Re^3, M^{-1} d\xi)} = (\psi_\alpha, D_t f_0) = 0, \tag{2.5}$$

where the ψ_α are the collision invariants.

The five equations

$$(\psi_\alpha, D_t f_0) = 0 \tag{2.6}$$

are nothing but the Euler equations, such that we can solve Eq. (2.4) provided M has parameters consistent with these hydrodynamical equations. However, f_1 is not completely known, because only the part of f_1 that is orthogonal to the invariants is determined by Eq. (2.4).

Next we analyze Eq. $(2.2)_2$. Abbreviate the right-hand side by S_1; then

$$\mathcal{L}_M f_2 = S_1 \tag{2.7}$$

is solvable if

$$(S_1, \psi_\alpha) = 0.$$

As $\int Q\psi_\alpha = 0$, this is equivalent to

$$(D_t f_1, \psi_\alpha) = 0. \tag{2.8}$$

Now denote by

$$f_1 = f_1^1 + f_1^2 \tag{2.9}$$

the decomposition of f_1 into its projection onto the subspace orthogonal to the five invariants, and its orthogonal complement. By Eq. (2.8),

$$(\psi_\alpha, D_t f_1^2) = -(\psi_\alpha, D_t f_1^1). \tag{2.10}$$

The right-hand side of (2.10) is given by the previous step. Therefore, Eq. (2.10) is nothing but a system of nonhomogeneous hyperbolic linear partial differential equations that can (in principle) be solved.

In conclusion, the solvability condition for f_2 fully determines f_1. The procedure can obviously be iterated to determine all the f_js.

At this point, one should start an investigation of the convergence of the Hilbert expansion, but this is a formidable job to do, if possible at all. A better strategy is to truncate the expansion in the form

$$f(t) = \sum_{j=0}^{n} \epsilon^j f_j + \epsilon^l r_\epsilon(t), \tag{2.11}$$

and write an equation for the remainder r_ϵ. In Eq. (2.11), l is not necessarily $n+1$. Rather, l should be considered as a parameter that will be chosen later, and the choice should be such that r_ϵ is uniformly bounded with respect to ϵ (see Section 4). We remark that this does not imply the convergence of the Hilbert expansion.

By completing this program, one could obtain a derivation of the Euler equations from the Boltzmann equation for all times for which a smooth solution of the latter exists. We shall do this in Section 4. First, however, we will discuss the entropy method, which has the potential to give a kinetic description of solutions of the Euler equations even when these solutions eventually lose smoothness.

11.3 The Entropy Approach to the Hydrodynamical Limit

We repeat that the approach based on the Hilbert expansion relies on the existence of smooth solutions of the Euler equations. Such solutions are, for smooth data, known to exist up to some time T that is small relative to the size of the initial conditions (in other words, smooth solutions exist locally in time). The general theory of conservation laws (and physical intuition and experience) suggests that solutions of hyperbolic equations like the Euler equations develop shocks, i.e., discontinuities. One is forced to pass to a weak solution concept.

Our discussion in this section must remain academic, because it is not known whether the system (1.20) has, for general initial values, a global weak solution. Simpler scalar conservation laws, which also allow the formation of shocks, are typically solved by use of the weak solution concept. There is a well-established theory of scalar conservation laws, documented, e.g., in the monograph by Smoller [19]; this theory has been successfully generalized to systems of two conservation laws in one dimension by R. DiPerna[12], with the theory of compensated compactness as the main tool. This is, at present, the state of the art. Recently, approximation methods based on so–called kinetic approximation schemes, which emulate the underlying kinetic structure, have been used to prove existence results for such systems of conservation laws [16].

Usually, there are many weak solutions to the initial value problem, and the physically relevant one is chosen by means of an entropy condition, which we next explain. Consider, for example, a scalar conservation law

$$\partial_t u + \partial_x F(u) = 0 \tag{3.1}$$

where $u = u(x,t)$ is a real function and F is a nonlinear function of u. From the analysis of even the simple case $F(u) = u^2$, which is known as the Burgers equation and can be solved explicitly, we expect in general the development of shocks. Therefore, even for smooth initial values, Eq. (3.1) must for sufficiently large times be interpreted in the weak sense.

Let $\langle , \rangle$ be an abbreviation for the space-time scalar product, then a weak solution of (3.1) is a function u such that

$$\langle \partial_t \Phi, u \rangle + \langle \partial_x \Phi, F(u) \rangle = 0 \tag{3.2}$$

for all $\Phi \in C_0^\infty([0,\infty) \times \Re^1)$.

As mentioned earlier, we expect (because of examples) many weak solutions for the same initial value, such that we have to find a recipe that will enable us to look for the interesting one. The fundamental idea to this end is to invoke the underlying microscopic structure. On a shock, the second law of thermodynamics must be satisfied, i.e., the physical entropy must increase. Here and in the following, we shall consider kinetic entropies, which are convex functions of the thermodynamical parameters.

Let $S(u)$ be an arbitrary function of the solution u. If u is smooth, we have that

$$\partial_t S(u) + \partial_x G(u) = 0 \tag{3.3}$$

where G must satisfy the consistency condition

$$G'(u) = S'(u)F'(u). \tag{3.4}$$

For weak solutions, the identity (3.3) is not true any longer. We say instead that a weak solution u satisfies an entropy condition if there is at least one convex function S (appropriately called "entropy") such that

$$\partial_t S(u) + \partial_x G(u) \leq 0, \tag{3.5}$$

with G given by (3.4).

A weak solution satisfying (3.5) is called an entropy solution with entropy S. In the case of one-dimensional scalar conservation laws, it is well known that an entropy solution is unique and is identical with the solution found via the vanishing viscosity limit (see Refs. 14 and 19 and Problem 1). Similar results for special systems of two scalar equations of conservation type are also available [12].

The system (1.20), in which we are now interested, eludes such methods for the time being from a rigorous mathematical point of view. However, a formal transition of the ideas is no problem: Defining

$$U = \begin{pmatrix} \rho \\ \rho v \\ \rho(e + \frac{1}{2}v^2) \end{pmatrix}, \tag{3.6}$$

Eq. (1.20) becomes

$$\partial_t U + \text{div} F(U) = 0 \tag{3.7}$$

for some nonlinear function F.

Next, given a pair (S, G), where S is a real convex function $S(U)$ and G is a vector field, we say that such a pair is an entropy pair if

$$\partial_t S(U) + \text{div} G(U) \leq 0 \tag{3.8}$$

and

$$\nabla_U G = \nabla_U S \cdot \nabla_U F. \tag{3.9}$$

Suppose that (ρ, v, T), with $T = \frac{2}{3}e$, is a sufficiently smooth solution of (1.20) or, equivalently, (3.7). Define the "physical entropy" S by

$$S(\rho, T) = \rho \ln \frac{\rho}{T^{\frac{3}{2}}}$$

(this is the usual definition from thermodynamics, except for a minus sign in front—we neglect this minus sign because we want S to be convex, not concave). The Euler equations (1.21) then imply that

$$\partial_t S + \nabla_x \cdot (vS) = 0.$$

S is a convex function of its arguments and therefore qualifies for an entropy functional in the sense described earlier, i.e., we believe that nonsmooth solutions of the Euler equations should satisfy the entropy inequality

$$\partial_t S + \nabla_x \cdot (vS) \leq 0. \tag{3.10}$$

Let us now recall that for the particular case of the Euler equations (1.20), we have a natural underlying microscopic model given by the Boltzmann equation. Thus, beyond the purely nominalistic analogy between the entropy of the conservation laws and the entropy in kinetic theory, it is natural to try to interpret the entropy condition (3.10) in terms of the H-theorem.

We return to the kinetic picture as given by Eq. (1.1). Assume that our initial value $f^\epsilon(0)$ is in the limit $\epsilon \to 0$ approaching a local Maxwellian

$$f^\epsilon(0) \to M \tag{3.11}$$

with initial macroscopic fields ρ_0, v_0 and T_0. Let $f_\epsilon(t)$ be a solution for this initial value; then by the H-theorem (see Chapter 3 Section 4)

(3.12)
$$\epsilon(\partial_t \int f_\epsilon \log f_\epsilon d\xi + \nabla_x \int \xi f_\epsilon \log f_\epsilon d\xi) = -\frac{1}{4}\int dn \int d\xi d\xi_* h_\epsilon(x,\xi,\xi_*,n,t)$$

with

$$h_\epsilon(x,\xi,\xi_*,n,t) = B(|\xi-\xi_*|,n)\log\frac{f'_\epsilon f'_{\epsilon *}}{f_\epsilon f_{\epsilon *}}(f'_\epsilon f'_{\epsilon *} - f_\epsilon f_{\epsilon *}). \tag{3.13}$$

Assume that we consider the Boltzmann equation either on a torus (i.e., on a box with periodic boundary conditions) or in all of $\Re^3$, such that the drift term in Eq. (3.12) will vanish if we integrate over the spatial variable. Then, after integration in x and t, assuming that the system is confined to a three-dimensional torus,

$$\begin{aligned} &4\epsilon(\int\int dx d\xi f^\epsilon(0)\log f^\epsilon(0) - \int\int dx d\xi f_\epsilon(t)\log f_\epsilon(t)) \\ &= \int_0^t ds \int dx \int\int d\xi d\xi_* \int dn h_\epsilon(x,\xi,\xi_*,n,t). \end{aligned} \tag{3.14}$$

Notice that the left-hand side of Eq. (3.14) is $4\epsilon(H(0)-H(t))$. Let us assume that the convergence in (3.11) is so strong that $\epsilon H(0)$ will converge to zero in the limit $\epsilon \to 0$. Then in this limit, it follows that the (non-negative) function h_ϵ satisfies $h_\epsilon \to 0$ almost everywhere in x, ξ, ξ_*, n, t. Therefore, if there is a continuous function f such that $f_\epsilon \to f$ as $\epsilon \to 0$, f must necessarily be a local Maxwellian (cf. Chapter 3). On the other hand, we also have

$$(\psi_\alpha, D_t f) = 0 \tag{3.15}$$

which are then the compressible Euler equations.

Let us now analyze the entropy condition. By assumption, we have pointwise a.e.

$$\partial_t \int f\log f d\xi + \text{div}\int \xi f \log f d\xi \le 0. \tag{3.16}$$

Also, the entropy of the Maxwellian with fields ρ, u, and T is

$$H(M) = \int M\log M d\xi = -\frac{3}{2}\rho[1+\log(2\pi)] + \rho\log(\rho/T^{3/2}). \tag{3.17}$$

Let $S = \rho\log(\rho/T^{\frac{3}{2}})$; then clearly

$$H(M) = S - \frac{3}{2}\rho[1+\log(2\pi)]. \tag{3.18}$$

Inserting M into Eq. (3.16) and using the continuity equation, we get

$$\partial_t S + div(vS) \le 0, \tag{3.19}$$

i.e., we have retrieved (3.10) from the kinetic level. The inequality (3.19) is indeed an entropy inequality in the sense of conservation laws (see Problem 2). Equality holds for smooth solutions.

We have, at a formal level, succeeded in linking the underlying kinetic model and the corresponding conservation laws via entropy functionals. This program has actually been carried out rigorously for scalar conservation laws and suitable simpler kinetic models by Lions, Perthame, and Tadmor [16]. Bardos [2] has collected the many technical properties one would have to prove for solutions of the Boltzmann equation in order to put the above discussion on a more rigorous foundation.

Problems

1. Consider a viscous perturbation of Eq. (3.1),

$$\partial_t u^\epsilon + \partial_x F(u^\epsilon) = \epsilon \partial_{xx} u^\epsilon.$$

The viscous term $\epsilon\partial_{xx}u^\epsilon$ assures the existence of a global smooth solution. Assume that u^ϵ converges to a limit u and prove that this limit satisfies the inequalities (3.5). (Hint: Use that $S''(u^\epsilon)(\partial_x u^\epsilon)^2 \geq 0$, by the convexity assumption.)

2. Find the entropy pair (S, G) for equation (3.19).

11.4 The Hydrodynamical Limit for Short Times

In this section we shall present a rigorous analysis of the Hilbert expansion, which we introduced in Section 2. First we recall the general philosophy underlying this method.

We start from a local Maxwellian with macroscopic parameters ρ_0, T_0, and v_0. It is known (see, for example, Ref. 17) that for a short time $t < t^*$ (which depends on some norms of ρ_0, T_0, and v_0) there exists a classical smooth solution of the Euler equations. We denote this solution by $(\rho(t), T(t), v(t))$ and let $M(x, \xi, t)$ be the time-dependent local Maxwellian with these parameters:

$$M(x,\xi,t) = \frac{\rho(x,t)}{(2\pi T(x,t))^{3/2}} \exp - \frac{|\rho - v(x,t)|^2}{2T(x,t)}. \tag{4.1}$$

Consider now the Boltzmann equation

$$\begin{cases} D_t f = \dfrac{1}{\epsilon} Q(f,f) \\ f(0) = f_0 \equiv M(0). \end{cases} \tag{4.2}$$

$M(t)$ is not expected to be the solution of the initial value problem (4.2), but it is expected to be close to it. If a solution of (4.2) can be expressed in terms of the Hilbert expansion

(4.3)
$$f^\epsilon = \sum_{n=0}^{\infty} \epsilon^n f_n$$

then the term f_n can be calculated by the arguments discussed in Section 2.

Unfortunately, the series (4.3) is not expected to be convergent. Therefore, we replace it, as already suggested at the end of Section 2, by

(4.4).
$$f^\epsilon = \sum_{n=0}^{N} \epsilon^n f_n + \epsilon^m R.$$

Here, the terms f_n are those determined by the Hilbert expansion (such that they are independent of ϵ) from Section 2, and the remainder R is implicitly given by the condition that f^ϵ satisfy the Boltzmann equation. The coefficients N and m are dictated only by the technical considerations that we want to have control (estimates) over R. A possible choice is $m = 3, N = 6$.

Notice that for the truncated Hilbert expansion the last term $f_N (= f_6)$ is not completely known; only its projection onto the subspace orthogonal to the five invariants is determined by the Hilbert expansion! In writing Eq. (4.4), we have tacitly assumed that the other component is zero. This assumption has an effect on the equation for R, which we are now going to derive.

Inserting (4.4) in Eq. (4.2), we obtain the following equation for the remainder R:

(4.5)
$$D_t R = \frac{2}{\epsilon} Q(M, R) + \mathcal{L}_1 R + \epsilon^2 Q(R, R) + \epsilon^2 A,$$

where

(4.6)
$$\mathcal{L}_1 R = 2Q(\sum_{k=1}^{6} \epsilon^{k-1} f_k, R)$$

(4.7)
$$A = -\epsilon D_t f_6 + \sum_{i,j: i+j \geq 7} \epsilon^{i+j-6} Q(f_i, f_j).$$

We associate the initial value $R = 0$ with Eq. (4.5). Notice that Eq. (4.5) is only "weakly" nonlinear and nonhomogeneous because of the factors ϵ^2 in front of $Q(R, R)$ and A.

At first sight we seem to face an easy problem. The factor ϵ^{-1} occurs only in front of the linear Boltzmann operator $2Q(M, \cdot)$, which is "almost" negative definite and therefore not expected to cause problems. However, the operator, which is really negative, is $2M^{-1/2} Q(M, M^{-1/2} \cdot)$, such that if we write $R = hM^{1/2}$ (as we did in Chapter 7) we find, due to the inhomogeneity of M, an extra term

$$(M^{-1/2}D_tM^{1/2})h \tag{4.8}$$

that behaves like $|\xi|^3$ for large velocities. This is a term that is difficult to control. Actually, it has a size larger than $\nu(\xi) \approx |\xi|$ [see the formula (7.2.13)], and we cannot hope to control $2\epsilon^{-1}Q(M,\cdot)$ via the multiplication with ν, as is usually the case when perturbations of Maxwellians are considered (see Chapter 7).

To overcome the difficulty, we introduce another global Maxwellian

$$\hat{M} = \frac{1}{(2\pi\hat{T})^{d/2}}\exp(-\frac{\xi^2}{2\hat{T}}) \tag{4.9}$$

with $\hat{T} > \sup_{x,t} T(x,t)$. ($d$ is the dimension of the physical space, and $T(x,t)$ is the temperature distribution given from a smooth solution of the Euler equations). There is then a constant $c > 0$ such that $\hat{M} \geq cM$. We decompose R as

$$R = M^{1/2}g + \hat{M}^{1/2}h, \tag{4.10}$$

where:

$$\begin{aligned} D_tg &= \frac{1}{\epsilon}Lg + \frac{1}{\epsilon}\chi\sigma^{-1}\hat{K}h \\ D_th &= -\mu\sigma g - \frac{1}{\epsilon}(\nu - \bar{\chi}\hat{K})h + L_1(\sigma g + h) \\ &\quad + \epsilon^2\hat{\nu}\hat{\Gamma}(\sigma g + h, \sigma g + h) + \epsilon^2 a. \end{aligned} \tag{4.11}$$

In (4.11) we have used the following abbreviations:

$$\begin{aligned} Lf &= 2M^{-1/2}Q(M, M^{1/2}f) = (-\nu + K)f \\ \hat{L}f &= 2\hat{M}^{-1/2}Q(M, \hat{M}^{1/2}f) = (-\nu + \hat{K})f \\ \nu &= \int_{\Re^3}\int_{S^2_+} M_*|V\cdot n|d\xi_* dn \\ \hat{\nu} &= \int_{\Re^3}\int_{S^2_+} \hat{M}_*|V\cdot n|d\xi_* dn \\ L_1f &= 2\hat{M}^{-1/2}Q(f_1 + \epsilon f_2 + \epsilon^2 f_3, \hat{M}^{1/2}f) \\ \hat{\nu}\hat{\Gamma}(f,g) &= \hat{M}^{-1/2}Q(\hat{M}^{1/2}f, \hat{M}^{1/2}g) \\ \nu\Gamma(f,g) &= M^{-1/2}Q(M^{1/2}f, M^{1/2}g) \end{aligned} \tag{4.12}$$

and

$$\chi(\xi) = \begin{cases} 1 \text{ if } |\xi| < \xi_0 \\ 0 \text{ otherwise} \end{cases} \tag{4.13}$$

(ξ_0 will be fixed later),

(4.14)
$$\bar{\chi} = 1 - \chi$$

(4.15)
$$\sigma = \sqrt{\frac{M}{\hat{M}}}$$

(4.16)
$$\mu = \frac{1}{2} M^{-1} D_t M$$

(4.17)
$$a = M^{-1/2} A.$$

It is not hard to show that a solution (4.10) of the problem (4.11) solves Eq. (4.5) (see Problem 1). Moreover, the term μ arising from the inhomogeneity of M is controlled by σ, which decays exponentially in $|\xi|^2$. Roughly speaking, we decomposed our solution into a low-velocity part (g) and a high-velocity part (h). The latter is controlled by a global Maxwellian.

We will now make some restrictive assumptions that will enable us to rigorously analyze Eq. (4.11). First, suppose that the dimension of the physical space is $d = 1$, and complement (4.11) with periodic boundary conditions. Note that $d = 1$ means that we have one-dimensional symmetry, i.e., $\rho = \rho(x,t)$, $v = v(x,t)$, $T = T(x,t)$ depend only on one space variable. However, at the level of the microscopic (Boltzmann) description, the velocity ξ continues to be a three-dimensional vector.

To deal with (4.11), we also need some preliminary results, which we list here.

i) It is known (see Ref. 17) that if $v(0)$, $\rho(0)$, and $T(0)$ are in H^s for some sufficiently large s, then there is a time $t^* > 0$ such that there exists a unique classical solution of the Euler equations, with values in H^s, on $[0, t^*)$.

ii) It is not difficult to show, from the definition and properties of $\mathcal{L}_M^{-1}$ (recall that $\mathcal{L}_M = 2Q(M, \cdot)$, and see Section 2), that for all positive integers k

(4.18)
$$\sup_{i \le 6, \xi} f_i M^{-\frac{1}{2}} |\xi|^k \le C_k.$$

iii) From the analysis in Chapter 7 we can extract the following estimates. Let

(4.19)
$$\|f\|_r = \sup_\xi (1 + |\xi|)^r \left(\int |f(x,\xi)|^2 dx \right)^{1/2}$$

(4.20)
$$\|f\|_{r,s} = \sum_{n=0}^{s} \|\nabla^n f\|_r.$$

Then we have

$$\|Kf\|_r \leq C\|f\|_{r-1} \tag{4.21}$$

$$\|Kf\|_0 \leq C\|f\|_{L^2_{x,\xi}} \tag{4.22}$$

$$\|Kf\|_{L^2_{x,\xi}} \leq C\|f\|_{L^2_{x,\xi}} \tag{4.23}$$

$$\|\Gamma(f,g)\|_{r,1} \leq C\|f\|_{r,1}\|g\|_{r,1}. \tag{4.24}$$

We emphasize that (4.24) holds because we only consider $d = 1$ (otherwise we would need a norm $\|\cdot\|_{r,s}$ with $s > d/2$). Note that the same estimates (4.21–24) hold with K and Γ replaced by $\hat{K}$ and $\hat{\Gamma}$.

iv) Finally,

$$\|\frac{1}{\nu}L_1 f\|_r \leq C\|f\|_r. \tag{4.25}$$

This follows from (4.18). Observe that the extra $|\xi|$ produced by the collision operator is compensated by the ν in the denominator.

The preliminary results i)–iv) enable us to prove the following.

(11.4.1) Lemma. *Let b be a known source term, and consider the linear problem*

$$D_t g = \frac{1}{\epsilon}Lg + \frac{1}{\epsilon}\chi\sigma^{-1}\hat{K}h \tag{4.26}$$

$$D_t h = -\mu\sigma g - \frac{1}{\epsilon}(\nu - \bar{\chi}\hat{K})h + L_1(\sigma g + h) + \epsilon^2 b \tag{4.27}$$

for $h(t=0) = g(t=0) = 0$. For $r \geq 3$, the following estimates hold:

$$\sup_{t\leq t^*}\|g\|_{r,1} \leq \epsilon^{1/4}\sup_{t\leq t^*}\|\frac{b}{\nu}\|_{r+1,1} \tag{4.28}$$

$$\sup_{t\leq t^*}\|h\|_{r,1} \leq \epsilon^{5/4}\sup_{t\leq t^*}\|\frac{b}{\nu}\|_{r,1}. \tag{4.29}$$

We postpone the proof of this lemma until the end of this section. First, we show how the estimates (4.28–29) lead to our principal result. Indeed, abbreviate the nonlinearity and inhomogeneity in (4.11) by b, i.e.,

$$b = \hat{\nu}\hat{\Gamma}(\sigma g + h, \sigma g + h) + a.$$

The estimates (4.18), the fact that H^1 is a Banach algebra in one dimension (i.e., $\|fg\|_{H^1} \leq \|f\|_{H^1}\|g\|_{H^1}$), and the observation that σ is exponentially decaying as a function of ξ^2 together imply an estimate

$$\|\frac{1}{\nu}b\|_{r,1} \leq C\left(\|g\|_{0,1}^2 + \|h\|_{r,1}^2 + 1\right). \tag{4.30}$$

Combining this with (4.28) and (4.29), we find

$$\|g\|_{r,1} \leq C\epsilon^{1/4} \sup_{t\leq t^*}(1 + \|g\|_{0,1}^2 + \|h\|_{r+1,1}^2) \tag{4.31}$$

$$\|h\|_{r+1,1} \leq C\epsilon^{5/4} \sup_{t\leq t^*}(1 + \|g\|_{0,1}^2 + \|h\|_{r+1,1}^2). \tag{4.32}$$

We can now use (4.32) to estimate $\|h\|_{r+1,1}$ in terms of $\|g\|_{0,1}$. By inserting the result into (4.31), one arrives at an a priori bound for $\|g\|_{r,1}$. This done, we are in a position to prove the existence of a unique solution (g, h) of the problem (4.11), continuous on $[0, t^*]$, with values in the Banach space endowed with the norm $\|\cdot\|_{r,1}$. The result follows.

(11.4.2) Lemma. *Consider the initial value problem associated with Eqs. (4.11) with the initial value $g = h = 0$. Then there exists a unique solution $(g, h) \in C[0, t^*]$.*

The analysis outlined here permits us to prove boundedness of the remainder R in (4.4). Our main theorem in this section is then just a simple corollary of the last lemma.

(11.4.3) Theorem. *There exists some $\epsilon_0 > 0$ such that for all $\epsilon < \epsilon_0$, there is a unique solution f^ϵ of the Boltzmann equation (4.2) with*

$$\sup_{t\leq t^*} \|f^\epsilon - M(t)\|_{L^2_{x,\xi}} < C\epsilon. \tag{4.33}$$

Remarks.

i) The convergence expressed by (4.33) can be improved.
ii) The dependence on only one space variable enters only in the estimates (4.28) and (4.29), where we have to control only one derivative. To deal with the full three–dimensional problem, we would have to consider Sobolev spaces with higher indices, which would make the estimates for the linear problem (4.26–27) more involved.
iii) The technical difficulties of the approach presented here are contained in the proof of Lemma 11.4.1; we give a sketch of this proof at the end of this section.

iv) The choice of the parameters $m = 3$, $N = 6$ is dictated by technical reasons. One chooses the lowest numbers for which the linear problem is manageable.

v) The preceding discussion was first carried out by R. Caflisch [6], who was the first to obtain a rigorous result on the hydrodynamical limit of the Boltzmann equation for times of existence of smooth solutions of the Euler equations. This result has been generalized along the same lines by M. Lachowicz [15], who discussed the three–dimensional case and the initial layer. More specifically, this means that if the initial condition f_0 is not a Maxwellian, we expect the solution of the Boltzmann equation to approach a local Maxwellian on a shorter time scale proportional to ϵt; this local Maxwellian then evolves on a slower time scale according to the fluid dynamical equations.

vi) Other results concerning the fluid dynamical limit of the Boltzmann equation, in terms of the compressible Euler equations, are due to Nishida [18] and Ukai and Asano [21]. Nishida used the Cauchy–Kowalewskaya theorem to reach a result similar to the one discussed here. Ukai and Asano used spectral methods to generalize to a more abstract setting.

vii) An interesting but essentially open question is that of a rigorous description of a regime in which the compressible Navier–Stokes equations are obtained from the Boltzmann equation. We will return to this problem in the next section, in connection with the incompressible scaling.

We conclude the present section with a sketch of the proof of Lemma 11.4.1. For more details, see Ref. 6.

First, we integrate (4.27). Denoting by $\bar{\nu}$ a lower bound of ν, we can estimate

$$\begin{aligned} h(x,\xi,t) = \int_0^t e^{-\frac{\nu(\xi)}{\epsilon}(t-s)}[-\mu\sigma g - \frac{1}{\epsilon}\bar{\chi}\hat{K}h \\ + L_1(\sigma g + h) + \epsilon^2 b](x-(t-s)\xi,\xi,s)\,ds. \end{aligned} \tag{4.34}$$

This representation entails

(4.35)

$$\begin{aligned} \sup_{t\le t^*}\|h\|_r &\le \sup_{t\le t^*}\left(\epsilon\|\frac{\mu\sigma g}{\nu}\|_r + \|\frac{\bar{\chi}\hat{K}h}{\nu}\|_r + \epsilon\|\frac{L_1(\sigma g+h)}{\nu}\|_r + \epsilon^3\|\frac{b}{\nu}\|_r\right) \\ &\le \epsilon\sup_{t\le t^*}\|g\|_0 + \frac{1}{C\xi_0}\sup_{t\le t^*}\|h\|_r + \epsilon\sup_{t\le t^*}(\|g\|_0 + \|h\|_r) + \epsilon^3\|\frac{b}{\nu}\|_r. \end{aligned}$$

The last estimate follows from the decay properties of σ and (4.25). By choosing ξ_0 large enough, we can solve the last inequality to get

$$\sup_{t\le t^*}\|h\|_r \le C\left(\epsilon\sup_{t\le t^*}\|g\|_0 + \epsilon^3\sup_{t\le t^*}\|\frac{b}{\nu}\|_r\right). \tag{4.36}$$

Similarly, we can differentiate Eq. (4.27) with respect to x and rewrite the result in integrated form. Estimating as above we get

$$\sup_{t\le t^*}\|h\|_{r,1}\le C\left(\epsilon\sup_{t\le t^*}\|g\|_{0,1}+\epsilon^3\sup_{t\le t^*}\|\frac{b}{\nu}\|_{r,1}\right). \tag{4.37}$$

To estimate g, one proceeds as before. After repeating the estimates from above one finds

$$\sup_{t\le t^*}\|g\|_r\le\sup_{t\le t^*}\|Kg\|_r+\sup_{t\le t^*}\|\chi\sigma^{-1}\hat{K}h\|_r. \tag{4.38}$$

To continue, note that by (4.36) and for $r>0$

$$\begin{aligned}\|\sigma^{-1}\chi\hat{K}h\|_r&\le e^{c\xi_0^2}\|\hat{K}h\|_r\\&\le e^{c\xi_0^2}\|h\|_{r-1}\\&\le e^{c\xi_0^2}\left(\epsilon\sup_{t\le t^*}\|g\|_0+\epsilon^3\sup_{t\le t^*}\|\frac{b}{\nu}\|_{r-1}\right)\\&\le\epsilon^{\frac12}\sup_{t\le t^*}\|g\|_0+\epsilon^{\frac52}\sup_{t\le t^*}\|\frac{b}{\nu}\|_{r-1},\end{aligned} \tag{4.39}$$

for sufficiently small ϵ. Combining (4.38) and (4.39), we arrive at

$$\sup_{t\le t^*}\|g\|_r\le\sup_{t\le t^*}\|g\|_{r-1}+\epsilon^{\frac52}\sup_{t\le t^*}\|\frac{b}{\nu}\|_{r-1}. \tag{4.40}$$

For $r=0$ we use the continuity of $\hat{K}$ with respect to $\|\cdot\|_0$ and (4.22) to get

$$\sup_{t\le t^*}\|g\|_0\le\sup_{t\le t^*}\|g\|_{L^2_{x,\xi}}+\epsilon^{\frac52}\sup_{t\le t^*}\|\frac{b}{\nu}\|_0. \tag{4.41}$$

On the other hand, by the energy estimate, and using that L is negative,

$$\frac12\frac{d}{dt}\|g\|^2_{L^2_{x,\xi}}\le\frac{1}{\epsilon}\|\sigma^{-1}\chi\hat{K}h\|_{L^2_{x,\xi}}\|g\|_{L^2_{x,\xi}} \tag{4.42}$$

and by (4.39)

$$\|\sigma^{-1}\chi\hat{K}h\|_{L^2_{x,\xi}}\le\|\sigma^{-1}\chi\hat{K}h\|_2\le C\left(\epsilon\sup_{t\le t^*}\|g\|_0+\epsilon^3\sup_{t\le t^*}\|\frac{b}{\nu}\|_1\right),$$

from which

$$\frac{d}{dt}\|g(t,\cdot)\|_{L^2_{x,\xi}}\le C\left(\sup_{\tau\le t}\|g(\tau,\cdot)\|_0+\epsilon^2\sup_{\tau\le t}\|\frac{b(\tau,\cdot)}{\nu}\|_1\right). \tag{4.43}$$

Thanks to (4.41), the estimate (4.43) allows us to bound $\|g(t,\cdot)\|$ in terms of $\sup_{t\le t^*}\|\frac{b}{\nu}\|_1$. As a final result, we have

$$\sup_{t\le t^*}\|g\|_r \le \epsilon^{\frac{3}{2}}\sup_{t\le t^*}\|\frac{b}{\nu}\|_{r-1}. \tag{4.44}$$

In order to derive the estimates (4.28–29), one has to differentiate Eq. (4.26) and proceed similarly.

Problems

1. Check that equation (4.11) implies that R, defined in Eq. (4.10), satisfies (4.5).
2. Prove (4.18).
3. Derive the estimate (4.37).
4. Prove (4.44).

11.5 Other Scalings and the Incompressible Navier-Stokes Equations

At the end of Chapter 3 and in the previous sections we have seen how a pure space-time scaling of the variables in the Boltzmann equation leads naturally to the hydrodynamic regime described by the compressible Euler equations. However, other scalings are also possible. For example, if we denote by (q,τ) the microscopic space and time variables (those entering in the Boltzmann equation) and by (x,t) the macroscopic variables (those entering in the fluid dynamical description), we can study scalings of the kind

$$q = \epsilon^{-1}x \tag{5.1}$$

$$\tau = \epsilon^{-\alpha}t \tag{5.2}$$

where $\alpha\in[1,2]$. For $\alpha=1$, this reduces to the compressible scaling we have considered so far. If $\alpha>1$, we are looking at larger "microscopic" times. We now investigate the limiting behavior of solutions of the Boltzmann equation in this limit.

Notice first that the compressible Euler equations (1.21) are invariant with respect to the scaling $t\to\epsilon^{-1}t$, $x\to\epsilon^{-1}x$. To investigate how these equations change under the scalings (5.1–2), let

$$\begin{aligned} v^\epsilon(x,t) &= \epsilon^\gamma v(\epsilon^{-1}x,\epsilon^{-\alpha}t), \quad \gamma=\alpha-1\\ \rho^\epsilon(x,t) &= \rho(\epsilon^{-1}x,\epsilon^{-\alpha}t)\\ T^\epsilon(x,t) &= T(\epsilon^{-1}x,\epsilon^{-\alpha}t)\end{aligned} \tag{5.3}$$

where (ρ, v, T) solve the compressible Euler equations (1.21). We easily obtain

$$\partial_t \rho^\epsilon + \operatorname{div}(\rho^\epsilon v^\epsilon) = 0 \tag{5.4}$$

$$\partial_t v^\epsilon + (v^\epsilon \cdot \nabla) v^\epsilon = -\frac{\nabla p^\epsilon}{\rho^\epsilon} \cdot \epsilon^{2(1-\alpha)} \tag{5.5}$$

$$\partial_t T^\epsilon + (v^\epsilon \cdot \nabla) T^\epsilon + \frac{2}{3} T^\epsilon (\nabla \cdot v^\epsilon) = 0. \tag{5.6}$$

The scaling of the velocity field v^ϵ in (5.3) is done in a dimensionally consistent way.

We expect that the hydrodynamical limit of the Boltzmann equation under the scaling (5.1–2) will be given by the asymptotic behavior of $(\rho^\epsilon, v^\epsilon, T^\epsilon)$, satisfying (5.4–6), in the limit $\epsilon \to 0$. We will now investigate this limit from a formal point of view.

To this end, let $\eta = \epsilon^{2(\alpha-1)}$ and expand

$$\begin{aligned} v^\eta \equiv v^\epsilon &= v_0 + \eta v_1 + \eta^2 v_2 + \dots \\ \rho^\eta \equiv \rho^\epsilon &= \rho_0 + \eta \rho_1 + \eta^2 \rho_2 + \dots \\ T^\eta \equiv T^\epsilon &= T_0 + \eta T_1 + \eta^2 T_1 + \dots . \end{aligned}$$

If we collect the terms of order η^{-1} in (5.5), we have

$$\nabla p_0 = 0 \tag{5.7}$$

and the terms of order η^0 give

$$\begin{aligned} \partial_t \rho_0 + \operatorname{div}(v_0 \rho_0) &= 0 \\ \partial_t v_0 + (v_0 \cdot \nabla) v_0 &= -\frac{\nabla p_1}{p_0} \\ \partial_t T_0 + (v_0 \cdot \nabla) T_0 + \frac{2}{3} T_0 (\nabla \cdot v_0) &= 0. \end{aligned} \tag{5.8}$$

From (5.7) and the perfect gas law $p_0 = \rho_0 T_0$, which we assume to hold at zeroth order; it follows that $\rho_0 T_0$ is constant as a function of the spatial variable. The first and third equations in (5.8) now imply

$$\partial_t (\rho_0 T_0) + \operatorname{div}(\rho_0 T_0 v_0) = -\frac{2}{3} \rho_0 T_0 \operatorname{div} v_0. \tag{5.9}$$

As $\rho_0 T_0$ is only a function of t, say $A(t)$, Eq. (5.9) implies that

$$A'(t) = -\frac{5}{3} A(t) \operatorname{div} v_0. \tag{5.10}$$

Under suitable assumptions Eq. (5.10) implies that $A'(t) = 0$ and then $\mathrm{div} v_0 = 0$. This is the case, for example, if we are in a box with nonporous walls, because then the normal component of v_0 is 0 and we can use the divergence theorem. A similar argument applies to the case of a box with periodicity boundary conditions. Or, if we are in all of $\Re^3$ and v_0 decays fast enough at infinity.

Assuming that we have conditions that imply that $\mathrm{div} v_0 = 0$, we easily get from the continuity equation that ρ_0 will be independent of x if the initial value is, and the same for T_0. But with this knowledge Eqs. (5.8) actually entail that ρ_0 and T_0 are constant.

Therefore, if the initial condition is "well prepared" in the sense that $u^\epsilon(x, 0)$ is a divergence–free vector field and $\rho^\epsilon(x, 0)$, $T^\epsilon(x, 0)$ are constant, we expect as a first-order approximation for $\rho^\epsilon, v^\epsilon, T^\epsilon$ the solutions of the equations

$$\mathrm{div} v_0 = 0$$

$$\partial_t v_0 + (v_0 \cdot \nabla) v_0 = -\frac{\nabla p_1}{\rho_0}$$

which are the incompressible Euler equations. This limit, known as "low velocity limit," is well known at the macroscopic level. We refer to Majda's book [17] for references and a detailed discussion. The variable η^{-1} enters into the theory as the square of the speed of sound. If this parameter is large compared to typical speeds of the fluid, then the incompressible model is well suited to describe the time evolution, provided that the initial velocity field is divergence–free and the initial density and temperature are constant.

Let us come back to the kinetic picture as described by the Boltzmann equation. The above discussion suggests that if $\alpha \in (1, 2)$, then in the scaling (5.1–3) the solutions of the Boltzmann equation will converge to a local Maxwellian whose parameters satisfy the incompressible Euler equations. This assertion can actually be proved rigorously (we will give some references later).

For $\alpha = 2$ something special happens. Of course, the incompressible Euler equations are invariant under the scaling (5.1–3); however, for $\alpha = 2$ the incompressible Navier-Stokes equations

$$\partial_t v + (v \cdot \nabla) v = -\nabla p + \nu \Delta v \tag{5.11}$$

$$\nabla \cdot v = 0 \tag{5.12}$$

are also invariant under the same scaling. It is therefore of great interest to understand whether the Boltzmann dynamics "chooses" in this limit the Euler or the Navier-Stokes evolution.

It is remarkable that the answer is Navier-Stokes. In other words, considering larger times than those typical for Euler dynamics ($\epsilon^{-2} t$ instead of $\epsilon^{-\alpha} t$, $\alpha < 2$), dissipative effects become non-negligible. For illustration, consider two layers of fluid moving with velocity v and $v + \delta v$. Suppose that we want to decide whether there is any momentum transfer between these layers (which is expected for the Navier–Stokes equations, but not for the Euler equations). The momentum transfer can in principle be affected by the thermalization contained in the Boltzmann collision term, but heuristic scaling arguments show that it is proportional to $\epsilon^{\alpha-2}\Delta v$, such that it remains only relevant for $\alpha = 2$.

These considerations can be put on a rigorous basis, and the viscosity coefficient $\nu > 0$ can be computed in terms of kinetic expressions. The incompressible Navier-Stokes equations can be derived from the Boltzmann equation if the time interval is such that smooth solutions of the hydrodynamical equations exist. The tool that yields this result is a Hilbert expansion similar to that discussed in Section 4, and one gets local convergence for the general situation (see De Masi et al. [11]) or global and uniform convergence if the data are small in a suitable sense (see Bardos and Ukai [5]). The initial layer can also be included (see Asano [1]). A different approach, closer in spirit and methodology to the arguments of Section 3, has been proposed and followed by Bardos et al. [3,4] to show that the cluster points of the global solutions for the Boltzmann equation given by the DiPerna-Lions existence theorem (see Section 5.3) are, under the scaling (5.1–3), identifiable with the weak (Leray) solutions of the incompressible Navier-Stokes equations. Unfortunately, some gaps remain in the rigorous treatment of the latter problem, related to the lack of strong energy conservation in the present general existence theory of the Boltzmann equation.

The situation is much more complicated if we want to investigate the relationship between the Boltzmann equation and the *compressible* Navier–Stokes equations. The first problem we face is that this case is not naturally recovered by a simple space-time scaling. The dissipative term in the compressible Navier-Stokes equations is usually interpreted as a correction of order ϵ to the inertial terms in the compressible Euler equations. This is what is also obtained by a formal expansion due to Enskog, usually called the Chapman–Enskog expansion (see Refs. 7,8, and 9 for details). The idea behind this expansion is that the functional dependence of f on the local density velocity and internal energy can be expanded into a power series. Although there are many formal similarities with the Hilbert expansion, the procedure is rather different. As remarked by Cercignani [8] (see also Ref. 7), the Chapman-Enskog procedure seems to introduce spurious solutions, especially if one looks for steady states. This is essentially due to the fact that one really considers infinitely many time scales (of orders $\epsilon, \epsilon^2, \ldots, \epsilon^n, \ldots$). Cercignani[8] introduces only two time scales (of orders ϵ and ϵ^2) and is able to recover the compressible Navier-Stokes equations. In order to explain the idea, we remark that the Navier-Stokes equations describe two kinds of processes, convection and diffusion, which act on two different time scales. If we consider only the first scale we obtain the compressible Euler equations; if we insist on the second we can obtain the Navier-Stokes equations only at the price of losing compressibility. If we want both compressibility and diffusion, we have to keep both scales at the same time and think of f as

$$f(x, \xi, t) = f(\epsilon x, \xi, \epsilon t, \epsilon^2 t).$$

This enables us to introduce two different time variables $t_1 = \epsilon t$, $t_2 = \epsilon^2 t$ and a new space variable $x_1 = \epsilon x$ such that $f = f(x_1, \xi, t_1, t_2)$. The fluid dynamical variables are functions of x_1, t_1, t_2, and for both f and the fluid dynamical variables the time derivative is given by

$$\frac{\partial f}{\partial t} = \epsilon \frac{\partial f}{\partial t_1} + \epsilon^2 \frac{\partial f}{\partial t_2}. \tag{5.13}$$

In particular, the Boltzmann equation can be rewritten as

$$\epsilon \frac{\partial f}{\partial t_1} + \epsilon^2 \frac{\partial f}{\partial t_2} + \epsilon \xi \cdot \nabla_{x_1} f = Q(f, f).$$

If we expand f formally in a power series in ϵ, we find that at the lowest order f is Maxwellian. The compatibility conditions at the first order give that the time derivatives of the fluid dynamic variables with respect to t_1 is determined by the Euler equations, but the derivatives with respect to t_2 are determined only at the next level and are given by the terms of the compressible Navier–Stokes equations describing viscosity and heat conduction. The two contributions are, of course, to be added as specified by (5.13) in order to obtain the full time derivative and thus write the compressible Navier-Stokes equations.

However, this formal expansion has not been justified so far with rigorous arguments as those discussed in Section 4.

References

1. K. Asano, "The fluid dynamical limit of the Boltzmann equation," in preparation.
2. C. Bardos, "Une interpretation des relations existant entre les équations de Boltzmann, de Navier–Stokes et d'Euler à l'aide de l'entropie," *Math. Applic. Comp.* **6 (1)**, 97–117 (1987).
3. C. Bardos, F. Golse, and D. Levermore, "Fluid dynamical limits of kinetic equations, I: Formal Derivation," *J. Stat. Phys.* **63**, 323–344 (1991).
4. C. Bardos, F. Golse, and D. Levermore, "Fluid dynamical limits of kinetic equations, II: Convergence proofs," preprint (1992).
5. C. Bardos and S. Ukai, "The classical incompressible limit of the Boltzmann equation," preprint, Tokyo Institute of Technology (1992).
6. R. Caflisch, "The fluid dynamical limit of the nonlinear Boltzmann equation," *Commun. Pure Appl. Math.* **33**, 651–666 (1980).
7. C. Cercignani, *The Boltzmann Equation and Its Applications,* Springer-Verlag, New York (1988).
8. C. Cercignani, *Mathematical methods in kinetic theory,* Plenum Press, New York, 2nd ed. (1990).
9. S. Chapman and T. G. Cowling, *The Mathematical Theory of Nonuniform Gases,* Cambridge University Press, Cambridge (1952).

10. A. De Masi, N. Ianiro, A. Pellegrinotti, and E. Presutti, "A survey of the hydrodynamical behavior of many particle systems," *Studies in Statistical Mechanics* **11** (W. Montroll and J. L. Lebowitz, eds.), 123–294, North Holland (1984).
11. A. De Masi, R. Esposito, and J. L. Lebowitz, "Incompressible Navier–Stokes and Euler limit of the Boltzmann equation," *Commun. Pure Appl. Math.* **42**, 1189–1214 (1989).
12. R. J. DiPerna, "Convergence of approximate solutions to conservation laws," *Arch. Rat. Mech. Anal.* **82 (1)**, 27–70 (1983).
13. D. Hilbert, *Grundzüge einer allgemeinen Theorie der linearen Integralgleichungen,* Chelsea, New York (1953).
14. S. N. Krushkov, "First order quasilinear equations in several variables," *USSR Sb.* **10 (2)**, 217–243 (1970).
15. M. Lachowicz, "On the initial layer and the existence theorem for the nonlinear Boltzmann equation," *Math. Methods Appl. Sci.* **9 (3)**, 342–366 (1987).
16. P. Lions, B. Perthame, and E. Tadmor, "Formulation cinétique des lois de conservation scalaires," *C.R.A.S. Paris* **312, I**, 97–102 (1991).
17. A. Majda, *Compressible Fluid Flow and Systems of Conservation Laws in Several Space Variables,* Springer-Verlag, Appl. Math. Ser. 53, (1984).
18. T. Nishida, "Fluid dynamical limit of the nonlinear Boltzmann equation at the level of the compressible Euler equations," *Commun. Math. Phys.* **61**, 119–148 (1978).
19. J. Smoller, *Shock Waves and Reaction–Diffusion Equations,* Springer-Verlag (1982).
20. H. Spohn, *Large Scale Dynamics of Interacting Particles,* Springer-Verlag, New York (1991).
21. S. Ukai and K. Asano, "The Euler limit and initial layer of the nonlinear Boltzmann equation," *Hokkaido Math. J.* **12**, 303–324 (1983).
22. S. R. S. Varadhan, "Entropy methods in hydrodynamic scaling," in *Nonequilibrium Problems in Many-Particle Systems,* Lecture Notes in Mathematics **1551**, 112-145, Springer-Verlag (1993).

12
Open Problems and New Directions

The first eleven chapters of this book comprise a collection of much of what we (the authors) know about the Boltzmann equation for hard spheres. In this last chapter, we want to revisit some of the questions addressed in the earlier chapters and discuss some possible further developments.

The first problem that we point out here as a worthwhile direction of research is the validation (local or global) of the Boltzmann equation for repulsive interaction kernels other than hard spheres. There is an unpublished result about this question, due to F. King[17], but apart from the fact that this paper is not widely disseminated, it applies only to smooth repulsive potentials and contains additional restrictive assumptions. There is a good chance to do the validation of the Boltzmann equation for more general hard potentials following the procedure from Chapter 4. One of the first major obstacles to this end will be the fact that the BBGKY hierarchy will look different—multiple collisions can no longer be neglected from the very beginning, because the particles are now mass points interacting via a strong short-range potential; there will be a non–negligible probability to have three, four, or even more particles "in collision" at a given time, and the terms accounting for these multiple collisions are only expected to disappear in the Boltzmann–Grad limit.

For long-range potentials, the situation is even more complicated and challenging. At the level of the Boltzmann equation, the gain and loss terms of the collision operator do not even make sense separately (at least not without assuming an angular cutoff), and it would certainly be interesting to understand this property at the level of the underlying particle system.

A related question is that of a rigorous derivation of the Fokker–Planck equation from the Boltzmann equation for Coulomb interactions. The right formal scaling has been outlined by Degond [14], but a rigorous proof is still missing. We must however mention a result due to Arsen'ev,[3] where the problem has been solved for interaction potentials other than the Coulomb potential.

Returning to the validation problem for the Boltzmann equation, we also mention that we still have to fully understand how the Boltzmann–Grad limit (or other kinetic scalings) applies to quantum models. This problem is far from being academic: The quantum Boltzmann equation is used in applications, e.g., in the theory of semiconductors. A good reference for these problems is a recent article by Spohn [22].

We turn to existence and uniqueness questions. The biggest open challenge is undoubtedly the uniqueness question. The situation here is similar to the one for the Navier–Stokes equations. The problem is that the method used by DiPerna and Lions (presented here in Section 5.3) is based on compactness arguments and therefore is not constructive. A proof of uniqueness would in all likelihood also give automatic results on smoothness and qualitative properties of the solution. At present, the search for new analytical tools to this end is on, but no success is in sight.

The result of DiPerna and Lions described in Chapter 5 can be considered the starting point of many developments. We saw one of these in Chapter 9, where the result was extended to the case of a bounded domain with isothermal boundaries at rest. An interesting extension concerns nonisothermal and moving boundaries. The case of an inhomogeneous temperature along a boundary at rest has been treated by L. Arkeryd and C. Cercignani[1], as mentioned in Chapter 9, with the help of a cutoff for large velocities. The removal of this cutoff would provide a significant extension of the results available on the Boltzmann equation.

Another important extension concerns the case of $\Re^3$, originally considered by DiPerna and Lions, if instead of looking at functions of L^1 we look at functions that are only locally in L^1 in x, but tend to a Maxwellian as x tends to infinity. In fact an extension of this kind would pave the way to two further extensions, i.e. the case of the flow past an obstacle, that we considered in Chapter 9 in the particular case of solutions close to a Maxwellian, and the case of $\Re^3$ in which the solution tends not to a Maxwellian but to a general L^1_ξ function (with finite energy and entropy as x goes to infinity). These extensions, although far from trivial, do not appear to be out of reach of the available tools. As this book goes to print, the authors learned that P. L. Lions has made progress in this direction [19].

In the theory of the spatially homogeneous Boltzmann equation, one of the most promising developments is the introduction into the field of methodology from information theory, as successfully done in recent work by Carlen and Carvalho [10]; they succeeded in obtaining the best available decay estimates for the H-functional in the spatially homogeneous case.

An exciting new tool was introduced into kinetic theory by M. Bony in 1987 [8]; for discrete velocity models in one space dimension without boundary,

$$(\partial_t + c_i \partial_x) f_i = \sum_{j,k,l} A_{ij}^{kl} \left(f_k f_l - f_i f_j \right),$$

$i, j, k, l = 1, \ldots, n$, Bony saw that the functional

$$I[f](t) = \sum_{i,j} \int_x \int_{y<x} (c_i - c_j) f_i(x,t) f_j(y,t) \, dy \, dx$$

satisfies an inequality

$$I[f](t) - I[f](0) = -\int_0^t \int_{-\infty}^{\infty} \sum_{i,j} (c_i - c_j)^2 f_i(x,s) f_j(x,s) \, dx \, ds \tag{1.1}$$

and is therefore monotonically decreasing. The equality (1.1) follows by differentiation of $I[f](t)$ and skillful use of mass and momentum conservation. Since the left-hand side of (1.1) is evidently bounded by the mass conservation law, the right-hand side, which consists of terms closely resembling the collision terms, is also bounded. Bony was able to use this to give a very elegant proof of global existence, uniqueness, and the asymptotic behavior as $t \to \infty$; the same results were obtained earlier, with a more complicated method, by Beale [5]. Bony called the functional $I[f]$ "potential for interaction," and his proof made no use of the H-theorem or the energy conservation law. The idea was in a sense also an application of tools from another field to kinetic theory, because his "potential for interaction" is very similar to the potential for interaction introduced in the 1960s by Glimm [15] to solve scalar conservation laws by the famous random choice method. There is a simple analogue to the functional I for the full Boltzmann equation in one dimension, and Cercignani [11] has pointed out a way to combine the information given by the decay of this functional with the decrease of the H-functional to prove classical global existence for the Boltzmann equation in one space dimension. A nice summary of Bony's work can be found in Ref. 7.

The most exciting questions related to this approach are: Are there potentials of interaction in more than one space dimension, and what do they look like? What happens if the gas is confined to a bounded domain? We have to mention at this point that for discrete velocity models in more than one dimension the state of the art of the existence theory is even more primitive than for the full Boltzmann equation. Indeed, the velocity–averaging lemma, which played such a fundamental role in the DiPerna–Lions result, is not applicable to discrete velocity models, and there is no general global existence theorem for discrete velocity models if the dimension is larger than one. One can speculate that this difficulty is related to the pathological behavior of discrete velocity models demonstrated in Chapter 4 Appendix C (Uchiyama's example).

A wide open field for research was briefly touched at the end of Chapter 9, where we mentioned the few available results on solutions of boundary value problems for the steady Boltzmann equation if the boundary values are far from equilibrium. For the problem where the gas is confined to a slab $0 \leq x \leq a$ and we have spatial dependence on x only, we know that there are solutions[2] (in a measure sense) if all the collisions between particles for which at least one of the (pre- or post-collisional) velocities has a small x-component are disregarded. The necessity for such a truncation arises from the structure of the equation itself, namely,

$$\xi_1 \frac{d}{dx} f = Q(f, f).$$

Clearly, the problem has no natural scale. Even a narrow slab will seem large for particles moving with small horizontal speed ξ_1. Exempting such particles from collisions turned out to make the problem accessible to existing fixed point theorems [2].

However, this is clearly an unphysical truncation, and a number of questions arise: Are there ways to remove the truncation? Even with the truncation, is the solution unique? And if it is not, which solutions are physically reasonable? How do we find these?

Another extension goes in the direction of studying the layers that arise near the boundaries or inside the gas (shock layers) when the Knudsen number is small; then outside this layer some kind of fluid dynamic equations hold, as indicated in Chapter 11, but this is not true if gradients on the scale of the mean free path occur locally, as in the mentioned layers. This aspect of the theory of the Boltzmann equation is treated, for instance, by C. Cercignani [12], but rigorous results are lacking. Even the limiting problems that arise when trying to study these layers (for example, the Kramers problem, the problem of the shock structure), have not been studied exhaustively, in spite of the fact that their solutions depend on just one space variable (plus the velocity variables). The main difficulty is again related to the ξ factor in front of the space derivative. As for the case of a shock layer, the only rigorous work seems to be due to Nicolaenko[21], who dealt with the existence of weak shock waves, while the half-space problems that arise in connection with the Knudsen boundary layers have been treated at the linearized level by Bardos, Caflisch, and Nicolaenko[4], Cercignani,[13] and Maslova [20].

Numerical experiments giving at least some insight to possible answers to these questions are desirable but difficult. This brings us to our next group of problems.

The field of numerical simulation techniques for the Boltzmann equation has seen a lot of activity in the last ten years; this activity was largely spawned by the emerging availability of sufficient and inexpensive computing power, but also by the needs of the aerospace industries. At the same time, the connections between simulation schemes, their interpretation as stochastic processes, and the Boltzmann equation was intensively investigated. For the future, some of the more mathematical problems associated with numerical solutions of the Boltzmann equation are

A) the development of more accurate implicit schemes for the spatially homogeneous equation. In view of the result by Bogomolov [6] mentioned in Chapter 10, a more accurate approximation of the homogeneous equation can significantly improve the accuracy with which one can solve a full nonhomogeneous problem. We suggest that the idea of returning particles with their post–collisional velocities into the sample and letting them be candidates for a second collision (see the discussion of Bird's scheme in Section 10.3), improves this accuracy, but there is no proof for this. R. Caflisch and G. Russo [9] have suggested implicit and semi–implicit schemes for kinetic equations; the problems one encounters is that such schemes cannot be easily designed such that positivity of the solutions and the usual conservation laws are guaranteed.

B) to obtain error estimates for the available schemes (or, for that matter, any schemes).

C) to investigate the links between particle simulation and the validation of the Boltzmann equation as done in Chapter 4. The similarities in the derivation of the equation and the derivation of simulation schemes suggest that there are intimate links between the two problems. In fact, it may be possible to deduce convergence of certain schemes from validity theorems. Also, the emergence of molecular chaos happens on each level, and it seems desirable to obtain a more profound understanding of the meaning of molecular chaos for numerical simulation.

D) finally, to explore the possibilities of combining particle simulation schemes for the Boltzmann equation with other approximations, like, e.g., discrete velocity models. Tentative steps in this direction were taken in Ref. 16, where the concept of a "random discrete velocity model" was introduced and shown to be consistent with the Boltzmann equation. There may well be other more efficient ways to solve the Boltzmann equation numerically; the five-dimensional integral in the collision operator is, as described at the beginning of Chapter 10, a serious mathematical obstacle to the accurate numerical solution; at the same time, the many dimensions in the equation offer many ways to approximate the collision integral, with the final target of more efficient numerical methods in kinetic theory on one hand and of a unified treatment of various approximations on the other.

References

1. L. Arkeryd and C. Cercignani, "A global existence theorem for the initial-boundary value problem for the Boltzmann equation when the boundaries are not isothermal," *Arch. Rat. Mech. Anal.* **125**, 271–288 (1993).
2. L. Arkeryd, C. Cercignani, and R. Illner, "Measure solutions of the steady Boltzmann equation in a slab," *Commun. Math Phys.* **142**, 285–296 (1991).
3. A. A. Arsen'ev, and O. E. Buryak, "On the solution of the Boltzmann equation and the solution of the Landau–Fokker–Planck equation," *Math. USSR Sbornik* **69, 2**, 465–478 (1991).
4. C. Bardos, R. Caflisch, and B. Nicolaenko, *Commun. Pure Appl. Math.* **22**, 208 (1982).
5. T. Beale, "Large–time–behavior of discrete Boltzmann equations," *Commun. Math. Phys.* **106**, 659–678 (1986).
6. S. V. Bogomolov, "Convergence of the total–approximation method for the Boltzmann equation," *U.S.S.R. Comput. Math. Phys.* **28 (1)**, 79–84 (1988).
7. M. Bony, "Existence globale et diffusion en théorie cinétique discrète," in *Advances in Kinetic Theory and Continuum Mechanics*, R. Gatignol and Soubbaramayer eds., 81–90, Springer, Berlin (1991).
8. M. Bony, "Solutions globales bornées pour les modèles discrets de l'équation de Boltzmann en dimension 1 d'espace," *Actes Journées E.D.P. St. Jean de Monts* **16** (1987).
9. R. Caflisch and G. Russo, "Implicit methods for collisional kinetic equations," *Proceedings of the 18th Symposium on Rarefied Gas Dynamics,* Vancouver, 1992 (to appear).
10. E. A. Carlen and M. C. Carvalho, "Strict entropy production bounds and stability of the rate of convergence to equilibrium for the Boltzmann equation," *J. Stat. Phys.* **67**, 575-608 (1992).
11. C. Cercignani, "A remarkable estimate for the solutions of the Boltzmann equation."*Appl. Math. Lett.* **5**, 59–62 (1992).
12. C. Cercignani, *The Boltzmann Equation and Its Applications*, Springer, New York (1988).
13. C. Cercignani, "Half-space problems in the kinetic theory of gases," in *Trends in Applications of Pure Mathematics to Mechanics*, E. Kröner and K. Kirchgässner, eds., Lectures Notes in Physics **39**, 323 (1986).
14. P. Degond, and B Lucquin-Desreux, "The Fokker–Planck asymptotics of the Boltzmann collision operator in the Coulomb case," *Math. Mod. and Meth. Appl. Sci.* **2,2**, 167-182 (1992).
15. J. Glimm, "Solutions in the large for nonlinear hyperbolic systems of equations," *Commun. Pure Appl. Math.* **18**, 95–105 (1965).
16. R. Illner and W. Wagner, "A random discrete velocity model and approximation of the Boltzmann equation,"*Jour. Stat. Phys.* **70 (3/4)**, 773–792 (1993).
17. F. King, Ph. D. thesis, Berkeley 1976
18. O. Lanford III, "Time evolution of large classical systems." *Lecture Notes in Physics* **38**, Moser, E. J., ed., 1–111. Springer-Verlag, New York (1975).
19. P. L. Lions, "Conditions at infinity for Boltzmann's equation," *Ceremade* **9334**, 1993.
20. N. Maslova, "Kramers problem in the kinetic theory of gases," *U.S.S.R Comput. Math. Math. Phys.* **22**, 208–219 (1982).
21. B. Nicolaenko, "Shock wave solution of the Boltzmann equation as a nonlinear bifurcation problem for the essential spectrum," in *Théories Cinétiques Classiques et Relativistes*, G. Pichon, ed., 127–150, CNRS, Paris (1975).
22. H. Spohn, to appear in *Micro, Meso and Macroscopic Approaches in Physics*, Plenum Press, New York (1994).

Author Index

Subject Index

Applied Mathematical Sciences

(continued from page ii)

52. *Chipot:* Variational Inequalities and Flow in Porous Media.
53. *Majda:* Compressible Fluid Flow and System of Conservation Laws in Several Space Variables.
54. *Wasow:* Linear Turning Point Theory.
55. *Yosida:* Operational Calculus: A Theory of Hyperfunctions.
56. *Chang/Howes:* Nonlinear Singular Perturbation Phenomena: Theory and Applications.
57. *Reinhardt:* Analysis of Approximation Methods for Differential and Integral Equations.
58. *Dwoyer/Hussaini/Voigt (eds):* Theoretical Approaches to Turbulence.
59. *Sanders/Verhulst:* Averaging Methods in Nonlinear Dynamical Systems.
60. *Ghil/Childress:* Topics in Geophysical Dynamics: Atmospheric Dynamics, Dynamo Theory and Climate Dynamics.
61. *Sattinger/Weaver:* Lie Groups and Algebras with Applications to Physics, Geometry, and Mechanics.
62. *LaSalle:* The Stability and Control of Discrete Processes.
63. *Grasman:* Asymptotic Methods of Relaxation Oscillations and Applications.
64. *Hsu:* Cell-to-Cell Mapping: A Method of Global Analysis for Nonlinear Systems.
65. *Rand/Armbruster:* Perturbation Methods, Bifurcation Theory and Computer Algebra.
66. *Hlavácek/Haslinger/Necasl/Lovísek:* Solution of Variational Inequalities in Mechanics.
67. *Cercignani:* The Boltzmann Equation and Its Applications.
68. *Temam:* Infinite Dimensional Dynamical Systems in Mechanics and Physics.
69. *Golubitsky/Stewart/Schaeffer:* Singularities and Groups in Bifurcation Theory, Vol. II.
70. *Constantin/Foias/Nicolaenko/Temam:* Integral Manifolds and Inertial Manifolds for Dissipative Partial Differential Equations.
71. *Catlin:* Estimation, Control, and the Discrete Kalman Filter.
72. *Lochak/Meunier:* Multiphase Averaging for Classical Systems.
73. *Wiggins:* Global Bifurcations and Chaos.
74. *Mawhin/Willem:* Critical Point Theory and Hamiltonian Systems.
75. *Abraham/Marsden/Ratiu:* Manifolds, Tensor Analysis, and Applications, 2nd ed.
76. *Lagerstrom:* Matched Asymptotic Expansions: Ideas and Techniques.
77. *Aldous:* Probability Approximations via the Poisson Clumping Heuristic.
78. *Dacorogna:* Direct Methods in the Calculus of Variations.
79. *Hernández-Lerma:* Adaptive Markov Processes.
80. *Lawden:* Elliptic Functions and Applications.
81. *Bluman/Kumei:* Symmetries and Differential Equations.
82. *Kress:* Linear Integral Equations.
83. *Bebernes/Eberly:* Mathematical Problems from Combustion Theory.
84. *Joseph:* Fluid Dynamics of Viscoelastic Fluids.
85. *Yang:* Wave Packets and Their Bifurcations in Geophysical Fluid Dynamics.
86. *Dendrinos/Sonis:* Chaos and Socio-Spatial Dynamics.
87. *Weder:* Spectral and Scattering Theory for Wave Propagation in Perturbed Stratified Media.
88. *Bogaevski/Povzner:* Algebraic Methods in Nonlinear Perturbation Theory.
89. *O'Malley:* Singular Perturbation Methods for Ordinary Differential Equations.
90. *Meyer/Hall:* Introduction to Hamiltonian Dynamical Systems and the N-body Problem.
91. *Straughan:* The Energy Method, Stability, and Nonlinear Convection.
92. *Naber:* The Geometry of Minkowski Spacetime.
93. *Colton/Kress:* Inverse Acoustic and Electromagnetic Scattering Theory.
94. *Hoppensteadt:* Analysis and Simulation of Chaotic Systems.
95. *Hackbusch:* Iterative Solution of Large Sparse Systems of Equations.
96. *Marchioro/Pulvirenti:* Mathematical Theory of Incompressible Nonviscous Fluids.
97. *Lasota/Mackey:* Chaos, Fractals, and Noise: Stochastic Aspects of Dynamics, 2nd ed.
98. *de Boor/Höllig/Riemenschneider:* Box Splines.
99. *Hale/Lunel:* Introduction to Functional Differential Equations.
100. *Sirovich (ed):* Trends and Perspectives in Applied Mathematics.

(continued on next page)

Applied Mathematical Sciences

(continued from page ii)

101. *Nusse/Yorke:* Dynamics: Numerical Explorations.
102. *Chossat/Iooss:* The Couette-Taylor Problem.
103. *Chorin:* Vorticity and Turbulence.
104. *Farkas:* Periodic Motions.
105. *Wiggins:* Normally Hyperbolic Invariant Manifolds in Dynamical Systems.
106. *Cercignani/Illner/Pulvirenti:* The Mathematical Theory of Dilute Gases.